Springer Series in

MATERIALS SCIENCE

145

Springer Series in

MATERIALS SCIENCE

The Springer Series in Materials Science covers the complete spectrum of materials physics, including fundamental principles, physical properties, materials theory and design. Recognizing the increasing importance of materials science in future device technologies, the book titles in this series reflect the state-of-the-art in understanding and controlling the structure and properties of all important classes of materials.

Please view available titles in *Springer Series in Materials Science* on series homepage http://www.springer.com/series/856

Preface

In the course of a year or slightly more that passed since the discovery of a new class of high-temperature superconductors (HTSCs) in FeAs-based compounds [1], the world's community of physicists, chemists and technologists achieved a substantial progress in understanding the mechanisms and details of this superconductivity. The intensity of researches coming about is comparable only to that which accompanied the discovery of HTSCs in cuprates. However, the present scientific context is markedly different from that having existed twenty years back. In those times, the researchers moved on while blindly palpating the terrain. At present, they can rely on a rich accumulated experience of work with complex compounds; novel experimental methods and numerical calculation schemes have emerged; computational resources became by far much more powerful, and, last but not least, the physical ideas elaborated in the studies of cuprates could have been immediately adapted for the study of new HTSC compounds.

An unprecedentedly fast advance of researches on the FeAs compounds was helped by an instantaneous propagation of knowledge via electronic data archives. A markedly international character of studies is noteworthy; as a rule, the articles on FeAs systems are published by joint teams of distant lands and laboratories that boosts a rapid augmentation of knowledge about the properties of systems under study and thinking over the wealth of experimental data. During last years (2008–2010), more than few thousand publications within this domain have appeared. This means that every day brought about, on an average, 2–3 new papers deposited in electronic archives.

If the epic of HTSC study in cuprates demanded years for arriving at some understanding of these materials' nature, with respect to new class of materials one year was sufficient as a due time to make a primary overview of the results obtained. Within half a year after the discovery of HTSC in FeAs compounds, first three reviews appeared in the *Physics – Uspekhi* [2–4]. In the beginning of 2009, a special issues of *Physica C* [5] and *New Journal of Physics* [6] appeared with review articles by leading scientists on the basics of the physics of the FeAs compounds, which also summarized the bulk of results accumulated within a year.

This book seems to be the world's first monograph on the physics of FeAs systems. It outlines in a systematic way the results of researches done in the global scientific community throughout the whole period since the end of February

2008, as the high-temperature superconductivity has been discovered in a LaOFeAs system.

The first three chapters cover experimental investigations of all classes of the FeAs compounds in which superconducting state has been discovered. The fourth chapter is devoted to theory models of these compounds and to the discussion, on this basis, of experimental results. Differently from the reviews published in [5, 6], which specifically addressed various aspects of the physics of FeAs systems in some detail, we attempted here to cover, within a unique concept, the whole bulk of experimental and theoretical material on these systems by now available. The authors' hope is that the book be of use for a broad fold of readers: those who already immediately work in this problem and who would wish to enter it.

Russia
August 2010

Yu. A. Izyumov
E.Z. Kurmaev

Contents

Acronyms

1111	compounds of type LaOFeAs
122	compounds of type $BaFe_2As_2$
111	compounds of type LiFeAs
11	compounds of type FeSe
LDA	Local Density Approximation
LSDA	Local Spin Density Approximation
DMFT	Dynamical Mean Field Theory
LDA + DMFT	joint LDA and $DMFT$ computational scheme
RPA	Random-Phase Approximation
FLEX	Fluctuation Exchange Interaction
Folded BZ	Folded BZ
Unfolded BZ	Unfolded BZ
SDW	Spin Density Wave
CDW	Charge Density Wave
NMR	Nuclear Magnetic Resonance
STS	Scanning Tunneling Microscopy
PCAR	Point-Contact Andreev Reflection
PES	Photoelectron spectroscopy
ARPES	Angle Resolved Photoelectron Spectroscopy
RXES	Resonant X-Ray Emission Spectroscopy
ZBC	Zero-Bias conductance
HTSC	High Temperature Superconductivity
OP	Order Parameter
GF	Green Function
BCS	Bardin–Cooper–Schrieffer's theory

Chapter 1
Introduction

The first report of superconductivity in LaOFeAs appeared in 2006 [1]; however, the transition temperature was low, $T_c = 3.5\,\mathrm{K}$. Similarly, LaONiP have shown $T_c = 4.5\,\mathrm{K}$ [7]. The breakthrough occurred in February 2008, as Kamihara et al. reported a superconductivity with $T_c = 26\,\mathrm{K}$ in fluorine-doped compound $LaO_{1-x}F_x$ FeAs [8]. Immediately afterwards several Chinese groups, by substituting lanthanum with other rare-earth elements, achieved much higher T_c values, namely, 41 K in $CeO_{1-x}F_xFeAs$ [9], 52 K in $PrO_{1-x}F_xFeAs$ [10] and reached 55 K in $SmO_{1-x}F_xFeAs$ [11].

The pristine (undoped) compounds are antiferromagnetic (AFM) metals, in which the magnetic ordering comes about simultaneously with structural phase transition at the Neél temperature $T_N \approx 140\,\mathrm{K}$ (in LaOFeAs) from tetragonal to orthorhombic phase. On substituting oxygen with fluorine, T_N rapidly falls down as the F concentration increases, and at $x \simeq 0.1$ the long-range magnetic ordering disappears, and a superconducting state sets on. A typical phase diagram of this type of compounds is shown in Fig. 1.1 in the (T, x) plane [12].

The situation so far resembles the HTSC in cuprates, e.g., $(La_{1-x}Sr_x)_2CuO_4$ exhibits a similar phase diagram. The superconductivity appears there in compounds of the type La_2CuO_4, which are also AFM under stoichiometry, in the course of lanthanum being substituted by strontium. In both systems, the doping brings along charge carriers (electrons or holes) that suppresses the AFM ordering and creates conditions for forming the Cooper pairs. This analogy supported a suggestion that the high-T_c superconductivity in newly discovered FeAs-based systems is influenced by the system's closeness to a magnetic phase transition, so that high T_c values are due to the carriers pairing mechanism via spin fluctuations.

An analogy between FeAs systems and cuprates becomes more apparent if we compare their crystal structures. The FeAs-based systems are built by stapling of the FeAs planes, intermediated by the LaO layers, similarly to how in cuprates the stacked CuO_2 planes are separated by the La- or Y-Ba layers. By force of their layered structure, both types of systems are strongly anisotropic, and electronic states therein are quasi two-dimensional.

Closely following the *Re*OFeAs compounds (with *Re* being a rare-earth element), the compounds of the type AFe_2As_2, (A = Ba,Sr,Ca) emerged, whose peculiarity is that a repeated unit in them contains a doubled FeAs plane, similarly to

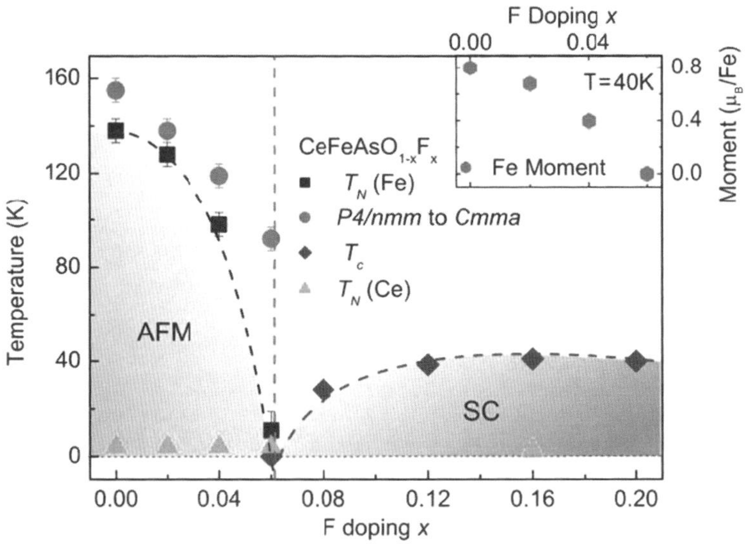

Fig. 1.1 Phase diagram in the (T, x) plane for the $CeO_{1-x}F_xFeAs$ compound

doubled layers in cuprates $YBa_2Cu_3O_6$. In doped AFe_2As_2, the superconductivity was immediately found with $T_c = 38$ K [13]. Further on, another class of FeAs-based systems has been discovered, the LiFeAs compound in which the FeAs planes are separated by the layers of lithium. It is remarkable that in this compound superconductivity with $T_c = 18$ K appears without any doping [14, 15].

A similar property is revealed by yet another structural type, namely, FeSe, FeS and FeTe, which are quite resembling the compounds of the FeAs group. These novel compounds are built from iron–chalcogen planes, in which, like in the FeAs compounds, the iron atoms form a squared lattice, each atom being surrounded by an octahedron of chalcogens. Here, no intermediary layers are present. In one such compound, FeSe, under pressure of ~1.5 GPa a superconductic transition with $T_c = 27$ K has been detected [16].

Therefore as of now we are aware of three classes of compounds build of the FeAs layers: these are LaOFeAs, AFe_2As_2, LiFeAs and moreover a similar structure type of FeSe in which the superconductivity with high T_c was detected. Physical properties of these compounds have many similarities and are dominated by the influence of a common planar structural element. More precise analysis of physical properties confirms this suggestion.

Calculations on electron–phonon coupling in these compounds have shown [17, 18] that the standard electron–phonon coupling mechanism cannot account for such high T_c values.

A similarity in physical properties of the FeAs-compounds with those of high-temperature superconducting cuprates puts forward a question about a role of electron correlations in these new materials. It is known that in the materials on the basis of transition-metal and rare-earth elements, such correlations do often play a primary role – see, e.g., a monograph by Fulde [19]. Another important question is that concerning the role of degenerate 3d orbitals of the Fe ions in the formation of electronic structure near the Fermi level in the FeAs-compounds, and about the spin state of the Fe ions in the compound [20]. Both these important questions will be addressed in the book from both experimental and theoretical viewpoints.

Chapter 2
Compounds of the ReOFeAs Type

2.1 Crystallochemistry and Basic Physical Properties of Doped Compounds

2.1.1 Crystal Structure

The highest values of T_c have been achieved in the row of *Re*OFeAs doped compounds, where *Re* stands for a rare-earth element (Table 2.1). All these compounds possess, at room temperature, a tetragonal structure with the $P4/nmm$ space group. Their crystal structure is formed by repeated FeAs layers, interlaced by the LaO layers. The FeAs layer is, in fact, created by three closely situated atomic planes: the middle one is a quadratic lattice of Fe atoms, sandwiched between two quadratic lattices of As, so that each atom of iron is surrounded by a tetrahedron of arsenic atoms. In other words, the FeAs layer is, in fact, formed by $FeAs_4$ complexes. The FeAs and LaO layers are separated by 1.8 Å.

The crystal structure of LaOFeAs is shown in Fig. 2.1. Lattice parameters for the *Re*OFeAs compounds are given in Table 2.1. As is seen, the tetragonal unit cell is strongly elongated, which explains a strong anisotropy of all its properties and a quasi-bidimensional nature of electronic states. The closest to each Fe atom are those of As, which underway to the next Fe neighbours, so that the electron transfer processes over the Fe sublattice are mediated by the Fe-As hybridization, and the exchange interaction between Fe atoms is of indirect character via the As atoms.

Crystallochemical properties of LaOFeAs compounds are determined by the configuration of the outer electron shells: Fe($4s4p3d$), As($4s4p$), La($6s5d4f$), O($2s2p$). The formal valences of ions are as follows: $La^{3+}O^{2-}Fe^{2+}As^{3-}$.

2.1.2 Electron Doping

On substituting an oxygen atom by fluorine, an extra electron goes into the FeAs layer; such situation is commonly referred to as electron doping. A substitution of lanthanum by, say, strontium, the LaO layer would lack one electron, which can be

Table 2.1 Maximal temperatures of superconducting transitions obtained by doping of the *Re*OFeAs compounds. In the last two lines, the lattice parameters of undoped compounds are given

ReOFeAs	*La*	*Ce*	*Pr*	*Nd*	*Sm*	*Gd*
T_c, K	41	41	52	51.9	55	53.5
Reference	[27]	[9]	[10]	[28]	[11]	[29]
a, *Å*	4.035	3.996	3.925	3.940	3.940	
c, *Å*	8.740	8.648	8.595	8.496	8.496	

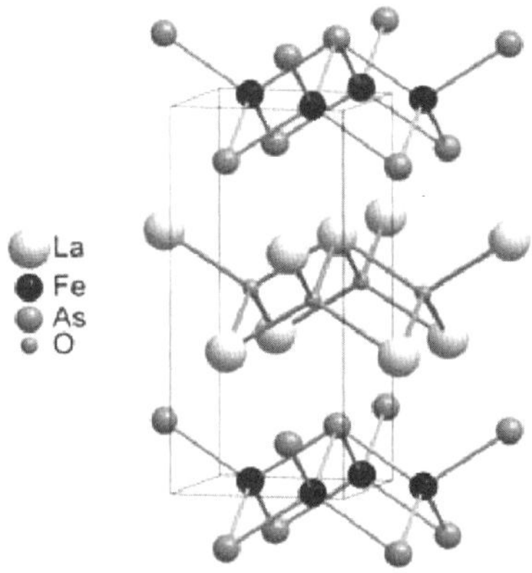

Fig. 2.1 Crystal structure of LaOFeAs

borrowed from the FeAs layer, leaving behind a hole. This would correspond to a hole doping. A re-distribution of electrons between the doped LaO and FeAs layers gives rise to a resulting conductivity of a compound. The nature of carriers can be deduced experimentally from the sign of the Hall constant R_H.

The measurements of the Hall effect have been done on a compound $LaO_{0.9}F_{0.1}$ FeAs [21] with $T_c = 24$ K soon after the discovery by Kamihara et al. [8] a superconductivity with $T_c = 26$ K on this very compound. In [21], it was concluded that R_H is negative and roughly independent on temperature up to 240 K. This indicates that the conductivity is dominated by electron carriers. From the Hall coefficient measured at $T \approx 100$ K, the carrier concentration was deduced to be $9.8 \cdot 10^{20}$ cm^{-3}. The authors of [22] confirmed these estimates. A measurement of R_H on a different sample $LaO_{0.89}F_{0.11}FeAs$ with $T_c = 28.2$ K, done at a temperature slightly superior to T_c, has shown that the concentration of electron carriers $n \approx 1 \cdot 10^{21}$ cm^{-3} [23] does, in fact, coincide with the results of [21, 22] (Fig. 2.2). In the inset of this figure, a temperature behaviour of the Hall coefficient R_H, throughout negative, is shown.

Another example of electron doping is given in Fig. 2.3 [24], where doped and undoped compounds are compared. In both cases, the Hall coefficient is negative. Compounds with other rare-earth elements, e.g. $NdO_{0.82}F_{0.18}FeAs$ [25], well indicate an electron nature of carriers.

A remarkable fact was a discovery of high-T_c superconductivity in the compounds *Re*OFeAs without fluorine doping, but under oxygen deficiency. Thus, [26]

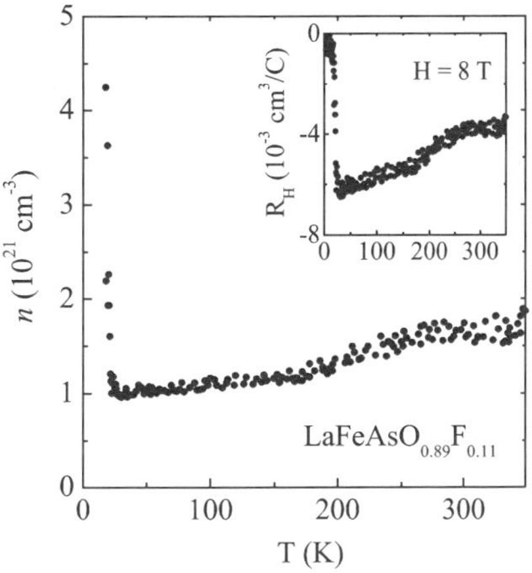

Fig. 2.2 Variation of the number of carriers and the Hall coefficient for the $LaO_{0.89}F_{0.11}FeAs$ compound [23]

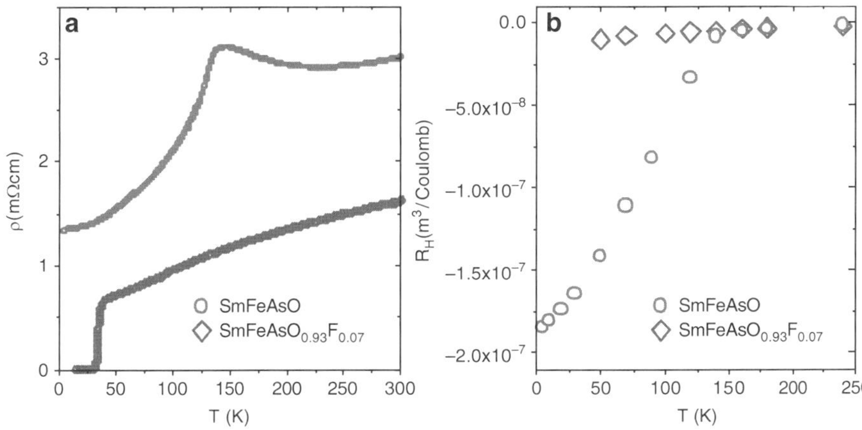

Fig. 2.3 Temperature dependence of resistivity and Hall coefficient for nondoped SmOFeAs and fluorine-doped $SmO_{0.93}F_{0.07}FeAs$ compound [24]

reports a detection of high T_c values in $LaO_{0.6}FeAs$ ($T_c = 28$ K), $LaO_{0.75}FeAs$ ($T_c = 20$ K), and $NdO_{0.6}FeAs$ ($T_c = 53$ K).

In [9–11, 27–29], the data are given concerning the compounds $ReO_{1-\Delta}FeAs$ with Re = Sm,Nd,Pr,Ce,La. Among them, the $SmO_{1-\delta}FeAs$ system indicated the highest $T_c = 55$ K. Hence, the fluorine doping and the oxygen deficiency produce similar effects in the initial stoichiometric compounds: they create electron carriers, suppress antiferromagnetic (AFM) ordering and result in the formation of a superconducting state.

Let us now consider the effect of substitution of a rare-earth element by a heterovalent dopant. A replacement of trivalent Re^{3+} by a quatrovalent substituent results in electron doping. For example, we take a system $Gd_{1-x}Th_xOFeAs$, where Gd^{3+} is substituted by Th^{4+}. At $x \approx 0.1$, a superconductivity with $T_c = 55$ K has been reported by [30]. Another example of electron doping is $Tb_{1-x}Th_xOFeAs$, where a substitution of Tb^{3+} by Th^{4+} results in $T_c = 52\,K$ [31].

2.1.3 Hole Doping

A completely different situation arises on substituting an Re^{3+} ion by a divalent element. On substitution of La^{3+} in LaOFeAs by Sr^{2+}, we deal with hole doping. The resulting compound, $La_{1-x}Sr_xOFeAs$, at $x = 0.13$ becomes superconducting with $T_c = 25\,K$ [32]. This was the first superconductor in the FeAs-row, obtained by hope doping, as has been confirmed by measuring the Hall coefficient R_H, which turned out in this system to be positive [33].

Apparently, an increase in strontium concentration suppresses the conventional AFM ordering in the pristine compound, and already at $x = 0.03$ the doped state becomes superconducting. T_c grows along with x and at $x \approx 0.11$–0.13 reaches the value of $T_c = 25\,K$. On substituting oxygen by fluorine, $T_c = 26\,K$. We can note a peculiar electron-hole symmetry: on doping a pristine compound by either electrons or holes the T_c grows in roughly similar way. There is, however, a certain difference between two situations. On doping with Sr, a rise of T_c is accompanied by a monotonous increase of the lattice parameters a and c, whereas fluorine doping reduces the lattice parameters [33].

A system $Pr_{1-x}Sr_xOFeAs$ offers another example of the hole doping, on substituting Pr^{3+} by Sr^{2+} [34]. A superconductivity of $T_c = 16.3\,K$ was achieved at the Sr concentration $x \approx 0.20$–0.25. Figure 2.4 shows a temperature dependence of the Hall coefficient, which is, below the room temperature, throughout positive. A similar result occurs in an Nd-based compound on substitution of the latter element with Sr. In an $Nd_{1-x}Sr_xOFeAs$ sample ($0 < x < 0.2$), $T_c = 13.5\,K$ has been achieved at $x \approx 0.2$ [35]. It should be noted that in difference from electron-doped compounds such as $ReO_{1-x}F_xFeAs$ where an increase of x the magnetic ordering is gradually suppressed and superconductivity occurs already at $x < 0.1$, in hole-doped systems

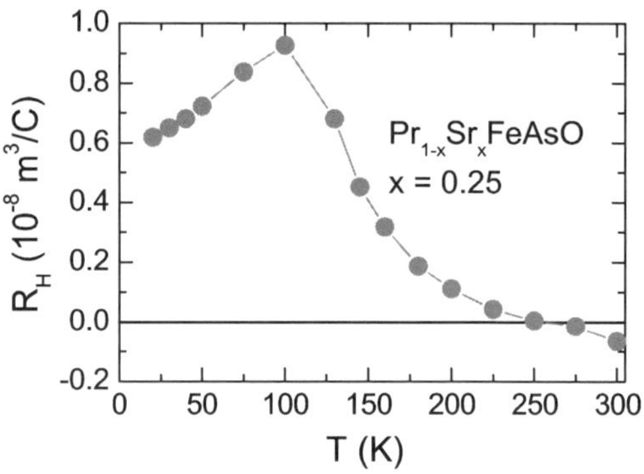

Fig. 2.4 Temperature dependence of the Hall coefficient R_H for the hole-doped $Pr_{0.75}Sr_{0.25}OFeAs$ compound [34]

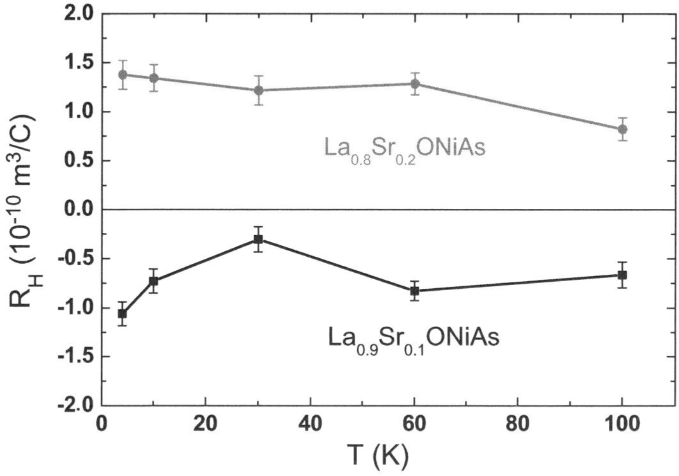

Fig. 2.5 Hall coefficient for two samples of the $La_{1-x}Sr_xONiAs$ compound [36]

an onset of a superconducting state demands higher concentrations of dopant. In this sense, the "electron-hole symmetry" does not hold.

The above discussion of main physical properties in LaOFeAs systems remains valid for those where Ni takes place of Fe, only that the T_c in such systems is much lower. It is noteworthy that the nature of carriers – are they holes or electrons – may vary depending on the dopant concentration (Fig. 2.5).

The LaOFeAs compound can be doped not only by fluorine which substitutes oxygen, but also by elements taking place of lanthanum, e.g. potassium. A doping with potassium adds hole carriers, rather than electrons. In [37], a new method of synthesis of superconducting compounds on the basis of LaOFeAs was suggested, allowing a simultaneous doping with fluorine and potassium. A synthesized compound $La_{0.8}K_{0.2}O_{0.8}F_{0.2}FeAs$ had $T_c = 28.5$ K.

The examples discussed above show that the superconductivity in FeAs systems might be induced either by electron doping (substituting oxygen by fluorine or due to the presence of oxygen vacancies), or by hole doping (via substituting La by Sr). These tendencies are maintained throughout the whole class of the *Re*OFeAs systems.

2.1.4 Substitutions on the Fe Sublattice

In earlier stages of studying the *Re*FeAsO system, it was shown that the superconductivity is induced by doping on either oxygen or rare-earth sublattice, which both are beyond the FeAs layers. Due to either substitution of oxygen by fluorine, or oxygen deficiency, the FeAs layers are infiltrated by charge carriers, that suppresses antiferromagnetic order of the pristine compound and leads to superconductivity.

In this sense, the new superconductors resemble the cuprates, where substitutions occur outside the CuO_2 planes.

A substitution of Fe atoms in the FeAs layers by Co does as well result in suppression of antiferromagnetism and appearance of superconductivity already at low concentrations of dopant. This feature makes a marked differences of new superconductors from cuprates, in which any intrusion into the CuO_2 planes suppresses superconductivity. In several works appeared simultaneously, astonishing results have been reported on a number of samples of $LaOFe_{1-x}Co_xAs$ [12, 38, 39]. At $x = 0.05$, the antiferromagnetism was suppressed, and at $x \approx 0.1$ a superconductivity with $T_c \approx 10$ K emerged, which further on disappeared at $x > 0.15$. This is confirmed by temperature dependencies of electrical conductivity at different x (Fig. 2.6) [38].

The phase diagram of this system in the (T, x) axes is shown in Fig. 2.7 [39]. It is shown that in the x range corresponding to superconductivity, for $T > T_c$ first a semiconductor-type behaviour is observed, which is followed at $T \approx 100$ K by metallic conductivity. Similar results were obtained for $SmOFe_{1-x}Co_xAs$ [39].

It turns out therefore that Co is an efficient dopant for inducing superconductivity. It is astonishing that superconductivity persists at quite high degree of disorder (broad interval of x) that apparently is an argument in favour of a non-standard symmetry of the order parameter, which is not sensitive to magnetic impurities [40]. It is interesting to note that for $x = 1$ the system becomes ferromagnetic with $T_c \approx 56$ K [38].

Note that electronic structure calculations for $LaOFe_{1-x}Co_xAs$ have appeared [41], which show that the Co doping displaces the Fermi level from its position at the slope of the partial density of Fe3d states into a more flat region. This circumstance explains a suppression of the SDW transition in the initial LaOFeAs on doping of its Fe sublattice.

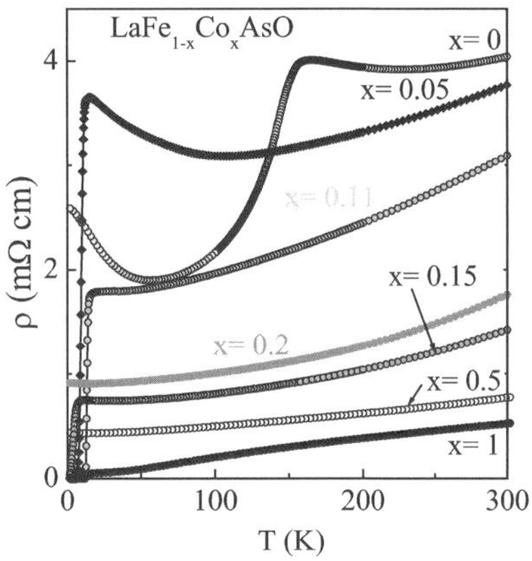

Fig. 2.6 Temperature dependence of electrical resistivity of the $LaOFe_{1-x}Co_xAs$ compound at different cobalt concentrations [38]

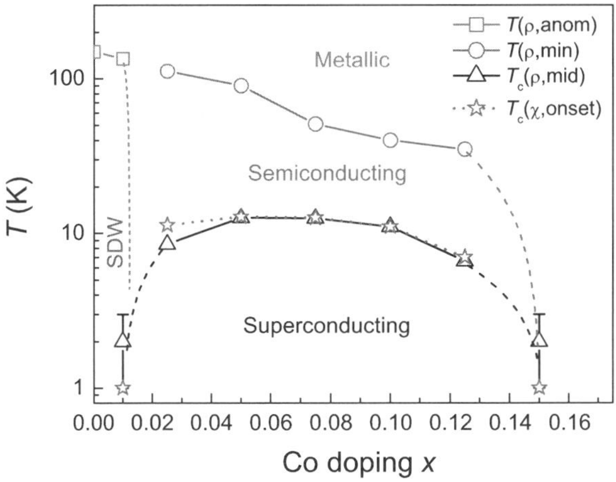

Fig. 2.7 Phase diagram of $LaOFe_{1-x}Co_xAs$ in the (T, x) plane [39]

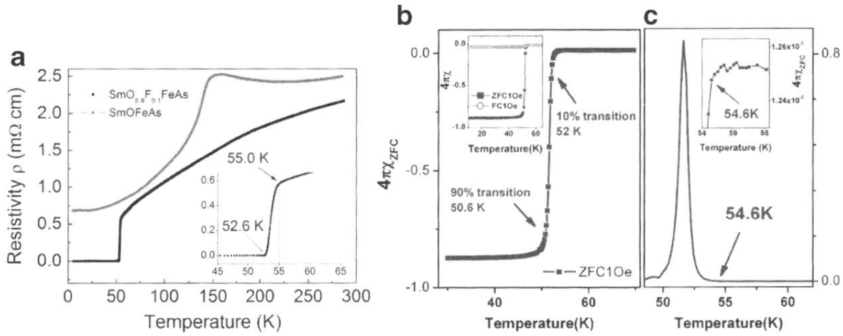

Fig. 2.8 Manifestation of superconducting state in the $SmO_{0.9}F_{0.1}FeAs$ compound as revealed by temperature dependence of (**a**) electrical resistivity, (**b**) magnetic susceptibility χ, (**c**) the derivative of χ in temperature [11]

2.1.5 Superconducting Transition Temperature

Now we turn to a more detailed description of superconducting properties in FeAs systems. How does a superconducting state in a doped material reveal itself in experiment? Let us take as an example the $SmO_{1-x}F_xFeAs$ compound in which at $x = 0.1$ the highest so far value of $T_c = 55$ K has been obtained [11]. Figure 2.8 shows the results of three different measurements: a sharp drop of electrical conductivity on lowering the temperature, a sharp appearance of diamagnetic response χ in applied magnetic field, and a sharp peak in the $d\chi/dT$ derivative. All three anomalies occur near the same temperature, which is, accordingly, the superconducting transition temperature. The curves as in Fig. 2.8 are typical for all superconducting systems on the FeAs basis. For comparison, corresponding curves for a group of

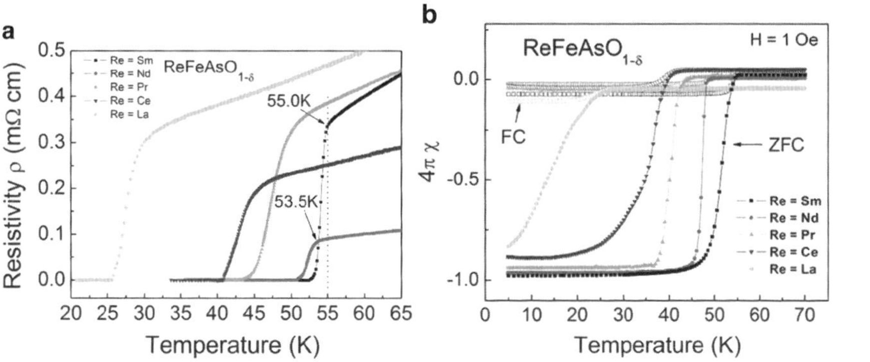

Fig. 2.9 Electrical resistivity (**a**) and magnetic susceptibility (**b**) of superconducting $Re\mathrm{O}_{1-x}\mathrm{FeAs}$ compounds with oxygen deficiency, as function of temperature [42]

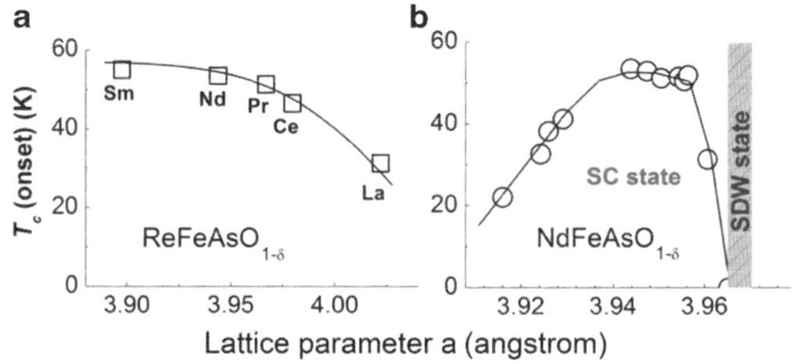

Fig. 2.10 T_c in the row of $Re\mathrm{O}_{1-\delta}\mathrm{FeAs}$ compounds as function of the lattice parameter a [42]

*Re*OFeAs compounds with different rare-earth constituents are shown in Fig. 2.9 [42]. The behaviour of electrical conductivity and magnetic susceptibility in the vicinity of T_c is similar between different systems. We note that, differently from the SmOFeAs system of Fig. 2.8 which was doped with fluorine, all superconducting compounds collected in Fig. 2.9 are deficient in oxygen. Despite different nature of dopants – fluorine substitution or oxygen vacancies – the manifestation of superconducting state in the temperature dependence of electrical conductivity and magnetic susceptibility is identical for both systems. It is instructive to compare the superconducting transition temperatures throughout a row of compounds with different rare-earth elements and hence lattice parameter a (Fig. 2.10). We see that T_c decreases with the rise of a (due to the increase of the element's ionic radius). For a given rare-earth constituent, T_c depends on the number of oxygen vacancies δ. In a synthesized compound, the vacancy concentration is revealed by the lattice parameter a (Fig. 2.10b).

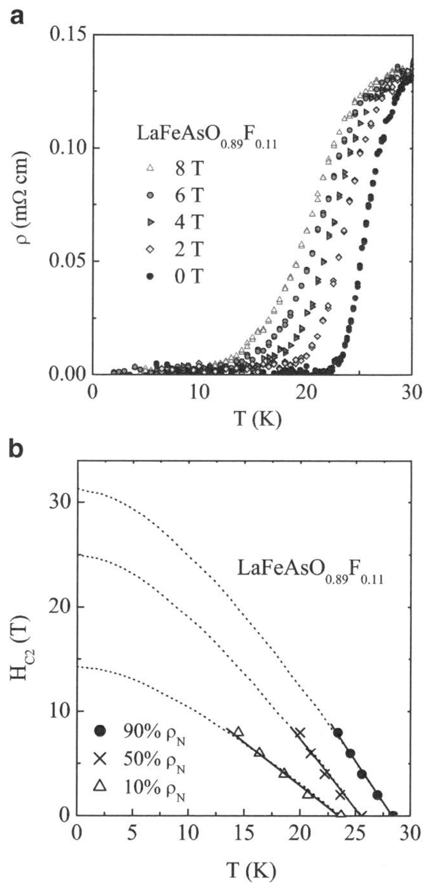

Fig. 2.11 Temperature dependence of (**a**) electrical resistivity in different magnetic fields, and (**b**) the values of H_{c2} extracted from these data, for $LaO_{0.89}F_{0.11}FeAs$ [23]

2.1.6 *Critical Fields*

Besides high transition temperatures, the FeAs-type compounds possess very high critical fields values. Consider, as an example, a study of the upper critical field H_{c2} in polycrystalline sample of $LaO_{0.98}F_{0.11}FeAs$ with $T_c = 28.2$ K [23].

H_{c2} is estimated from the data on the temperature dependence of electrical resistivity in magnetic field. In Fig. 2.11a, such data in the field range up to 8 T are given, and in Fig. 2.11b, the $H_{c2}(T)$ results extracted from the latter. It is seen from Fig. 2.11a that the interval of the drop in resistivity shifts towards lower temperatures on applying the field, that is typical for II order superconductors. The superconducting transition temperature $T_c(H)$ is defined by the condition that $\rho(T_c, H)$ equals a certain fraction (percentage) of resistivity ρ_N in the normal phase, for a given field magnitude H. The thus defined values of $T_c(H)$ for $\rho = 10$, 30 and 90% of ρ_N are shown in Fig. 2.11b along with the critical fields $H_{c2}(T)$. In all cases, $H_{c2}(T)$ exhibit linear dependence without any tendency towards saturation.

The slope $dH_{c2}/dT\ |_{T=T_c}$ equals −0.87 T/K for $\rho = 10\%\rho_N$, −1.41 T/K for $\rho = 50\%\rho_N$ and −1.59 T/K for $\rho = 90\%\rho_N$. In the BCS theory, the H_{c2} is linear in T in the vicinity of T_c and saturates towards $T = 0$. According to the Werthamer–Helfand–Hohenberg formula [43],

$$H_{c2}(0) \approx -0.693\, T_c \left.\frac{dH_{c2}}{dT}\right|_{T=T_c}. \tag{2.1}$$

The dashed lines in Fig. 2.11b are extrapolations of linear experimental curves towards the thus calculated values of $H_{c2}(0)$. For $\rho = 90\%\rho_N$, the $H_{c2}(0)$ exceeds 30 T. From the known Ginsburg–Landau formula for the correlation length $\xi(0) \approx (\Phi_0/2\pi H_{c2})^{1/2}$, where Φ_0 is a flux quantum, an estimation follows: $\xi(0) \approx$ 48 Å for H_{c2} (10% ρ_N), $\xi(0) \approx 36$ Å for H_{c2} (50% ρ_N), $\xi(0) \approx 33$ Å for H_{c2} (90% ρ_N). These values are comparable to those measured in cuprates for corresponding values of T_c.

Measurements of the Hall constant on the same sample revealed its negative sign (that corresponds to electron carriers) and carriers concentration of $\approx 1.7{\cdot}10^{21}$ cm^{-3} at room temperature and $\approx 1{\cdot}10^{21}$ cm^{-3} at a temperature just above T_c (assuming a single carriers band).

The above data concerning the sample studied of $LaO_{0.89}F_{0.11}FeAs$ are quite representative for the whole series of superconducting compounds $Re$$O_{1-x}F_xFeAs$. Thus for $NdO_{0.82}F_{0.18}FeAs$ with $T_c = 51$ K [44], $H_{c2}(48\text{ K}) = 13$ T has been measured, and the critical field $H_{c2}(0)$ estimated after (2.1) turned out to be within 80–230 T. Measurements on a single crystal of the same composition [45] revealed a large anisotropy of H_{c2}.

The critical fields estimated after (2.1) for the field directions in the basal plane (ab) and along the tetragonal axis (c) are: $H_{c2}^{ab}(0) \approx 304$ T and $H_{c2}^{c}(0) \approx 62$–70 T. The measurements on a Sm-based compound confirmed high values of H_{c2}. Thus for a sample $SmO_{0.85}F_{0.15}FeAs$ with $T_c = 46$ K, the measurements of specific heat in the fields of up to 20 T gave [46] $[dH_{c2}/dT]_{T=T_c} = -5$ T/K, that according to (2.1) gives an estimate $H_{c2}(0) = 150$ T. For another sample, $SmO_{0.7}F_{0.3}FeAs$ [47] with $T_c = 54.6$ K, the estimated $H_{c2}(0)$ is even higher: $H_{c2} \approx 200$ T. A detailed review of (H, T) phase diagrams of FeAs compounds can be found in [48].

2.1.7 Effect of Pressure on the T_c

Soon after the discovery of superconductivity in $LaO_{1-x}F_xFeAs$ it was reported that in the compound with $x = 0.11$, the T_c increases under applied pressure and reaches the maximum value of 43 K at 4 GPa [49]. It was suggested that the lattice compression is responsible for this effect. Indeed, in *Re*OFeAs compounds the atoms of rare-earth element have smaller radius than La, and T_c in these compounds is markedly higher, exceeding 50 K. In a subsequent work [50], the measurements of electrical conductivity in LaOFeAs under high pressures, up to 29 GPa, have

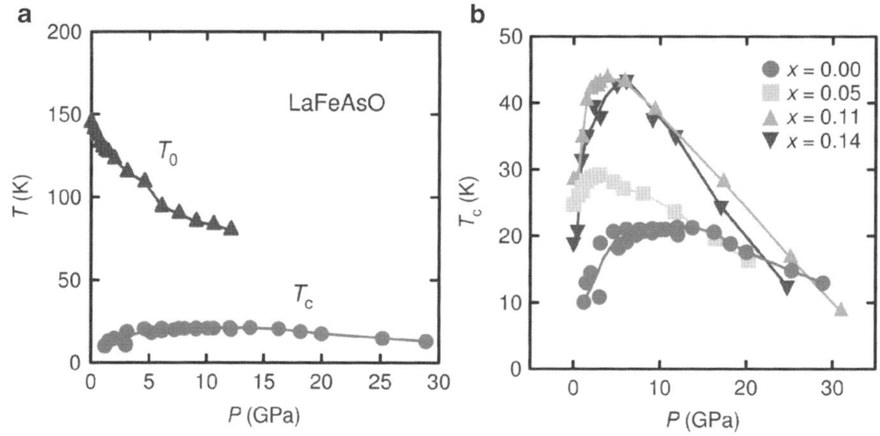

Fig. 2.12 (**a**) (T, P) phase diagram for LaOFeAs, obtained from the measurements of electrical resistivity at different pressures [50]; (**b**) variation of T_c with pressure in $LaO_{1-x}F_xFeAs$ compounds. The data for doped compounds are taken from [49,51]

been done. The results concerning the variation of the temperatures of structural (magnetic) phase transition T_0 and of the superconducting transition temperature, extracted from the raw data on $\rho(T)$ at different pressures, are shown in Fig. 2.12.

The (P, T) phase diagram shown in Fig. 2.12a resembles the (x, T) phase diagram for doped $LaO_{1-x}F_xFeAs$ compounds [8]. This similarity may be explained by an observation that the oxygen substitution with fluorine, beyond modifying the carriers density, results in reducing the lattice constant. Thus, as $x = 0.05$ the unit cell squeezes from 0.14186 nm^3 that is accompanied by an appearance of superconductivity with $T_c = 24$ K [8]. According to the variation of the unit cell volume under pressure [51], the above variation corresponds to a pressure of ~0.3 GPa. Correspondingly, in the context of merely changing the volume, the substitution of oxygen with fluorine is more efficient in suppressing structural and magnetic phase transitions and the onset of superconductivity that an effect of external pressure.

As is seen from Fig. 2.12b, the maximal $T_c = 21$ K in stoichiometric compound LaOFeAs is achieved at the pressure of ~12 GPa. As regards the variation of T_c with pressure in doped compounds, it first rises with pressure, passes through maximum and falls down. A similar behaviour of T_c under pressure is observed in $LaO_{1-x}F_xFeAs$ of a different composition, as well as in oxygen-deficient LaOFeAs compounds (Fig. 2.13). The latter have maximal $T_c \sim 50$ K at the pressure of 1.5 GPa [52].

A relation between the changes of T_c under pressure and variation of the lattice parameter is shown in Fig. 2.14. At high pressures ($P > 10$ GPa), the lattice parameters and T_c in the compound investigated $LaO_{0.9}F_{0.1}FeAs$ do decrease linearly [53].

A similar behaviour of T_c under pressure was observed in another compound type, LaOFeP. At the ambient pressure, T_c in doped $LaO_{1-x}F_xFeP$ compounds is

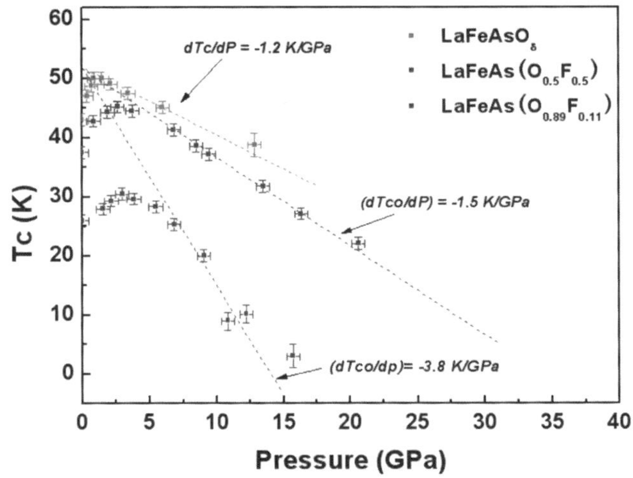

Fig. 2.13 Variation of the superconducting transition temperature with pressure for two $LaO_{1-x}F_xFeAs$ compounds and a $LaO_\delta FeAs$ compound with oxygen vacancies [52]

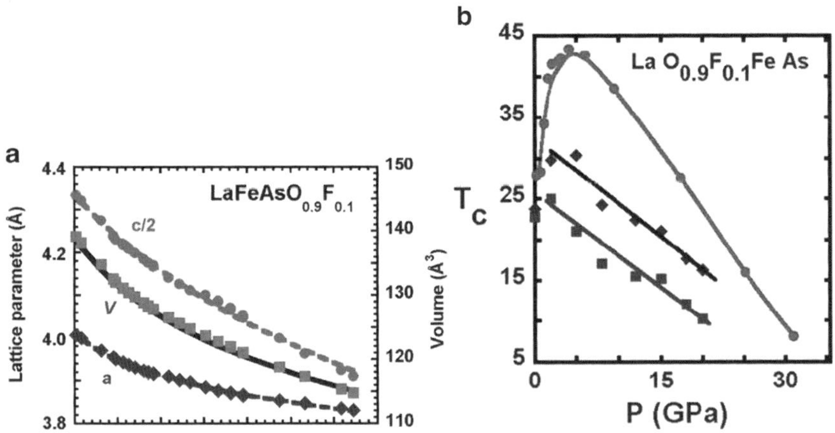

Fig. 2.14 Variation under pressure of (**a**) lattice parameters, and (**b**) superconducting transition temperature, for the $LaO_{0.9}F_{0.1}FeAs$ compound [53]

2–7 K. On applying the pressure, T_c rises rapidly, achieving 8.8 K already at $P = 0.8$ GPa, after which it falls down at the rate $dT_c/dP > 4$ K/GPa [54].

Finally, we discuss an aspect of chemical pressure which occurs on a substitution of an ion in the compound by another ion of a smaller radius. In this case, the shrinking of the lattice parameter is observed as the dopant concentration grows. This situation is illustrated by Fig. 2.15, taken from [55].

Yttrium has smaller ionic radius than lanthanum, therefore replacing the latter by the former reduces the lattice parameter a. As is seen from the figure, T_c grows with the yttrium concentration, whereas T_N decreases. This trend is common for all

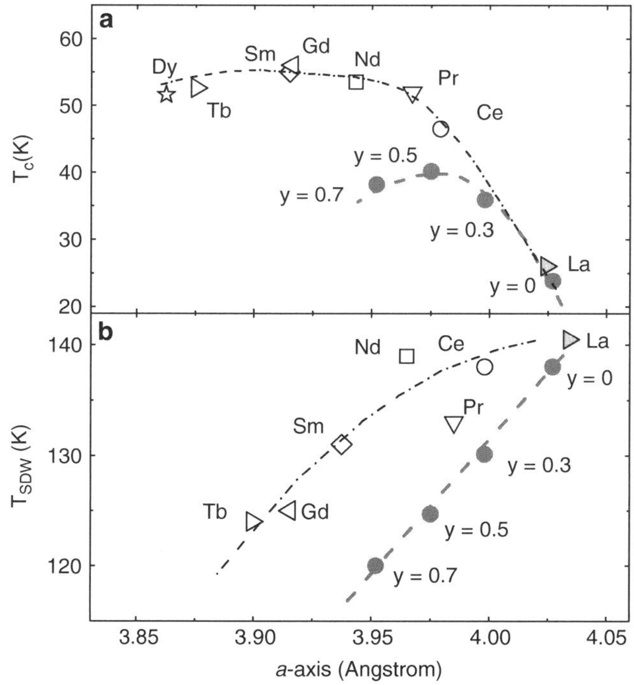

Fig. 2.15 (**a**) T_c and (**b**) T_N for *Re*OFeAs compounds as functions of the lattice parameter. *Black dots* indicate the values of T_c and T_N for the $La_{1-y}Y_yF_{0.15}FeAs$ compound at the levels of yttrium concentration $y = 0; 0.3; 0.5; 0.7$ [55]

$ReO_{1-x}F_xFeAs$ compounds, which exhibit a maximum T_c value at some optimal fluorine doping x.

A doping of the stoichiometric compound LaOFeAs with yttrium up to $y \leq 0.7$ does not lead to superconductivity, because an effect of chemical pressure is by far weaker than that of the fluorine doping. On yttrium doping of already superconducting compound $LaO_{1-x}F_xFeAs$, T_c increases from 24 to 40 K. Obviously, this happens not so much due to a decrease of the lattice parameter (chemical pressure) as because of adding new carriers to the compound, as La is partially replaced by Y. Therefore, the role of chemical pressure at the onset of superconductivity in FeAs-based compounds is considerably limited, in comparison with the effect of doping by heterovalent elements.

A particularly interesting behaviour of superconductivity under pressure is observed in Ce-containing compounds. In general, compounds with Ce do often exhibit anomalies induced by the Kondo screening of localized moments of the $4f$ shell of Ce atoms. In metallic Ce, an isostructural $\alpha \rightarrow \Gamma$ phase transition under pressure occurs, whose nature is purely electronic one, induced by a change of the Ce valence [56].

In [57], based on a thorough study of transport properties and X-ray absorption spectra under pressure, a competition of superconductivity and Kondo screening was found in the $CeO_{0.7}F_{0.3}FeAs$ compound. On increase of pressure, the superconducting transition temperature is gradually decreasing, and from $P = 8.6$ GPa on it drops abruptly, reaching zero at $P = 10$ GPa. XAS studies show a re-distribution of statistical weight from the main absorption line towards the satellite, indicating an appearance of the $4f^0$ states in the main bulk of the $4f^1$ states of Ce ions. A similar behaviour is observed in metallic Ce at the $\alpha \rightarrow \Gamma$ transition, see [56].

Therefore, the X-ray absorption spectra reveal the Kondo screening of the localized moments at Ce ions, caused by pressure. The spectra of the superconducting state are similar to the XAS of pure Ce. Guided by this analogy, the authors of [57] arrived at a conclusion that the reason of the suppression of T_c by pressure in $CeO_{0.7}F_{0.3}FeAs$ is in an emergence of a state with Kondo singlets, that expels the state with the Cooper pairs. Therefore, a quantum phase transition under pressure takes place, driven by a screening of localized moments of the Ce$4f$ shell by the Fe$3d$ electrons.

We mention in this relation the work [58], in which, in non-superconducting $CeOFeAs_{1-x}P_x$, two quantum critical points have been found, under the variation of the phosphorus content x. In the (T, x) phase diagram, for $x < 0.37$ an antiferromagnetic phase was detected with an ordering of localized moments at Fe and Ce sites. Further on, in the $0.92 < x < 1.00$ range a non-magnetic state with heavy fermions come about, induced by the Coulomb screening.

2.2 Magnetic Properties

2.2.1 Magnetic Structure

Stoichiometric *Re*OFeAs compounds are antiferromagnetics. The first indications of a possibility of magnetic ordering in LaOFeAs stem from measurements of temperature dependences of electrical conductivity and magnetic susceptibility, which exhibited anomalies near $T \approx 150$ K. At this temperature, a structure transition from tetragonal into orthorhombic phase have been detected. It was initially suggested that the magnetic ordering occurs at the same temperature. By now, full neutron diffraction studies done at a nuclear reactor in Oak Ridge clarified the situation [59]. At $T \approx 155$ K, indeed, a structural transition occurs with changing the symmetry from tetragonal space group $P4/nmm$ to monoclinic $P112/n$ at lower temperatures (in some cases a transition into orthorhombic phase *Cmma*) has been detected).

It turned out that magnetic phase transition happens at a lower temperature, $T_N = 137$ K. In neutron diffraction patterns, magnetic reflects (103), corresponding to doubling the primitive cell along the c axis, have been found. The main result of the study of LaOFeAs is shown in Fig. 2.16, where points and squares mark the

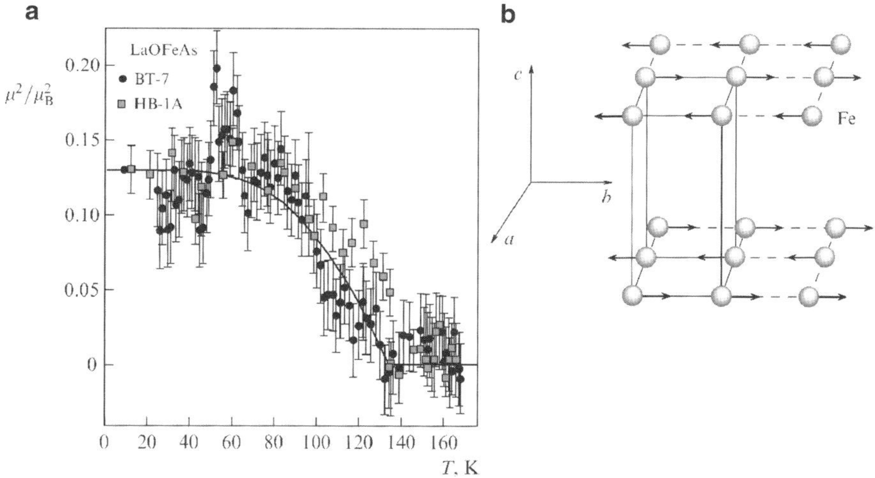

Fig. 2.16 (**a**) Intensity of the magnetic Bragg peak in LaOFeAs according to neutron diffraction data, depending on temperature [59]. Experimental *dots* and *squares* correspond to measurements at different diffractometers. (**b**) Magnetic ordering in the Fe sublattice

temperature dependence of the magnetic reflex, which scales as the square of the magnetic order parameter. The inset at the top shows the magnetic structure formed in the Fe sublattice.

As is shown from the Fig. 2.16, the resulting magnetic structure is an antiferromagnetic alteration of ferromagnetic chains (stripes) in the basal plane. This structure quite agrees with theory predictions following from band structure calculation of this compound [60], even as the compound is in fact metallic. In this sense, the magnetic ordering in FeAs compounds is drastically different from that in cuprates. Cuprate compounds at the stoichiometry exhibit rather conventional antiferromagnetic ordering in basal planes, whereby magnetic moments of Cu atoms are set antiparallel to the moments of their nearest neighbours in the copper sublattice. Stoichiometric cuprates are Mott insulators.

As was earlier pointed out, a doping of cuprates destroys long-range magnetic ordering and results at the onset of superconducting state. The situation is similar in the FeAs systems. Neutron diffraction studies of doped superconducting compound $LaO_{1-x}F_xFeAs$ ($T_c = 26$ K) showed an absence of magnetic order. Therefore, similarly to how it is in cuprates, the superconductivity occurs in the vicinity of magnetic phase transition that indicates an important role of antiferromagnetic fluctuations in electron pairing. Magnetic ordering as shown in Fig. 2.16 has been also found in another stoichiometric compound, namely in NdOFeAs with $T_N = 141$ K [61].

In both these compounds (with *Re* = La, Nd), the magnetic moment per Fe atom, at low temperatures, is anomalously small: 0.36 μ_B for the La-based and 0.25 μ_B for the Nd-based compound, whereas according to theory predictions it is expected to be $\sim 2\,\mu_B$ [62, 63]. This discrepancy marks a thus far persisting problem. Even as it is evident that an occurrence of frustrations (two antiferromagnetic interactions between Fe atoms in the basal plane) might reduce the mean magnetic moment somehow, a reduction by a factor of five is difficult to understand. We note, however, an existence of calculations of electron density done within the LDA+U scheme with negative U value (to be understood as $U_{eff} = U - J$), which result in a substantial reduction of the magnetic moment [12].

Neutron diffraction studies of $CeO_{1-x}F_xFeAs$ [9] did in part confirm an antiferromagnetic structure of other compounds. More precisely, the spins of Fe atoms in the basal plane do build ferromagnetic stripes, alternating antiferromagnetically as shown in Fig. 2.16, and this magnetic structure is repeated or alternate along the c axis. Presumably, it is related to a special role of Ce atoms in compounds, because of the tendency of valence electrons of cerium to easily hybridize with other electrons. It is noteworthy that the Fe magnetic moments in this compound are higher, of about 1 μ_B. A study of a fluorine-doped compound revealed phase diagram rather similar to that of cuprates.

Besides the above cited publication, one should mention the studies of magnetic ordering in the SmOFeAs by magnetic measurements [64]. The stoichiometric compound SmOFeAs shows an anomaly in magnetic susceptibility as $T \approx 140$ K due to an onset of antiferromagnetic order. At $T \approx 6$ K, another peak in $\chi(T)$ is detected, which reveals an antiferromagnetic ordering established in the Sm sublattice. In the doped compound $SmO_{0.85}F_{0.15}FeAs$ the antiferromagnetic ordering in the FeAs planes is suppressed, whereas the ordering in the Sm sublattice persists even as the material becomes a superconductor with $T_c = 52$ K. Therefore, we face a situation when superconductivity and magnetic ordering so coexist, even if they manifest themselves on different electron subsystems, belonging to Fe and Sm atoms, respectively. It is interesting to note that the magnetic susceptibility of the SmOFeAs system has a Curie–Weiss contribution from Fe atoms, from which the magnitude of magnetic moments at the latter can be estimated to be about 1.4 μ_B.

2.2.2 Theoretical Explanation of Long-Range Magnetic Ordering in ReOFeAs

Shortly after clarifying the magnetic structure of the LaOFeAs compound in a neutron diffraction experiment [59], its theoretical explanation was given by Yildirim [65]. This explanation is based on the LDA total energy calculations for possible frustrated magnetically ordered phases of the compound in question, along with an idea that frustrations can be removed by crystal lattice distortions.

An initial suggestion was that the J_1 exchange interaction between the nearest Fe atoms on the square lattice of undoped LaOFeAs and the J_2 interaction between next-nearest neighbours, is antiferromagnetic. This immediately puts forward two schemes of antiferromagnetic ordering, AFM1 and AFM2, shown in Fig. 2.17. An analysis of the exchange energies expressed consistently with the localized Heisenberg model of classical spins does easily result in a conclusion that the AFM1 is stable for $J_1 \gg J_2$ and AFM2, on the contrary, for $J_2 \gg J_1$. This result

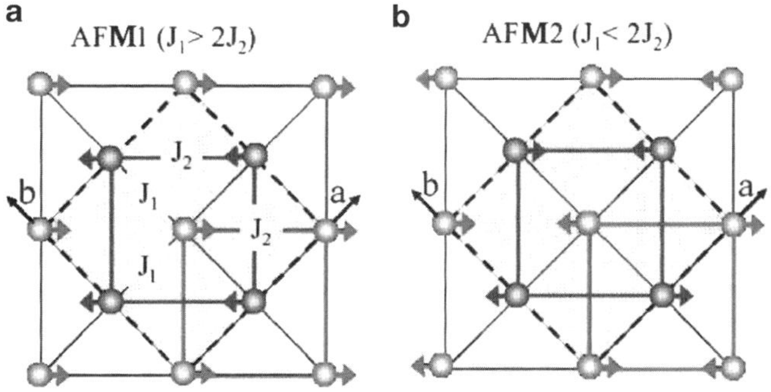

Fig. 2.17 Two possible types of antiferromagnetic ordering under antiferromagnetic exchange interactions J_1, J_2 [65]

is understandable, because in case of large J_1 the highest priority is to satisfy the antiferromagnetic ordering between nearest neighbours, leaving the weaker J_2 interaction frustrated. In the opposite case of large J_2, the AFM2 structure assures antiferromagnetic spin setting between the next-nearest neighbouring Fe atoms, at the expense of leaving the J_1 interactions frustrated.

In both cases, the antiferromagnetic structure of the whole plane can be considered as two interlaced antiferromagnetic sublattices. A further difference between the two structures is the following. In the AFM2 structure, each Fe atom is surrounded by other Fe nearest neighbours, so that its spin finds itself in zero exchange field, hence the mutual orientation of spins in two sublattices is not fixed, and we deal with fully frustrated situation. It is known that in frustrated magnetic systems, a frustration in nearly all cases is lifted either by structure distortions, or by thermic or quantum fluctuations [66, 67]. Therefore if the AFM2 ordering gets materialized, one could expect ordered structure distortions of the tetragonal lattice that was indeed detected experimentally [59]. Such simple yet fruitful considerations was put forward on the basis of the localized model, despite the fact that the latter cannot be, strictly speaking, attributed to the LaOFeAs compound, which is not an insulator but a metal. Therefore, it was essential to probe energies of emerging magnetic structures by calculations done within the itinerant model as well. In [65], total energies of four different phases have been calculated: the nonmagnetic (NM) and ferromagnetic (FM) ones, along with the antiferromagnetic AFM1 and AFM2. The calculations have been done in the LDA. The energies calculated for these four phases in dependence on the fixed mean moment, m, of the spin projection $\langle S^z \rangle$ at the Fe atom onto the quantization axis have shown that the AFM2 structure has indeed the lowest energy, which is achieved at $m = 1$. The calculations in which the mean spin was not fixed revealed the minimum at somehow smaller value of $m = 0.87\,\mu_B$. We note in passing that this value does still exceed the experimental value of $0.36\,\mu_B$ by a factor of nearly two.

A comparison of calculated total energy values with the formulas following from the localized model with two exchange interactions, J_1 and J_2, permits to extract the

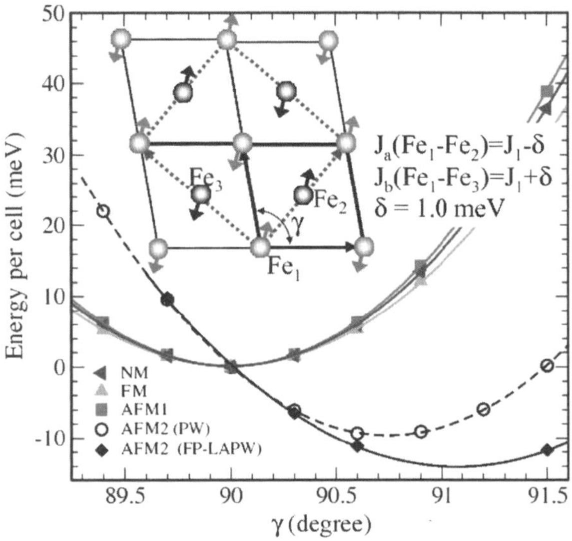

Fig. 2.18 Total energy per elementary unit depending on the angle γ for four phases: NM, FM, AFM1, and AFM2. In the inset, a distorted magnetic structure with the AFM2 magnetic ordering is shown [65]

values of the latter. Both interactions turned out to be antiferromagnetic and of the same order of magnitude, whereas however $J_2 > 1/2\,J_1$, confirming the preference towards the AFM2 type of ordering.

Let us now discuss the effect of lattice distortions onto the magnetic structure. Orthorhombic distortions of the tetragonal lattice, like those found in LaOFeAs simultaneously with the onset of a magnetic ordering, can be realized by varying the angle Γ (Fig. 2.18) in either sense away from 90° [65]. In this case, the distance between Fe_1 and Fe_2 atoms increases, and between Fe_1 and Fe_3 decreases. The exchange integral J_1 gets smaller between atoms situated along the a direction and larger along the b direction. This can be casted into the formula $J_a = J_1 - \delta$, $J_b = J_1 + \delta$, whereby the net exchange energy of crystal is decreasing, exploiting the changes in exchange interaction between spins along the a direction and along the b direction. The energies of all four phases, calculated in the LDA depending on the angle Γ, are shown in Fig. 2.18. It is shown that the AFM2 ordering has the lowest energy for $\Gamma = 91°$. The experimental value is 90.3°, whereby the calculated value of the mean spin $m = 0.48\,\mu_B$ is reasonably close to the experimental one, $0.36\,\mu_B$. The structure distortion reduces the magnitude of the local moment in the AFM2 ordering from 0.87 to $0.48\,\mu_B$. The density of states at the Fermi level in distorted phase gets almost doubled. Therefore, a competition of two antiferromagnetic interactions J_1 and J_2 in pristine LaOFeAs compound is resolved by an orthorhombic lattice deformation, which stabilizes the AFM2 magnetic structure. This picture yields the lattice distortion parameters and mean magnetic moment values, which are close to the experimental data.

A problem of anomalously small values of mean magnetic moments on Fe atoms was discussed in [68, 69]. In [68], the methods of Mössbauer spectroscopy and muon spin rotation (μSR) were applied to minutiously measure the temperature dependence of the sublattice magnetization in undoped LaOFeAs and the SDW order parameter. From these two measurements, a mean magnetic moment per Fe site has been extracted to be 0.25 μ_B. Calculations which used the four-band model of an antiferromagnet in the mean-field approximation resulted in a higher value, 0.33 μ_B, – a qualitatively expected trend, due to the fact that the mean-field theory neglects the fluctuations. In [69], calculations of the electronic structure of LaOFeAs and of the magnetic moments on Fe atoms were done under assumption of different AFM structures, within both LDA and GGA. Specifically, three structures have been inspected, AFM-1, AFM-2 and AFM-3, different in the orientation of magnetic moments of adjacent Fe atoms within the basal plane and between neighbouring planes. One of the structures, AFM-3 with the magnetic cell doubled along the c axis, was that deduced from the neutron diffraction studies. The calculations of electronic structure and mean moments on Fe atoms for each of the three AFM structures have been done for different volumes of the primitive cell, i.e. for different ambient pressures.

The results were unexpected. At fixed cell volume, the mean magnetic moments turned out to be very sensitive to the type of AFM structure. In the vicinity of the equilibrium volume, the magnetic moment decreased strongly, especially in the AFM-1 structure. For the really existing AFM-3 structure, the magnitude of the mean magnetic moment changed by a factor of three within the pressures range of ± 5 GPa. At the pressure of 5 GPa, the calculated moments agree with the experimentally measured ones. Through the pressure range of -10 to $+10$ GPa, the Fe magnetic moment falls down from the maximal possible one (2 μ_B) to nearly zero. Therefore, an increase of negative pressure (i.e. lattice dilatation) is expected to induce a large rise of the magnetic moment. This effect could have been checked on hydrogenated samples, where the hydrogen absorption permits to imitate a negative pressure.

It is remarkable that the electronic structure does not substantially change within broad range of pressures, whereas the magnetic moment varies considerably. This indicates that the magnetic state of the compound (its magnetic structure and the magnitude of mean magnetic moment) results from a delicate equilibrium between the kinetic energy, which determines the electronic structure within the LDA, and inter-electron interaction. Therefore, a description of FeAs-type compounds within a purely localized spins model, like the Heisenberg one, might be quite problematic: these compounds are with certainty rather itinerant magnets.

The main problem concerning the nature of magnetic properties of the FeAs-compounds is explaining the smallness of the mean magnetic moment at Fe atoms in the magnetically ordered SDW state, and finding out the structure of magnetic excitations spectrum in the doped compounds.

In relation with the question of magnetic moments, we draw attention to [70] in which, within the self-consistent spin waves theory, an attempt has been done to calculate the mean moment $\langle S^z \rangle$ at the Fe atoms, proceeding from the fully localized Heisenberg model. Hereby, four different exchange interactions have been considered. Two of them are between nearest neighbours, J_{1a} along the chain of identically magnetized atoms and J_{1b} – between those nearest neighbours, which belong to the chains of opposite spin direction. Further on, J_2 is between next-nearest neighbours in the ab plane, and J_c between the nearest neighbours along the c direction.

In [70], via the Green's function of spin waves, self-consistent equations for $\langle S^z \rangle$ have been obtained. From these equations, in the limit $\langle S^z \rangle = 0$, a formula for the Neél temperature T_N in terms of the four exchange interaction parameters is derived. The authors of [70] proceeded from the experimental value $T_N = 138$ K for the LaOFeAs compound and, making use of the self-consistent equation for $\langle S^z \rangle$, estimated the most probable magnitudes of exchange interactions: $J_{1b} = 50 \pm 10$ meV, $J_{1a} = 49 \pm 10$ meV, $J_2 = 26 \pm 10$ meV, $J_c = 0.020 \pm 0.01$ meV. A solution of the equation for $\langle S^z \rangle$ for these parameter values yielded the temperature dependence of $\langle S^z \rangle$ throughout the whole temperature interval, up to T_N. For $T = 0$, it turned out that $\langle S^z \rangle = 0.7$. In this calculation, the value of atomic spin $S = 1$ at the Fe atom has been taken, which most closely corresponds to the value of the magnetic moment calculated from first principles.

Therefore, the spin fluctuations do decrease, for chosen values of exchange parameters, the magnitude of $\langle S^z \rangle$ as in the SDW ground state from $\langle S^z \rangle = 1$ to $\langle S^z \rangle = 0.7$. This spin contraction is much weaker than that in fact observed in LaOFeAs and other compounds of this type. An additional reduction of the mean moment at Fe atoms is, most probably, a consequence of a more itinerant character of magnetism in these systems.

In what concerns the magnetism of rare-earth ions in *Re*OFeAs systems, a question remains open: why, throughout the series of superconducting $Re\text{O}_{1-x}\text{F}_x\text{FeAs}$ compounds with localized moments (*Re* = Ce, Pr, Nd, Sm), the T_c values stay higher than in the La-based compound? Systematic studies of rare-earth magnetism in these systems were performed in [71] using μSR experiments and a symmetry analysis. Different combinations of magnetic orderings in rare-earth and Fe sublattices have been taken into consideration. A strong influence of the magnetic ordering in the *Re*-sublattice onto the magnetism of the Fe-sublattice has been demonstrated, for the case of different symmetries of the magnetic order parameter over these sublattices. The symmetry analysis revealed that in *Re*OFeAs systems, there is no Heisenberg exchange between the spins of *Re* and Fe ions. The authors conclude that the magnetic *Re*–Fe interaction cannot be a substantial reason of the observed enhancement of T_c in the $Re\text{O}_{1-x}\text{F}_x\text{FeAs}$ compounds with magnetic *Re* ions.

2.2.3 Phase Diagrams

On doping of stoichiometric compounds possessing an SDW-type magnetic structure, their magnetic ordering temperature T_N decreases gradually, and a superconducting state sets on. The magnetic ordering is accompanied by structural distortion of the pristine (ideal) tetragonal phase, which turns into an orthorhombic one. It is of utter interest to trace the boundaries of all three phases – magnetic, structural and superconducting ones – in the (T, x) plane. In a rough approximation, the magnetic and the orthorhombic phases are coinciding on this plane, but it is difficult to explore

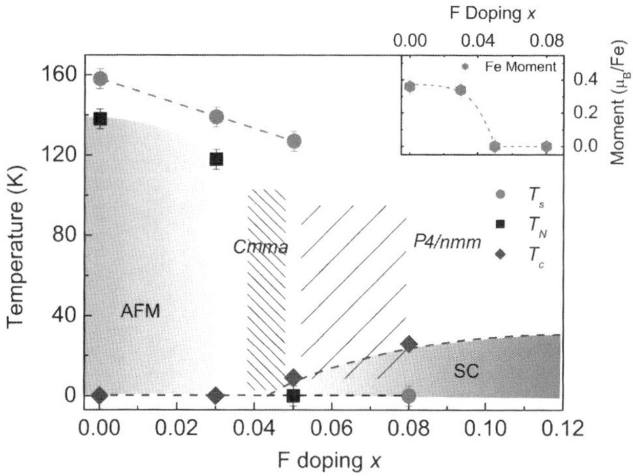

Fig. 2.19 Structural and magnetic phase diagram of $LaO_{1-x}F_xFeAs$, as determined from neutron experiments on the samples with $x = 0$; 0.03; 0.05; 0.08 [72]. In the inset, the x dependence of the mean magnetic moment per Fe atom, measured at 4 K, is shown

their borders in detail. Local methods, such as μSR or Mössbauer spectroscopy, provide information about long-range magnetic order, but they are not sensitive to structure distortions. On the other side, X-ray structure analysis accounts for the latter but neglects the magnetic ordering. There is only the neutron diffraction which allows to probe the both factors at the same time.

In [72], a detailed phase diagram of $LaO_{1-x}F_xFeAs$ has been obtained, based on the neutron diffraction analysis in combination of the traditional methods to study the superconducting state (Fig. 2.19). This publication extends the data obtained in [59, 73, 74] by neutron scattering, X-ray structure analysis, transport properties and μSR.

It is seen from Fig. 2.19 that magnetic and structural phase transitions occur in a narrow region of concentration x (shaded) around $x = 0.04$. A careful analysis shows that the superconductivity starts at x values still within the orthorhombic phase, in which, however, the long-range magnetic order does not exist anymore. The larger part of the concentration domain of existence of superconductivity does, however, fall onto the tetragonal phase. This observation is typical for many *Re*OFeAs compounds, whereas the fine details of the (T, x) phase diagram differ over systems. For example, Fig. 1.1 shows the phase diagram of $CeO_{1-x}F_xFeAs$ [12]. Different from the $LaO_{1-x}F_xFeAs$ system, the disappearance of magnetic ordering is gradual, and its existence region, along with that of the orthorhombic phase, does not overlap with the region of superconductivity. The phase diagram in question resembles those of cuprates.

A more delicate situation takes place in $SmO_{1-x}F_xFeAs$ [75–77]. In an early work [75], the phase diagram shown in Fig. 2.20 has been obtained. The temperature of SDW-type ordering does rapidly fall down with doping. A superconducting state

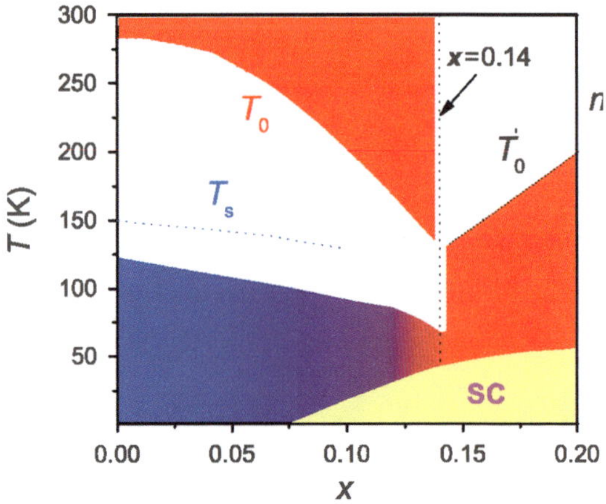

Fig. 2.20 Phase diagram of the $SmO_{1-x}F_xFeAs$ system [75]

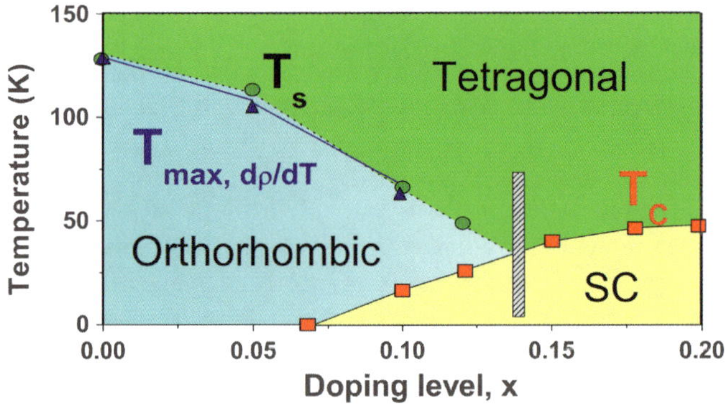

Fig. 2.21 Refined phase diagram of the $SmO_{1-x}F_xFeAs$ system [77]

appears at $x \sim 0.10$, passes through maximum of the transition temperature $T_c = 54$ K at $x = 0.20$ and further on does not change much. There exists a concentration region $0.05 < x < 0.15$ in which superconductivity and the SDW-type ordering do coincide. A more detailed study of the border of the region where both order parameters coincide needs yet to be better studied near the crossing of the lines $T_s(x)$ (dashed line in Fig. 2.20) and $T_c(x)$.

A possible coexistence of magnetic and superconducting order parameters in $SmO_{1-x}F_xFeAs$ has been addressed in [78]. The arguments in its favour come from μSR measurements for $x = 0.05$–1.0, in which region the $T_s(x)$ and $T_c(x)$ curves do overlap. This coexistence occurs at nanoscale, e.g. the domains of magnetic and superconducting phases do mutually disperse one into the other, having size of about 2 nm. This is exactly a typical coherence length of ξ in this compound. Such small-grain disperse coexistence does probably reveal a competition between magnetism and superconductivity in this compound.

In Fig. 2.21, a refined phase diagram of this compound is shown, from which it is clearly seen that superconductivity starts to grow inside the orthorhombic phase.

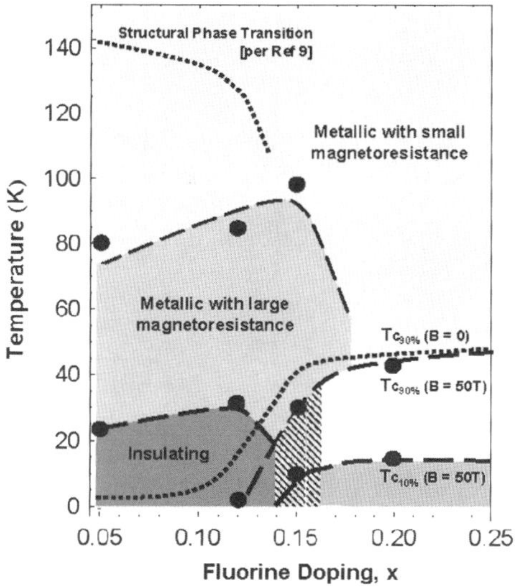

Fig. 2.22 Phase diagram of the $SmO_{1-x}F_xFeAs$ system in magnetic field of 50 T [76]

A shaded region around $x = 0.14$ marks the boundary between two different types of behaviour in the temperature dependence of electrical conductivity [77].

A further level of phase diagram detailing has been achieved in [76]. Figure 2.22 resumes the data on magnetoresistance; a region with large magnetoresistive effect is marked as shaded. At low levels of doping, a metallic phase with high magnetresistance persists, which, however, is replaced at low temperatures by a "dielectric" phase. The quotes mean that the electrical resistivity in this region varies with temperature as $\ln T$. The narrow shaded region at $0 < x < 0.14$ corresponds to the "dielectric" to metal transition. In the same region, a quite abrupt transition from orthorhombic into tetragonal phase takes place, in agreement with Fig. 2.21.

2.2.4 *Magnetic Fluctuations*

Fluctuations of the magnetic order parameter have effect onto the behaviour of magnetic susceptibility. We will consider here only the statical susceptibility $\chi(T)$ i its dependence on temperature. In Fig. 2.23, $\chi(T)$ over broad temperature range is shown for stoichiometric LaOFeAs [79]. Beyond the magnetic phase transition, $\chi(T)$ grows linearly, and notably this dependence is typical for FeAs-based systems. In the inset, the temperature dependence of the magnetic contribution in heat capacity, defined as $\partial(\chi_T)/\partial T$. As is seen, this latter property shows anomalies near the temperatures of magnetic (T_N) and structural (T_s) transitions. Particularly strongly do magnetic fluctuations reveal themselves near structure phase transition at $T_s = 156$ K (a sharp maximum in the $\partial(\chi_T)/\partial T$ curve).

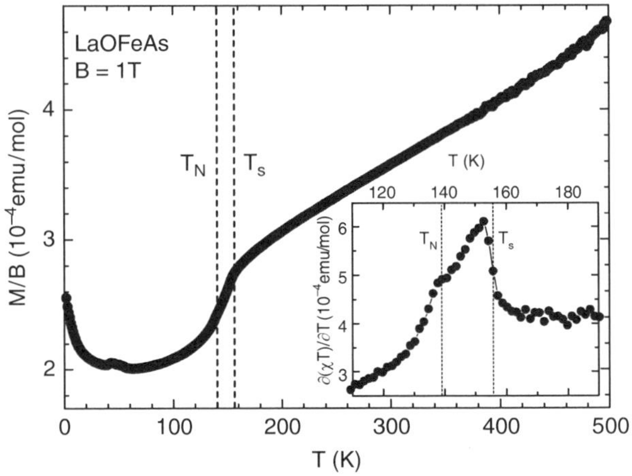

Fig. 2.23 Temperature dependence of magnetic susceptibility for LaOFeAs. In the inset: temperature dependence of the property $\partial(\chi_T)/\partial T$ in the vicinity of magnetic and structural phase transition [79]

In doped compounds, $\chi(T)$ does not change much up to concentrations $x < 0.04$, when superconductivity sets on, and for $T < T_c$ the magnetic susceptibility $\chi(T)$ exhibits a typically diamagnetic behaviour. Within the region of existence of superconductivity, up to $x < 0.125$, the slope of $\chi(T)$ in the normal (metallic) phase is only weakly x-dependent, i.e. $\chi(T)$ only slightly increases with temperature.

2.3 Electronic Structure

2.3.1 Stoichiometric Compounds

First-principles electronic structure calculation of the LaOFeP compound, in which superconductivity ($T_c = 4$ K) had been first found, was done prior to the discovery of high T_c values in this class of compounds [80]. Calculations for LaOFeAs have been done in fact independently by several groups [17, 18, 62, 63, 81–85]. We report here the results by Singh and Du [81], obtained by the augmented plane waves method (LAPW) within the local density approximation (LDA). In the calculation, the experimental values of the lattice parameters for LaOFeAs have been used: $a = 4.03552$ Å, $c = 8.7393$ Å. Internal coordinates of La and As atoms in the plane formed by Fe atoms were determined by minimization of total energy calculated in the LDA, yielding $z_{La} = 0.1418$, $z_{As} = 0.6926$. This resulted in the following inter-atomic distances: Fe–Fe 2.854 Å, As–As 3.077 Å, Fe–As 2.327 Å.

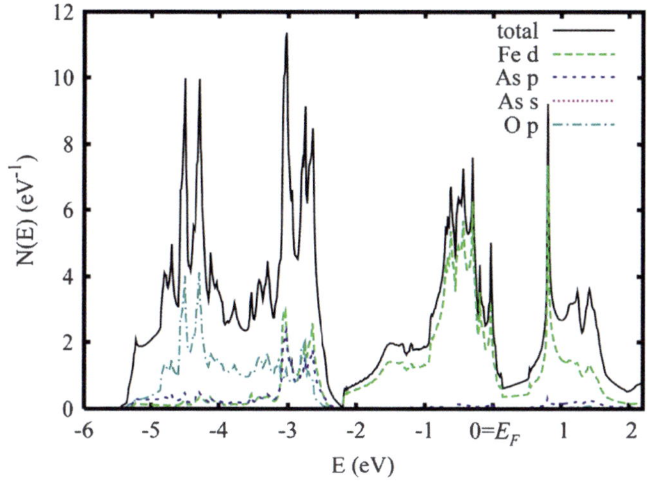

Fig. 2.24 Calculated density of states in LaOFeAs (*solid line*) and partial contributions to it from the orbitals of Fe, As and O (*dashed lines*) [81]

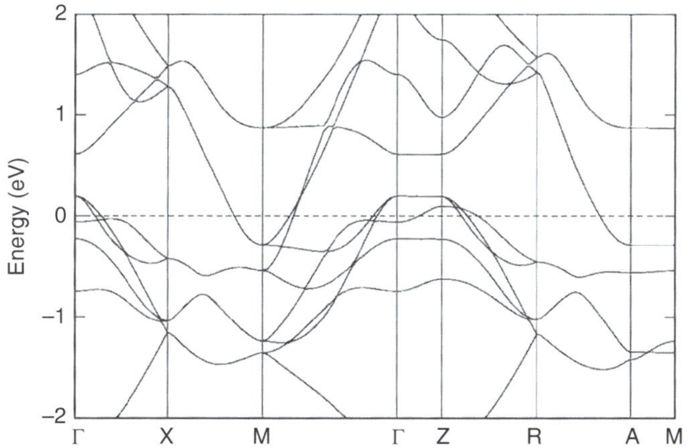

Fig. 2.25 Band structure of LaOFeP [80]

The calculated density of states (DOS) is shown in Fig. 2.24, where, along with the total one over unit cell, the partial contributions from Fed states, Op and Asp states are shown.

The Fermi energy cuts the region of Fed states which occupy the range from roughly -2 to $+2$ eV (assigning zero energy to the Fermi level). A broad region below the d states is formed by p states of As and O. The states of the La atoms fall into the range above the Fermi energy.

The band structure is shown in Fig. 2.25 taken from an earlier work by Lebeque [80] for the sole reason that the corresponding figure in [81] is complicated by

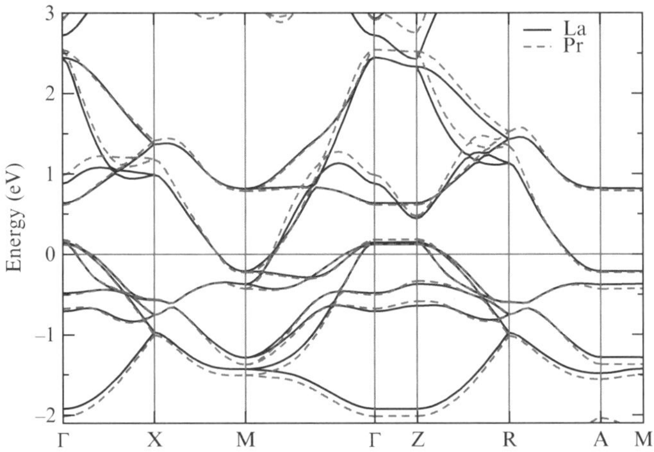

Fig. 2.26 Band structure of LaOFeAs (*solid line*) and PrOFeAs (*dashed line*), after [82]

additional data related to the variation of As coordinates in the lattice, whereas otherwise the results of [80] for LaOFeP are identical to those for LaOFeAs.

For a comparison, we show in Fig. 2.26 band structures calculated for two compounds, LaOFeAs and PrOFeAs, whose crystal structures are identical and lattice parameters are close to those of LaOFeP [82]. We see that the dispersion curves for the three compounds shown in Figs. 2.25 and 2.26 are qualitative identical, and quantitative differences tiny. A similarity between calculation results, even zoomed in a fine energy scale around the Fermi level, is maintained throughout the whole class of ReOFeAs-like compounds, so that Figs. 2.24 and 2.25 do faithfully represent all compounds of this type. In [82], a calculation through the series with Re = La, Ce, Sm, Nd, Pr, Y has been done and shown that there is practically no difference in the total DOS as well as in the details of the partial Fe3d states distribution.

The details of the dispersion of the bands crossing the Fermi level do determine the Fermi surface, which is multi-sheet in LaOFeAs compounds. The Fermi level crosses two hole bands centred at Γ and two electron bands emerging from M. Noteworthy is a flat character of curves along the $\Gamma{-}Z$ direction, i.e. a weak dependence of hole quasiparticles' energy on the k_z momentum, so that the Fermi surface around Γ has cylindrical shape. The same observation applies to the sheets of the Fermi surface in the vicinity of M (as follows from the flat dispersion of the electron bands along the $M{-}A$ line).

Therefore, the Fermi surface of the LaOFeP compound contains two hole cylindrical sheets with the axis along $\Gamma{-}Z$ and two electron ones along $M{-}A$. This reveals a quasi-two-dimensional character of electronic states, formed by d_{xz} and d_{xy} orbitals. Beyond the said four cylindrical sheets, a three-dimensional hole pocket is present, centred at Z (see Fig. 2.27) and formed by the Fed states, hybridized with p states of As and La.

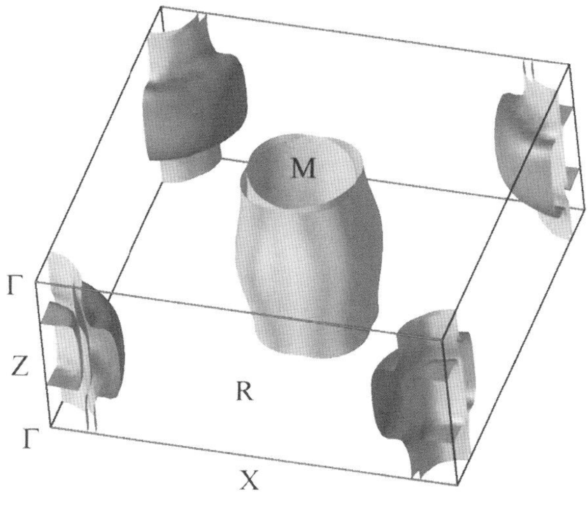

Fig. 2.27 Fermi surface of LaOFeAs [81]. Symmetry points of the Brillouin zone: Γ (0,0,0), *Z* (0,0,1/2), *X* (1/2, 0,0), *R* (1/2.0,1/2), *M* (1/2,1/2,0) in the units of $2\pi/a$

Three hole sheets, taken together, make 80% of the state density $N(E_F)$ at the Fermi level. The average electron velocity on them, taken in the xy plane, is $0.81 \cdot 10^7$ cm s^{-1}. The corresponding numbers for two electron sheets are $2.39 \cdot 10^7$ and $0.37 \cdot 10^7$ cm s^{-1}. From here, a high anisotropy of conductivity (of about 15) follows, which again underlines a quasi-two-dimensional character of electron states in this material.

The summary volume confined within two electron cylinders (and equal to the volume within the hole cylinders) is 0.13 electrons per formula unit. The value of the density of states (per formula unit and both spin components) is $N(E_F) = 2.02\,(\text{eV})^{-1}$. Hence, LaOFeAs is a conductor with low carrier concentration and relatively high density of states on the Fermi level, quite differently from the situation in cuprates.

Numerous calculations of electronic structure for different LaOFeAs-type compounds give quite similar results: flat regions in the dispersion curves around the Fermi level along Γ–Z and M–A, which yield hole and electron cylindrical sheets of the Fermi surface and reveal quasi-two-dimensional character of electronic states in the FeAs layers.

Let us now turn more attentively to the structure of Fe3d states, which play a major role in the formation of electronic properties of FeAs-type compounds, since, namely, these states are pronounced at the Fermi level. In a free Fe ion, the fivefold degenerate 3d term includes five orbitals – d_{xz}, d_{yz}, d_{xy}, $d_{x^2-y^2}$ and $d_{3z^2-r^2}$, – whose wave functions are shown in Fig. 2.28. The d_{xz} and d_{yz} orbitals each have four lobes, positioned in the xz and yz planes, correspondingly. The d_{xy} and $d_{x^2-y^2}$ orbitals are in the xy plane, with the difference that the lobes of $d_{x^2-y^2}$ are directed along the x and y axes, whereas in the d_{xy} – along the diagonals of the quadrants,

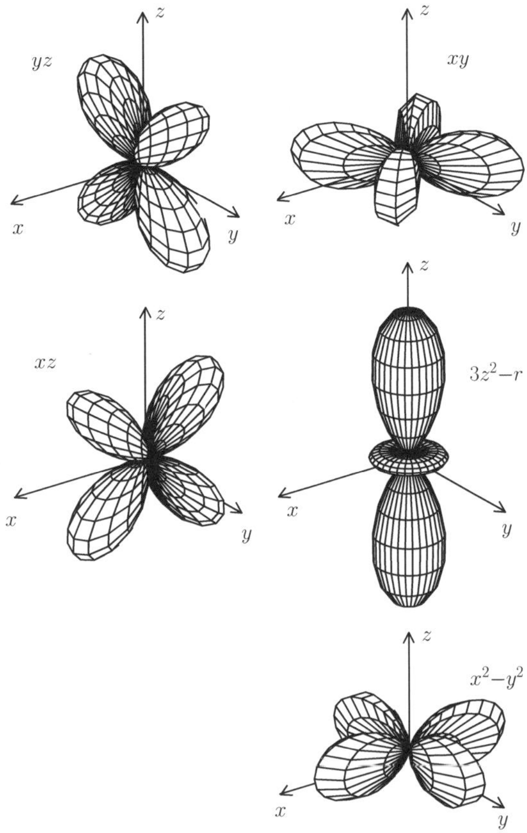

Fig. 2.28 Graphical representation of five degenerate orbulals for $3d$ electrons of the Fe ion

i.e. rotated by 45°. Finally, the $d_{3z^2-r^2}$ orbital has one lobe directed along the z axis, complemented by a z-axial symmetric structure in the xy plane.

In cubic-symmetry crystal field, the fivefold degenerate $3d$ level splits into the e_g doublet and t_{2g} triplet, which correspondingly include the following orbitals:

$$e_g : \left(d_{3z^2-r^2},\, d_{x^2-y^2}\right); \quad t_{2g} : \left(d_{xz},\, d_{yz},\, d_{xy}\right). \tag{2.2}$$

In the structure of LaOFeAs, each Fe ion is surrounded by four As atoms, which form a distorted tetrahedron. In tetrahedral crystal field, the t_{2g} level is situated lower in energy than the e_g one.

For a more convenient discussion on the LaOFeAs crystal structure, it is more convenient to rotate the coordinate system by 45° around the z axis, which results in interchanging the $d_{x^2-y^2}$ and d_{xy} orbitals, so that the attribution into e_g and t_{2g} groups becomes as follows:

$$e_g : \left(d_{3z^2-r^2},\, d_{xy}\right); \quad t_{2g} : \left(d_{xz},\, d_{yz},\, d_{x^2-y^2}\right). \tag{2.3}$$

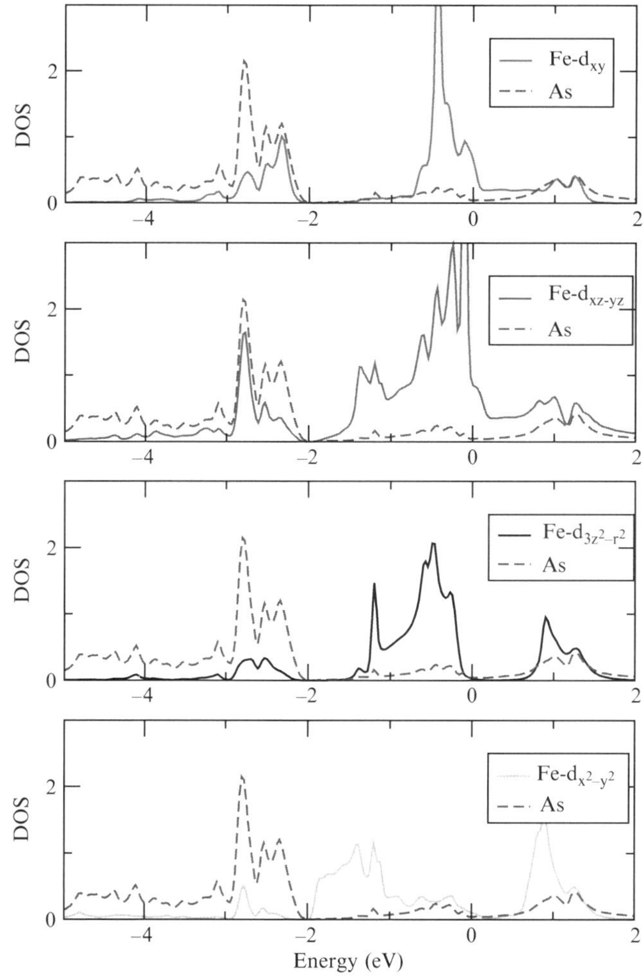

Fig. 2.29 Partial densities of states of the Fe-d orbitals (*solid lines*) and As-p orbitals (*dashed lines*), after [85]

This is the setting traditionally used in the description of electronic properties of the FeAs-type compounds.

Partial contributions of different orbitals into the total Fe3d density of states were calculated in [85] for LaOFeAs, see Fig. 2.29. It follows from Fig. 2.29 that the d_{xz}, d_{yz} and d_{xy} orbitals have a considerable weight in the energy range from −3 to −2 eV, where the contributions of the Asp orbitals are concentrated. An overlap of the said Fed orbitals with the Asp orbitals results in their large hybridization, which ensures an efficient hopping of d electrons over Fe sublattice to next-nearest neighbours, via intermediary As ions.

2.3.2 The Role of Magnetic Ordering and Doping

The results shown in Figs. 2.24–2.29 refer to paramagnetic stoichiometric compounds. At low temperatures, the SDW-type ordering takes place which influences on the electronic structure, – dispersion curves and density of states. The other relevant problem is the calculation of the electron spectrum of doped systems. It can be a priori expected that the electron doping is about to enlarge the electron pockets of the Fermi surface, and the hole doping – the hole pockets. However, to figure out how precisely does this happen, calculations are needed for compounds with a given (controllable) level of doping.

Doped systems, e.g. $LaO_{1-x}F_xFeAs$, are disordered in what regards the distribution of F ions over the LaO sublattice. A modelling of such systems poses a theoretical problem, which demands to adopt some approximative treatment. In practice, two approaches are largely accepted: the virtual crystal approximation and the supercell approach, i.e. a substitution of a disordered crystal, at some special level of doping, by an ordered one, immediately including dopant at some sites. For example, the situation $x = 0.125 = 1/8$ can be imitated by an $8\times$ enlarged supercell, in which one of eight oxygen atoms is replaced by fluorine. The translation symmetry thus restored, a calculation can be done by a standard band-structure technique, in an assumption that its results would faithfully enough reproduce the expected behaviour of the disordered structure for the x given. The above value of $x = 0.125$ is indeed very common for many doped LaOFeAs systems, since it is close to this level of doping that the superconducting state sets on.

Let us discuss the first works in which the both problems – the SDW-type ordering and doping – are addressed simultaneously [86]. Figure 2.30 shows the band structure of stoichiometric LaOFeAs along with that doped with fluorine ($x = 0.05$), for non-magnetic case. The curves 1, 2 and 3 correspond to the hole-type spectrum of quasiparticle states, and the curves 4 and 5 to the electron-type one. It can be seen from the figure that the fluorine doping decreases the hole pockets and

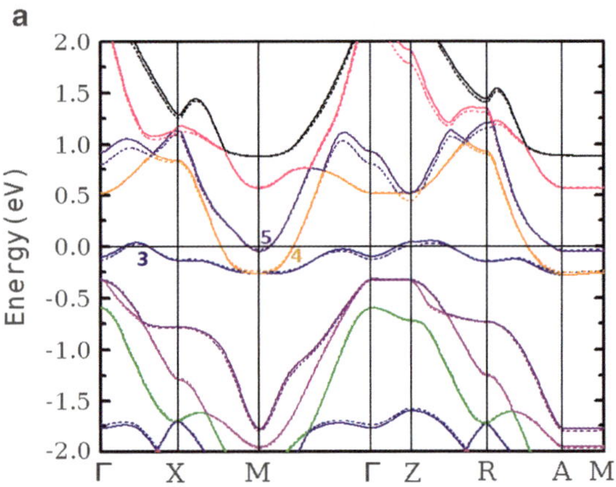

Fig. 2.30 Calculated band structure of the nonmagnetic compound $LaO_{1-x}F_xFeAs$ for stoichiometric composition (*solid line*) and under fluorine doping (*dotted line*) ($x = 0.05$) [86]

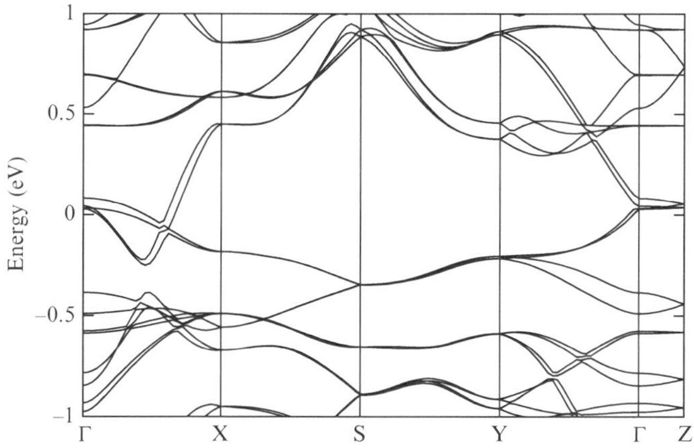

Fig. 2.31 Band structure of LaOFeAs in its magnetically ordered phase with the wave vector $\mathbf{q}_M$, taking the orthorhombic distortion into account. Symmetric points of the Brillouin zone: Γ (0,0,0), S (1/2,1/2,0), Y (0,1/2,0) and Z (0,0,1/2) [88]

increases the electron ones. Therefore, the fluorine doping does, indeed, correspond to electron doping that is confirmed by the sign of the Hall coefficient. It is reported in [86] that an increase of x leads to a further extension of the electron pockets.

The band structure of magnetically ordered phase has been addressed in [84, 86–88]. It was shown that both the antiferromagnetic (AFM) and the SDW phases are energetically preferable over the nonmagnetic one. In the last of these works [88], the true SDW-type magnetic ordering found in the FeAs-type compounds has been considered, namely a stripe magnetic structure with antiferromagnetic alteration of chains in the basal plane and along the z axis, described by the wave vector $\mathbf{q}_M$.

In Fig. 2.31, calculated energy dispersion curves along symmetry directions of the orthorhombic Brillouin zone are shown after [88]. The resulting picture is very close to that obtained for the SDW-type structure with ferromagnetic ordering of chains, instead of their alternation, along the z axis, so that the third component of the wave vector $\mathbf{q}_M$ be zero. This similarity is explained by the smallness of exchange interaction between adjacent FeAs layers, due to a quite large separation of the latter.

As can be seen from a comparison of Fig. 2.31 with Fig. 2.25 for non-magnetic case, the magnetic ordering does substantially alter the band spectrum. A broad gap between the valence and the conduction bands appears over almost the whole Brillouin zone, and is also revealed as a pseudogap in the density of states. The Fermi level is crossed by four dispersive bands close to Γ, at some points along the Γ–X line. The hole sheets have the shape of narrow cylinders due to the presence of a flat region in the dispersion curves along the Γ–Z. The electron sheets are more deformed along the c axis.

The density of states for magnetically ordered crystal is shown in Fig. 2.32, where along with the results for the stoichiometric compound, the calculations for two

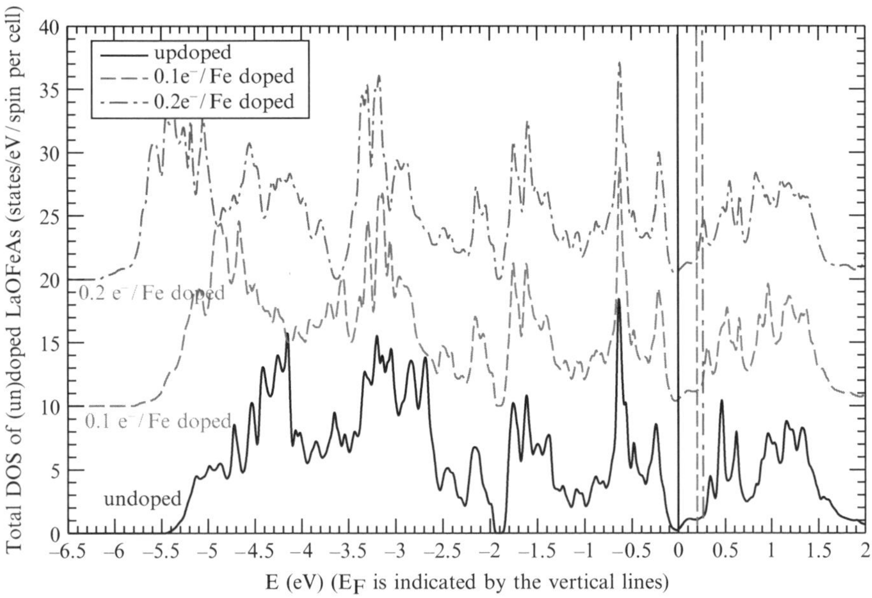

Fig. 2.32 Density of states in magnetically ordered LaOFeAs compound, nondoped and electron-doped to 0.1 and 0.2 electrons per Fe atom. For convenience, the *curves* are shifted one with respect to the other by 10 units along the ordinate axis [88]

electron-doped ones are presented [88]. It is seen that the Fermi level falls into the pseudogap; however, in the range where the density of states is dominated by Fe*d*, it is hardly affected by doping. This is a consequence of the virtual crystal approximation used, i.e. in fact, the rigid-band approximation. In reality, as is known from experiment, at such level of doping as $x = 0.2$, the SDW-type ordering does not survive. To grasp this effect, it is necessary to go beyond the LDA and take electron correlations into account, as will be described in Chap. 4.

An effect of external pressure on the electronic structure is shown in Fig. 2.33. Along with the changes in the energy dispersion curves, the density of states is affected. Specifically, the first peak above E_F shift towards the Fermi level, but the peak below E_F, in the interval 0.1 eV downwards, rests practically unchanged. Consequently, the pressure must induce changes in superconducting properties of electron-doped materials, whereas the expected effect onto the properties of hole-doped ones should be less important.

The results of the LDA calculations are shown in Fig. 2.33 used relative positions of As atoms (i.e. z_{As} parameters) as obtained by total energy minimization (relaxation procedure). The importance of this procedure is illustrated by Fig. 2.34 and Table 2.2 [88].

At all values of z_{As}, the total energy of the $\mathbf{q}_M$ (SDW)-phase is the lowest one in relation to the ferromagnetic and antiferromagnetic phases. The magnitude of the magnetic moment in the SDW phase is substantially lower at the optimized value

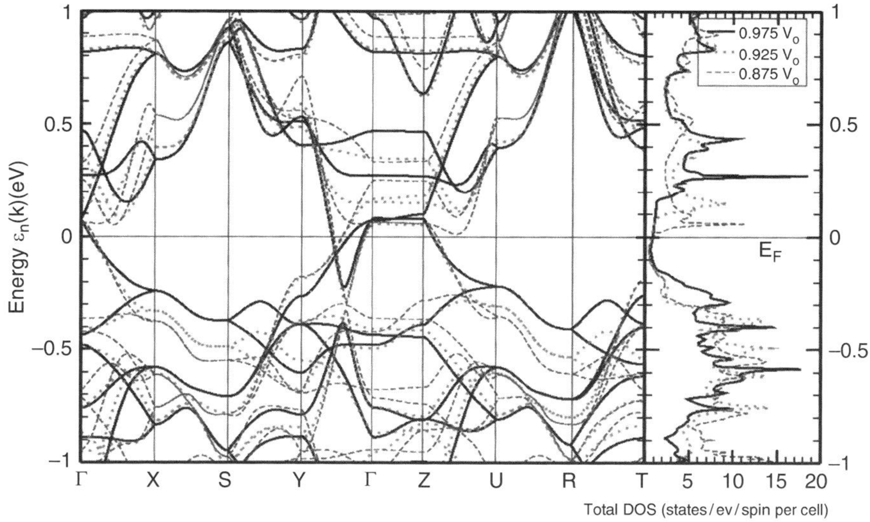

Fig. 2.33 Band structure of magnetically ordered LaOFeAs compound, calculated at different volumes: 0.975 V_0, 0.925 V_0 and an applied pressure. In the *right panel*, the density of states at different volumes is shown [88]

Table 2.2 Calculated electronic structure of compounds. The sheets of the Fermi surface are indicated: $h(\Gamma)$ and $h(Z)$ – the hole ones, centred at Γ and Z points, respectively; $e(M)$ – electron ones, centred near the M point of the Brillouin zone

Compound	Fermi surface	References
LaOFeAs	$2h(\Gamma), h(Z), 2e(M)$	[17, 18, 37, 62, 81, 84, 85]
LaOFeAs	$h(Z)$ absent, *APM*	[63]
LaOFeP	$2h(\Gamma), h(Z), 2e(M)$	[80]
LaOFeP	*ARPES*:$h(\Gamma), e(M)$	[92]
NdOFeAs	*ARPES*:$h(\Gamma), e(M)$	[93]
LaONiP	$h(X), 3e(M)$	[93]
LaONiAs	$h(X), 2e(M)$	[37]
*Re*OFeAs *Re* = La,Ce,Sm,Nd,Pr,Y)		[82]

of z_{As} than in experiment. The differences of the values listed in the last column for spins ↑ and ↓ do exactly yield the magnitude of the magnetic moment shown in the first column.

The comparison of the second and the third lines in the Table 2.2 reveals that the numerical results are quite sensitive to the calculation scheme adopted. It is moreover noteworthy that in the above cited work [88] further factors affecting the calculation results are cited. Moreover, this work covers the results related to a substitution of As by other elements (P, Sb, N), as well as La substitution by other rare-earth elements.

A comparison of results obtained by different groups and with different calculation techniques does often reveal some disagreements – minute ones in what

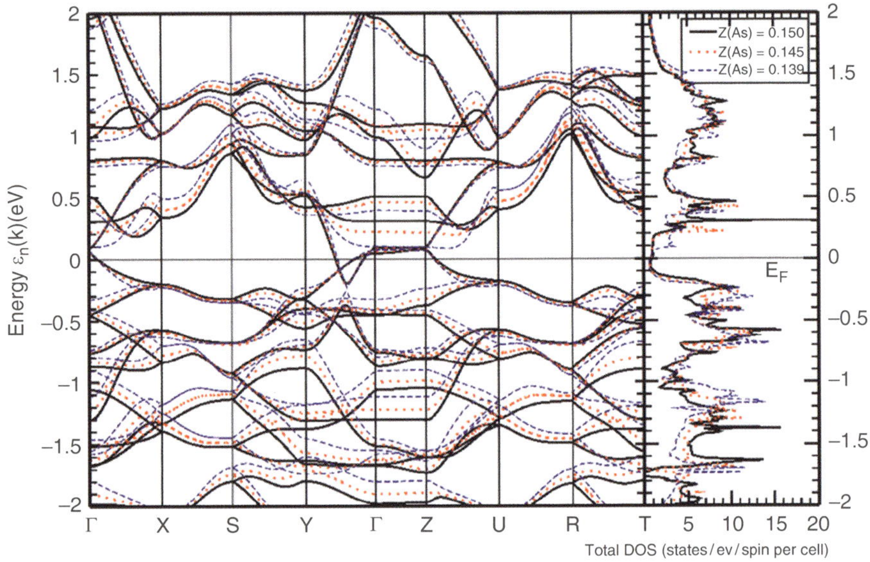

Fig. 2.34 Band structure of magnetically ordered LaOFeAs, calculated for different values of the z_{As} parameter: 0.150, 0.145, and 0.139 [88]

regards the band structure and the Fermi surface, yet sometimes contradictory in relation to magnetic properties and, in particular, the magnetic ground state of the FeAs-type systems. Since the discovery of superconductivity, calculations of the pristine compound LaOFeAs have been reported, which predict both ferromagnetic and antiferromagnetic structures, moreover the latter one of both chessboard-like and stripes-like types, as the ground state. The smallness of the energy differences between the structures is due to itinerant character of magnetism in the FeAs-type systems, extremely sensitive to the details of their electron and crystalline structure and, in particular, to the precise positioning of the As atoms, given by the z_{As} parameter. In [90], four different methods (two all-electron ones and two pseudopotential ones) have been used to calculate all electronic properties of the LaOFeAs compound, and the conclusions drawn about the abilities and accuracy of each method. Thus, for the analysis of magnetic properties, e.g. the calculation of magnetic phase diagrams, an LDA calculation with theoretically determined z_{As} parameters could have been recommended, how it has been used, e.g. in [91].

2.3.3 Experimental Studies of the Fermi Surface

There are two methods of experimental determination of the Fermi surface in metals, one based on the de Haas–Van Alphen (dHvA) effect, the other on the angle-resolved photoelectron spectroscopy (ARPES). In the first case, an information

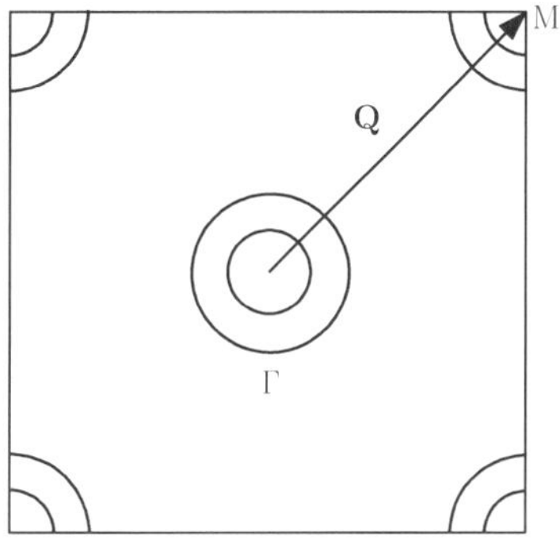

Fig. 2.35 Schematic representation of the Fermi surface in the *Re*OFeAs compounds

about cross-sections of the Fermi surface is extracted from detected fluctuations of magnetization as function of the magnetic field. This method is very accurate but it does not relate the measured cross-section to its actual placement in the **k**-space of the Brillouin zone. ARPES has inferior accuracy but permits a direct recovery of the Fermi surface in the **k**-space. Hence, both methods have advantages and shortcomings and are, in practice, complementary.

LDA calculations of ReOFeAs compounds and other isomorph compounds, e.g. LaOFeP, exhibit a common structure of the Fermi surface. It consists of two cylindrical hole pockets, which include the Brillouin zone centre, two electron pockets centred at the Brillouin zone corners and, moreover, a three-dimensional hole pocket around Γ. A $k_z = 0$ section of such multi-sheet Fermi surface is shown in Fig. 2.35.

The **Q** vector connecting the Γ and M points is close to the nesting vector which connects congruent points of the hole and electron pockets of the Fermi surface, since the sizes of these pockets, according to LDA calculations, are almost identical. The nesting determines the details of magnetic susceptibility of these compounds, in particular, the SDW magnetic structure, as will be shown in Chap. 4.

An experimental verification of these conclusions has been done with the use of ARPES [92, 93]. Thus, in [93] the ARPES spectra of the $NdO_{1-x}F_xFeAs$ single crystal have been measured, which indicated the pockets of the Fermi surface around Γ and M points of the Brillouin zone (Fig. 2.36), in agreement with first-principles calculations.

In [92], the measured PES of the LaOFeP compound, integrated over angles, enabled to recover the density of states over the broad energy interval, which comes out in agreement with numerical calculations. Angle-resolved measurements of photoemission detected two sheets of the Fermi energy around the Γ point along with a further sheet (apparently a degenerate one) around the M point. The works cited let us to conclude that the main features of calculated electronic structure of the LaOFeAs-type compounds have found experimental confirmation.

In parallel, first work on the study of the Fermi surface using the dHvA effect have appeared [94, 95]. The study was done for the LaOFeP compound whose crystal structure is the same as of LaOFeAs, and the superconducting transition temperature is $T_c \sim 7\,\text{K}$. This choice was motivated by a need of a high-purity

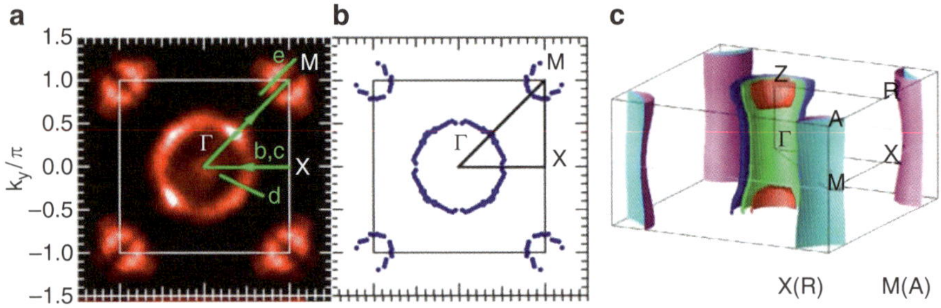

Fig. 2.36 Pockets of the Fermi surface, reconstructed from ARPES measurements on an $NdO_{1-x}F_xFeAs$ single crystal (**a**–**b**), and their comparison with the Fermi surface calculated within the LDA (**c**) [92]

single crystal sample with not so high value of the upper critical field ($H_{c2} \approx 0.68$ T for $B_{\parallel c}$ and 7.2 T for $B_{\perp c}$), to be able to suppress superconducting state by an experimentally readily accessible field. The results have shown that this compound has two cylindrical hole surfaces, centred at Γ, and two electron ones, centred at M. Hence, a full agreement with the LDA results by Lebeque [80] was found, albeit with higher effective masses, of 1.7–2.1 m_e (m_e: free electron mass), instead of about 0.8 m_e as calculated. Therefore, the main conclusions of the LDA calculations for FeAs-type systems found experimental verification on the basis of different methods.

A detailed comparison of LDA calculation results with the ARPES data revealed certain discrepancies. Thus, in [96] a thorough analysis of electron spectra of two stoichiometric compounds LaOFeP and LaOFeAs has been done (Fig. 2.37). In the (a) panel, around Γ two hole pockets Γ_1 and Γ_2 are seen, and an electron pocket – near the M point. In LaOFeP, they include, correspondingly, 1.94 holes, 1.03 holes and 0.05 electrons. For LaOFeAs, the corresponding numbers are 1.86 and 1.18 (Γ_1 / Γ_2); the electron M pocket is difficult to measure because of its peculiar cross-like shape. We note in addition that the inner (Γ_1) pocket in both compounds is doubly degenerate, as well as the M-centred electron one. Therefore, the Fermi surface consists of five sheets ($2 \times \Gamma_1 + \Gamma_2 + 2 \times M$), exactly as the LDA calculations for both compounds predict it to be.

A detailed comparison of LDA calculations with ARPES data [96] reveals a good agreement between theory and experiment, which is further confirmed by recent studies of the dHvA effect [95] on LaOFeP. For the other compound, LaOFeAs, the agreement between ARPES and LDA calculations is worse (a presence of the cross-shaped pocket at M, not predicted in the calculations). As of now, the reason for such disagreements is not clear. Taking into account the situation with other compounds of the ReOFeAs group, one can conclude about the agreement between LDA calculations and experimental data only in what regards the main result, the number and approximate size of electron and hole pockets; the fine details of electronic structure may differ.

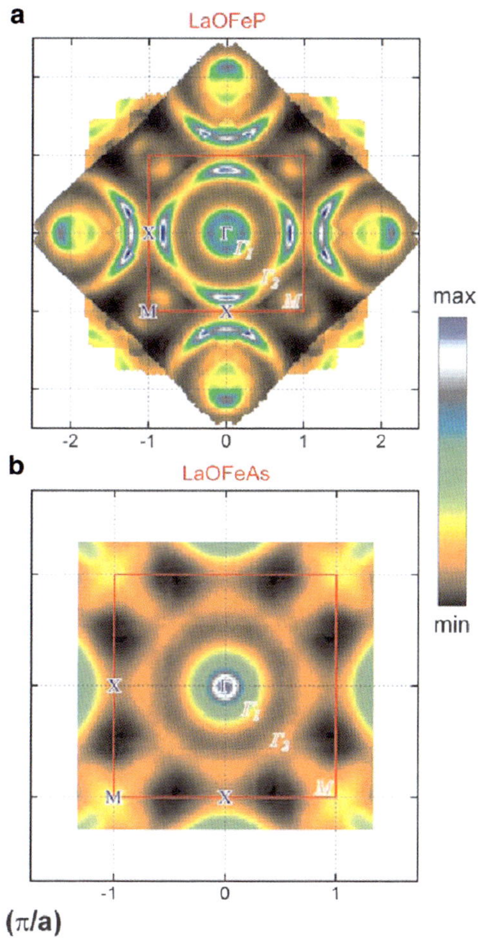

Fig. 2.37 Map of the Fermi surface of two compounds LaOFeP (**a**) and LaOFeAs (**b**), obtained from symmetrized ARPES data [96]

2.4 Symmetry of the Superconducting Order Parameter

2.4.1 Experimental Methods of Determining the Order Parameter

A knowledge of the symmetry and the shape, in the momentum space, of the superconducting order parameter (gap in the electron spectrum at the Fermi surface) is of particular importance, since it allows to make conclusion about the pairing mechanism. Several scenarios have been suggested for explaining the mechanism of superconductivity in the new class of superconductors, corresponding to different predictions for the symmetry of the order parameter: s, p or d type. There are several ways to determine experimentally the superconducting gap, and, consequently, to study the symmetry of the superconducting order parameter.

One of such methods is nuclear magnetic resonance (NMR), in which the Knight shift and the spin–lattice relaxation rate are measured. From the temperature dependence of the one and the other, information about superconducting order parameter can be gained.

Particularly informative in this sense are spectroscopic methods, in which the current I is measured as function of the voltage applied to the sample and the conductance dI/dV is determined; further on, the superconducting gap is found by adjustment of experimental curves to theoretical ones. The methods which fall within this group are: scanning tunnel spectroscopy (STS), photoelectron spectroscopy (PES), angle-resolved photoelectron spectroscopy (ARPES), point contacts with Andreev reflection (PCAR). In these methods, the density of quasiparticle states in a superconductor is directly measured. Immediately close to these methods is the Josephson contact spectroscopy.

There are also methods of different types: from measurement of the temperature dependence of the penetration depth of magnetic field into a superconductor, $\lambda(T)$. Moreover, certain information about the symmetry of superconducting order parameter can be gained from temperature dependence of electronic specific heat $C_v(T)$.

The whole spectrum of these methods has been applied to *Re*OFeAs compounds doped with different elements yielding them superconducting. The results of these studies are summarized in Table 2.3. In the following, we describe these results in detail, grouping them according to the methods used.

2.4.2 Nuclear Magnetic Resonance

Measurements of the resonance frequency and linewidth of the NMR at ^{57}Fe and ^{75}As nuclei make it possible to extract important information on electronic and magnetic properties of the FeAs-type systems. This information is contained in two characteristics immediately measurable in the NMR: the Knight's shift K and the spin–lattice relaxation rate $1/T_1$. The latter is expressed from the imaginary part of the dynamic electron susceptibility $\chi(\mathbf{k}, \omega)$ via a well-known relation:

$$1/T_1 \sim |A_{nf}|^2 \sum_{\mathbf{k}} \frac{\mathrm{Im}\,\chi(\mathbf{k}, \omega_0)}{\omega_0}, \tag{2.4}$$

where ω_0 is the NMR frequency, and A_{nf} – the constant of hyperfine interaction, coupling the nuclear spin of an isotope to the conductivity electrons.

Since ω_0 is small against characteristic electron-related frequencies, including kT_{c}, the $1/T_1$ property is determined by low-frequency density of states in the spectrum of spin fluctuations in the electron system. Due to a presence of a gap on the Fermi surface of a superconductor, $\mathrm{Im}\,\chi(\mathbf{k}, \omega_0)$ is exponentially small for $T < T_{\mathrm{c}}$, which in its turn leads to an exponential dependence of $1/T_1$ upon temperature in

Table 2.3 Superconducting order parameter in the *Re*OFeAs compounds, according to the data obtained by different methods.

Compound	T_c, K	Experiment	Order parameter	Reference
$LaO_{0.7}FeAs$	28	NMR	$1/T_1 \sim T^3$	[99]
$LaO_{0.92}F_{0.08}FeAs$	23	NMR	d or s, $\Delta_1 = 4kT_c$, $\Delta_2 = 1.5, kT_c$	[100]
$LaO_{0.89}F_{0.11}FeAs$	28	NMR	Pseudogap	[101]
$LaO_{1-x}F_xFeAs$ $0.04 \le x \le 0.14$	–	NMR	$1/T_1 \sim T^3$, pseudogap	[102]
$LaO_{1-x}F_xFe_{0.95}$ $Co_{0.05}As$	–	NMR	Differs from $1/T_1 \sim T^{2.5-3}$ and exponent	[103]
$LaO_{0.9}F_{0.1}FeAs$	26	NMR	$1/T_1 \sim T^3$, pseudogap	[98]
$LaO_{1-x}F_xFeAs$	27	PCAR	$\Delta = 2.8$–4.6 meV,s-type, pseudogap	[126]
$LaO_{0.9}F_{0.1}FeAs$	28	PCAR	$\Delta = 3.9$ meV, *ZBP*	[104]
$SmO_{0.85}F_{0.15}FeAs$	42	PCAR	$\Delta = 6.67$ meV, s-type	[105, 106]
$SmO_{0.85}FeAs$	52	STS	$\Delta = 8$ meV, *ZBP*	[108]
$SmO_{0.9}F_{0.1}FeAs$	51.5	PCAR	$\Delta_1 = 10.5$ meV, $\Delta_2 = 3.7$ meV, *ZBP*	[109]
$SmO_{0.8}F_{0.2}FeAs$	49.5	TRS	$\Delta = 8$ meV, pseudogap $\Delta_{GP} = 61 \pm 9$ meV	[112]
$NdO_{0.9}F_{0.1}FeAs$	–	PCAR	$\Delta_1, \Delta_2.ZBP$	[113]
$NdO_{0.9}F_{0.1}FeAs$	–	$\lambda(T)$	Δ, s-type	[114]
$NdO_{0.9}F_{0.1}FeAs$	53	ARPES	$\Delta = 15$ meV, s-type	[111]
$NdO_{0.86}F_{0.14}FeAs$	48	STS	$\Delta = 9.2$ meV, pseudogap, s-type	[110]
$NdO_{0.85}FeAs$	52	PES	Gap on the electron M-surface	[115]
$NdO_{0.9}F_{0.1}FeAs$	43	$\lambda(T)$	Two gaps, $\Delta\lambda \sim T^2$	[107]
$PrO_{0.89}F_{0.11}FeAs$	45	NMR	$1/T_1 \sim T^3$, $\Delta_1 = 3.5kT_c$, $\Delta_2 = 1.1\,kT_c$	[97]
$PrO_{1-y}FeAs$	35	$\lambda(T)$	$\Delta/kT_c \ge 1$	[116]

the superconducting state and results in an appearance of the Hebel–Slichter peak in the vicinity of T_c.

The above is valid if the superconductor gap does not become zero on the Fermi surface, e.g. in case of s symmetry of the superconducting order parameter. If the superconducting gap has zeros in some points or along certain lines on the Fermi surface, then the $1/T_1$ property, and also electron specific heat, do exhibit temperature dependence in the form of power law:

$$1/T_1 \sim T^n, \tag{2.5}$$

with some n value. For a two-dimensional system in the case of, say, d symmetry of the order parameter, $1/T_1 \sim T^3$. From measurements of spin–lattice relaxation in a superconductor, one can judge about the symmetry of the order parameter or at least draw conclusions on whether the superconductor gap has zeros on the Fermi surface.

If the gap Δ has no zeros on the Fermi surface, it can be extracted from the experimental data on $1/T_1$ using the relation following of the BCS theory:

$$\frac{T_{1,N}}{T_{1,S}} = \frac{2}{kT} \iint N_S(E)N_S(E')[1 - f(E')]\delta(E - E')\,dE\,dE', \tag{2.6}$$

or a correspondingly more involved expression for the case when the superconductor has two gaps on different sheets of the Fermi surface. Here, $T_{1,S}$ and $T_{1,N}$ stand for spin–lattice relaxation times in superconducting and normal phases of a metal, correspondingly, and $N_S = \frac{E}{\sqrt{E^2-\Delta^2}}$ is density of states in the superconducting state.

An information concerning the gap can as well be extracted from the Knight shift which, in the BCS theory, can be expressed as follows [97]:

$$\frac{K_S}{K_N} = \int N_S(E)\frac{\partial f(E)}{\partial E}\mathrm{d}E. \tag{2.7}$$

In the last formula, $f(E) = (1 + e^{E/kT})^{-1}$ is the Fermi function. Hence from measurements of $1/T_1$ one can judge about the presence of zeros of the gap on the Fermi surface. If such zeros do not come about, a comparison of experimental data on the Knight shift with model calculations results permits to extract the gap value.

NMR studies of superconducting FeAs-type compounds have been done in a number of works [97–103], with the results collected in Table 2.3. It is seen that in many cases, a power-law behaviour is detected. It is moreover remarkable that in no system was the Hebel–Slichter coherent peak detected. Joint data on the temperature dependence of $1/T_1$ are shown in Fig. 2.38.

For most of the FeAs-compounds, the power law close to T^3 is observed, however in the $LaO_{0.89}F_{0.11}FeAs$ sample studied in [103], the behaviour of $1/T_1$ is

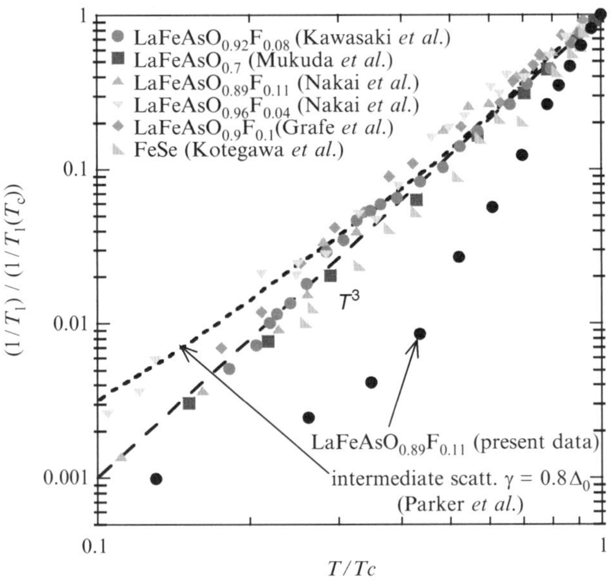

Fig. 2.38 Temperature dependence of the spin–lattice relaxation rate $1/T_1$ for a number of superconducting $LaO_{1-x}F_xFeAs$ compounds [103]

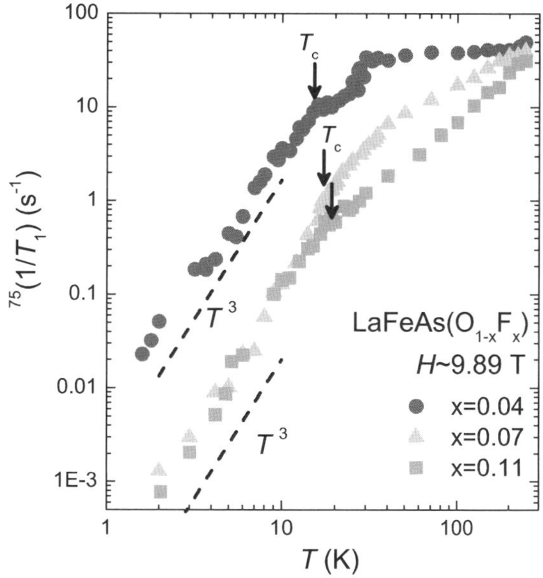

Fig. 2.39 Temperature dependence of $1/T_1$ in the $LaO_{1-x}F_xFeAs$ system at different levels of doping [102]

more complicated. At $T > 0.5\,T_c$, the $1/T_1$ follows the temperature rather as $\sim T^6$, and at even lower temperatures no power-law behavior is seen at all.

Systematical studies on the same system $LaO_{1-x}F_xFeAs$ over broad interval of doping ranges have shown that the power law stands – see results in [102] shown in Fig. 2.39. In the same work, devoted to measurements of $1/T_1$ over broad temperature ranges, for $x > 0.11$ a pseudogap Δ_{PG} of the order of kT_s, $T_s \sim 150$ K being about the temperature of magnetic transition in non-doped compound, has been detected above T_c. A presence of pseudogap in the spectrum of quasiparticle states in the normal phase of a superconductor is typical for cuprates where, as it is well established by now, it appears due to interaction of electrons with spin fluctuations. Among the FeAs-type systems, the presence of a pseudogap has been confirmed, by NMR and other methods, in a number of compounds – see Table 2.3.

Now, an important comment can be in place. An observation of the power-law behavior of $1/T_c$ does not yet mean that in a superconductor given, a gap is realized with zeros somewhere on the Fermi surface. One should take into account that in the FeAs-type systems, the Fermi surface is a multi-sheet one; hole-like around Γ and electron-like around M. On both hole and electron sheets, an s-type superconducting state (that without zero gap) can emerge; however, the signs of the gap function on these sheets may either coincide or be inverse. In case of sign coincidence one talks of extended s symmetry of the order parameter, whereas the opposite signs are referred to as the $s^{\pm}$ symmetry. In [90, 117], an idea has been put forward that in the FeAs-type systems, namely the $s^{\pm}$ symmetry of the order parameter is realized. This idea turned out to be very fruitful (see Chap. 4) and allowed to explain the power-law temperature dependence of $1/T_1$ in a different way.

When considering the gaps on the hole and electron sheets of the Fermi surface in the **k**-space, then obviously on a transition from hole to electron sheet, in case of $s^{\pm}$ symmetry, the gap must pass through zero; however, the zero lines are situated out of the Fermi surface, because the hole and electron pockets are separated. Moreover, it should be reminded that superconductivity occurs in doped, i.e. disordered systems, so that a scattering on non-magnetic impurities may transfer quasiparticles from the hole to electron sheet and back. Such scattering suppresses superconductivity, similarly to how a scattering on magnetic impurities in conventional superconductors works. However, a different issue is important in this context: a calculated spin–lattice relaxation rate $1/T_1$ in superconductors with $s^{\pm}$ symmetry of the order parameter, in the presence of non-magnetic impurities, changes its exponential temperature dependence into the power-law one, close to $1/T_1 \sim T^{2.5-3}$ [90]. Consequently, an observed power-law temperature dependence of $1/T_1$ might not necessarily imply a non-standard symmetry of the order parameter with gap zeros, but also an existence of coupling with the $s^{\pm}$ symmetry of the order parameter. This concept is supported by the study of superconductivity in FeAs-based systems of other classes, covered by Chaps. 2 and 3, as well as by discussions in the theory-related Chap. 4.

2.4.3 Point-Contact Andreev Reflection

In this method, the current is measured which flows through a point contact of a normal metal to a superconductor, N/S, as function of applied voltage. According to the Blonder–Tinkham–Klapwijk (BTK) theory [118], based on the BCS model with some phenomenological parameters added to account for the quasiparticles damping and the barrier characteristics, the current through the point contact is given by the formula (see [105]):

$$I_{\mathrm{NS}}(V) = C \int [f(E - eV) - f(E)]\,[1 + A(E) + B(E)]\,\mathrm{d}E. \qquad (2.8)$$

Here, $A(E)$ and $B(E)$ are functions determined via modified coherent factors,

$$\widetilde{U}^2 = \frac{1}{2}\left[1 + \frac{\sqrt{(E+i\Gamma)^2 - \Delta^2}}{E+i\Gamma}\right], \quad \widetilde{V}^2 = \frac{1}{2}\left[1 - \frac{\sqrt{(E+i\Gamma)^2 - \Delta^2}}{E+i\Gamma}\right],$$

and C is a constant sensitive to the contact properties on the surface of the superconducting sample. Specifically, it follows:

$$A(E) = |a|^2, \quad B(E) = |b|^2, \quad \text{with}$$
$$a = \widetilde{U}\widetilde{V}/\Gamma; \quad b = -(\widetilde{U}^2 - \widetilde{V}^2)(Z^2 + iZ)/\Gamma; \quad \Gamma = \widetilde{U}^2 + (\widetilde{U}^2 - \widetilde{V}^2)Z^2,$$

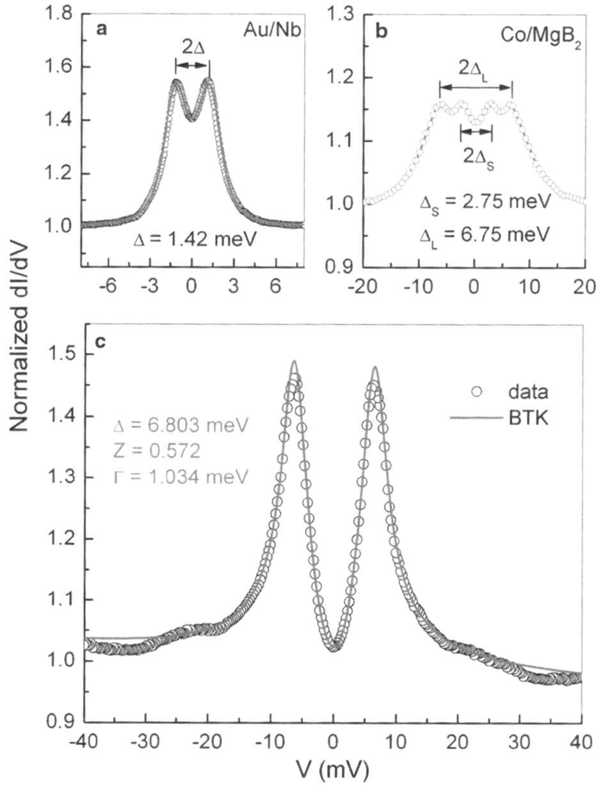

Fig. 2.40 PCAR spectra of superconductors: (**a**) Nb; (**b**) MgB_2; (**c**) $SmO_{0.85}F_{0.15}FeAs$ [105]. In the latter case, the parameters of the BTK model are indicated

which formulae transform into the known Dynes formulae [119] for the tunnel N/S current in the limit of $Z \to \infty$.

As follows from the general formula (2.8), for a conventional superconductor with s symmetry of the gap Δ, the current $I_{NS}(V)$ shows a two-peak structure around $V = 0$, whereby the distance between the spectral peaks equals 2Δ. In Fig. 2.40, an example of a spectrum for the Au/Nb contact is given, where the points mark experimental data and the continuous curve is a result of fitting, with some adjustable parameters added. In Fig. 2.40, the measurement results for the MgB_2 superconductor (the Co/MgB_2 contact) are given.

The observed four maxima reveal the existence of two superconducting gaps, Δ_S (the small one) and Δ_L (the large one). These examples confirm the efficiency of PCAR. Finally, in Fig. 2.40 a spectrum from an $SmO_{0.85}F_{0.15}FeAs$ compound is given, which indicates the presence of just one gap Δ [105]. This spectrum has been obtained for a given point contact on the superconductor surface. In [105], spectra have been collected over 70 point contacts, and their evaluation done by varying the adjustment parameters. The averaged Δ value, obtained with the best fit, makes $2\Delta = 13.34 \pm 0.3$ meV; hereby, taking into account the value $T_c = 42$ K, we arrive at the estimate $2\Delta/kT_c = 3.68$, quite close to 3.52 value of the BCS theory. Therefore, according to the data of this particular study it appears that the superconductor in question has one isotrop gap, i.e. it is an s-type superconductor. No gap zeros on the Fermi surface seem apparent, and the temperature dependence of $\Delta(T)$ is of conventional BCS type.

In another work [109], more detailed results have been obtained on a sample of $SmO_{0.9}F_{0.1}FeAs$ having $T_c = 51.5$ K. In Fig. 2.41, PCAR spectra are given, recorded in some points at the superconductor surface. In panel c, the results corresponding to two different contacts are shown together from which the presence of two gaps is obvious, $\Delta_1 = 10.5 \pm 0.5$ meV and $\Delta_2 = 3.7 \pm 0.4$ meV. Their temperature dependence is shown in the panel d. Both gaps disappear at the superconducting transition temperature T_c.

A remarkable result is shown in the f panel, where a three-peak structure of spectrum is seen. Beyond two conventional coherent peaks, the so-called zero-bias conductance (ZBC) peak is seen as $V = 0$. It appears due to the formation in the superconducting gap Δ of Andreev bound states, witnessing the existence of zero gap on the Fermi surface. We note that in [105] no such spectra were reported, from which a conclusion was done that the Sm-based superconductors do not have zeros in the superconductor gap.

Since in [109] a ZBP, peak has been detected, a conclusion has been done that the order parameter in the said superconductor has the d symmetry, and all theory curves shown in Fig. 2.41 have been calculated assuming the corresponding angular dependence of the order parameter, $\Delta = \Delta_0 \cos 2\theta$.

As is seen from Table 2.3, also the ZBP peaks were detected in an Nd-containing compound.

A possibility to identify the $s^{\pm}$ gap symmetry in FeAs-type systems with the help of the Andreev reflection was addressed in a row of theory publications [120–124]. In [120], an increase of the density of states at zero energy for an N/$s^{\pm}$ contact was demonstrated; however, this work was mostly numerical one, which made it difficult to establish a relation between the effect announced with a formation of the Andreev bound states in the contact plane. A more complete and physically transparent study was that reported in [123], where the authors generalized the BTK method [118] of phenomenological characterization of contact for the analysis of the Andreev reflection.

The coupled Andreev states appear for both $N/s^{\pm}$ and N/s_{++} contacts, where s_{++} stands for a two-band superconductor in which the signs of the superconducting order parameter coincide on both sheets of the Fermi surface. In Fig. 2.42, the calculated conductance for both cases is shown. The calculation is done for a tunnel contact with $Z = 10$ and the situation with two gaps on the sheets of the Fermi surface, $\Delta_2 = 2\Delta_1$. Moreover, different magnitudes have been considered of the α parameter which gives the ratio of probability amplitudes for an electron coming from a normal metal into superconductor to end up in a state on either electron, or hole surface: $\Psi = \Psi_n + \alpha\psi_e$.

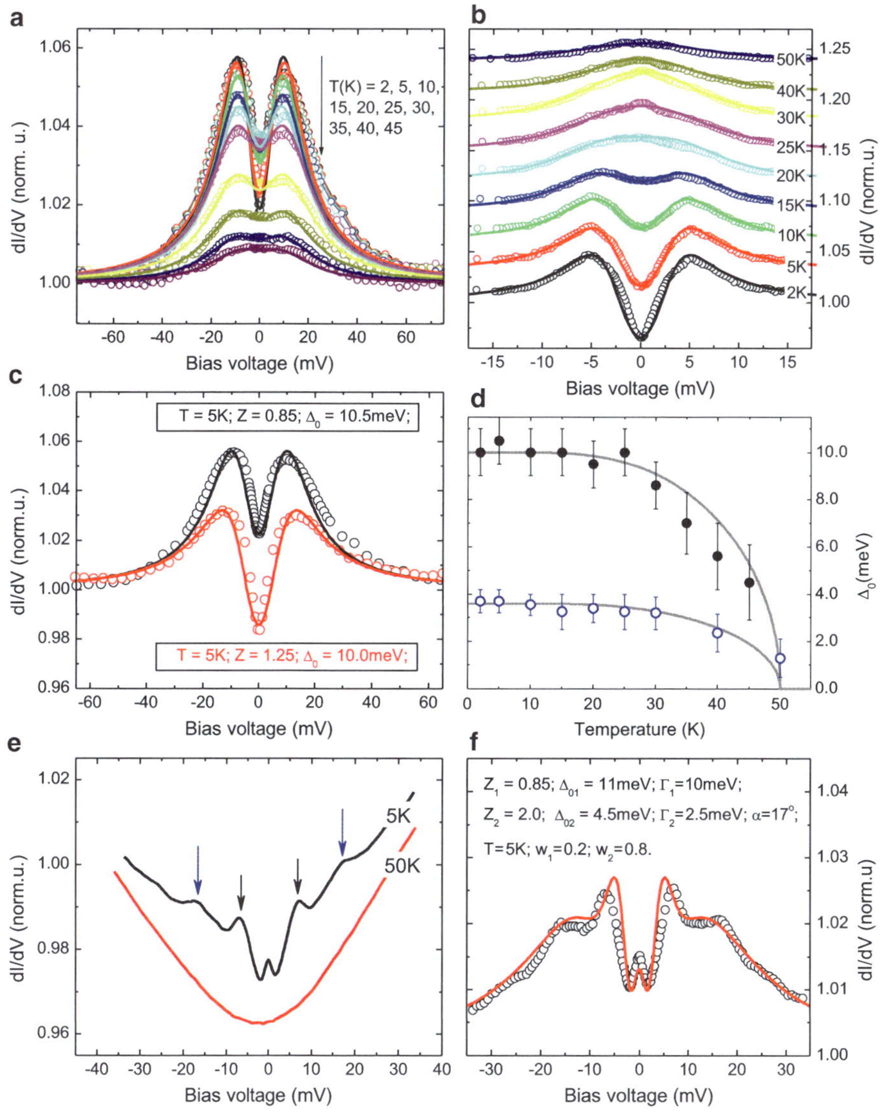

Fig. 2.41 PCAR spectra of superconducting $SmO_{0.9}F_{0.1}FeAs$ [109]. In the (**a**–**c**) and (**f**) panels, the experimental points and theory curves (*solid lines*) with the fitting parameters used are given. (**d**) Temperature dependencies of the gap values determined from fits as shown in (**a**) and (**b**)

As is seen from Fig. 2.42, in the case of $s^{\pm}$ symmetry the peaks in the gap (the smallest one of Δ_1 and Δ_2) due to the formation of bound Andreev states may appear at non-zero values of the potential V. In general, the Andreev bound states may exist in a broad interval of α values: $0 \leq \alpha^2 \leq \Delta_1/\Delta_2$, whereby the ZBP peaks (those at $V = 0$) appear at $\alpha^2 = \Delta_1/\Delta_2$. An emergence of peaks at $V = 0$

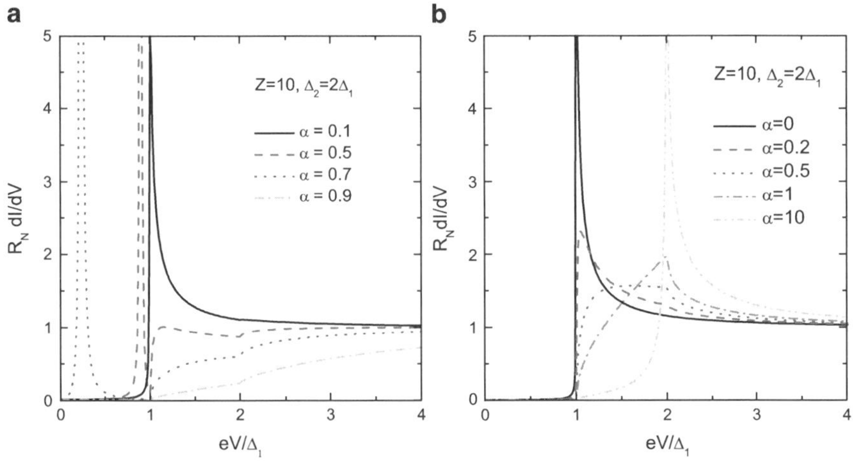

Fig. 2.42 Conductance in the low-transparency regime, (**a**) $N/s^{\pm}$ and (**b**) N/s_{++}, at different values of the α parameter [123]

does not yet mean zeros of the superconducting order parameter, because the $s^{\pm}$ symmetry may be also responsible for the appearance of exactly such peaks. On the other side, an extended s symmetry (the s_{++} one) does not yield bound states in the gap; however, they may exist outside of the latter, and can be erroneously taken for a signature of the multigap state. Summarizing, the observation of ZBP peaks in N/S contacts does not yet unambiguously indicate a presence of zeros in the superconducting order parameter.

Detailed PCAR studies on polycrystalline samples of two superconducting compounds *Re*OFeAs, *Re* = La,Sm, are outlines in [125]. Figure 2.43 shows the measured conductance for two-point contacts, which are characterized by resistivity of the normal metal, R_N, for $LaO_{1-x}F_xFeAs$ at 4.3 K. The experimental data are indicated by dots, whereas continuous curves give the fitting results after the BTK method with a single gap (dashed line) and two gaps (solid line). We see that the observed four-peak structure can be well mapped onto the adjustment curve corresponding to two gaps. Similar results have been obtained for another compound, $SmO_{0.8}F_{0.2}FeAs$, with $T_c = 51.5$ K.

An evaluation of data over numerous contacts results in the following gap values. In $LaO_{1-x}F_xFeAs$, $\Delta_1 \approx 3$ meV, $\Delta_2 \approx 8$–10 meV; in $SmO_{0.8}F_{0.2}FeAs$, $\Delta_1 \approx 6$ meV, $\Delta_2 \approx 19$–20 meV. Hence in both cases, the relation between the large and the small gaps is $\Delta_2/\Delta_1 \approx 3$.

It is remarkable that in no contacts where ZBP peaks detected, so that apparently no gap zeros exist on the Fermi surfaces of these two compounds. On the same contacts from which the data of Fig. 2.43 have been collected, the measurements of conductance have been done at different temperatures (Fig. 2.44), and the temperature dependencies of superconductor gaps $\Delta_1(T)$, $\Delta_2(T)$ extracted. Corresponding curves have also been extracted for the second compound studied, $SmO_{0.8}F_{0.2}FeAs$. Temperature dependencies of the gap for two compounds are markedly different.

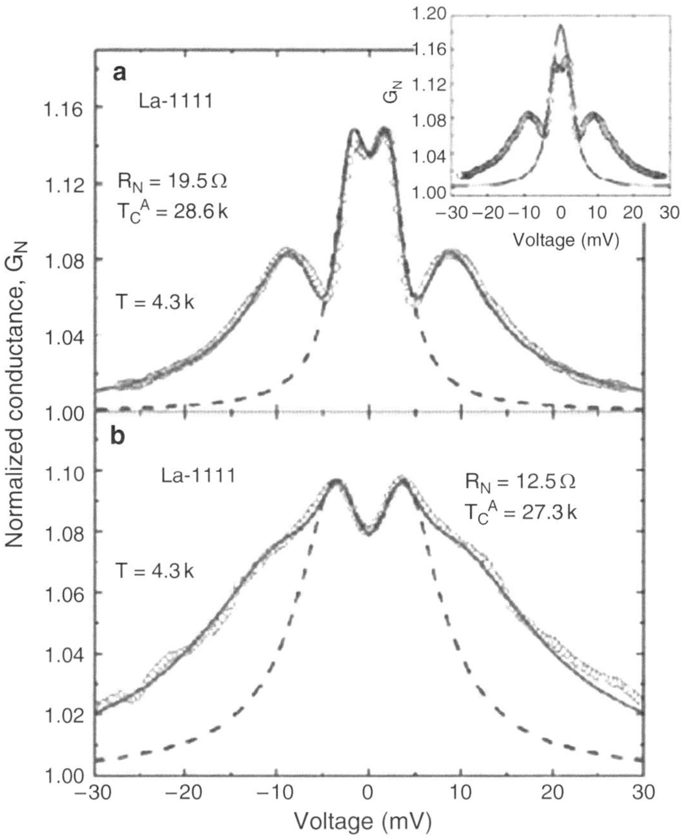

Fig. 2.43 PCAR measurements of conductance for two typical contacts of a $LaO_{1-x}F_xFeAs$ superconductor [125]. In the inset, an adjustment curve corresponding to a suggestion of the *d* symmetry of the superconducting order parameter is given

While in the Sm-containing compound, both gaps close at T_c, the situation in the La-containing compound is more complicated: the larger gap Δ_2 escapes detection already at $T \approx 0.8\,T_c$, whereas the smaller one is still non-zero at $T > T_c$. This situation remains so far obscure; it has been suggested that the Δ_2 may not be necessarily related to a superconducting state. It is interesting to note that in both compounds, the smaller gap is inferior to what could be expected from the BCS theory, namely, $2\Delta_1/kT_c = 2.2$–3.2, whereas the larger gap is substantially beyond the BCS value: $2\Delta_2/kT_c = 6.5$–9.

Even if a number of results obtained is not ultimately clarified, the conclusion remains beyond doubt that the superconductivity occurring in the *Re*OFeAs compounds with *Re* = La and Sm is characterized by the presence of two superconducting gaps, and the absence of gap zeros at the Fermi surface. This superconducting state has an extended *s* symmetry; however, it was not possible to relate the gaps

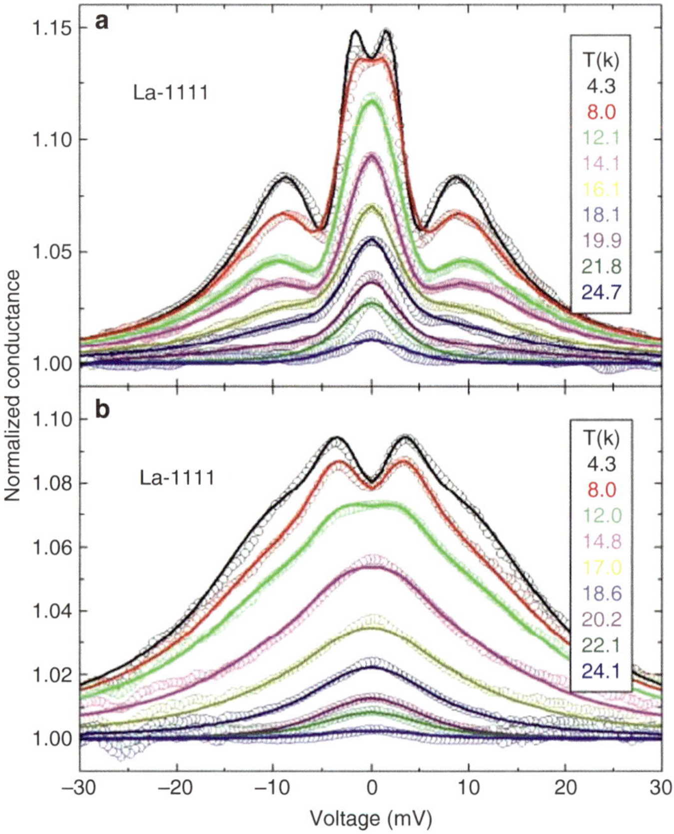

Fig. 2.44 Temperature dependence of conductance for the same point contacts of $LaO_{1-x}F_xFeAs$ which correspond to Fig. 2.43 [125]

Δ_1 and Δ_2 to the hole and electron sheets of the Fermi surface, correspondingly. Moreover, it was not possible to establish a phase relation between the Δ_1 and Δ_2 order parameters. The results so far obtained, however, do not contradict an idea of the $s^{\pm}$ symmetry of the superconducting order parameter in the *Re*OFeAs compounds. Very pronounced ZBP was discovered in $TbO_{09}F_{0.1}FeAs$ [127].

In another work [126], done on the $LaO_{1-x}Fe_xFeAs$ compound using the PCAR method, three energy gaps have been detected: two superconducting gaps Δ_1 = 2.8–4.6 meV and Δ_2 = 9.8–12 meV, which do not possess zeros at the Fermi surface, and a pseudogap which survives at temperatures by far exceeding T_c, up to $T^* \sim 140$ K, that is close to the Néel temperature for the undoped compound. This pseudogap is, probably, induced by spin fluctuations, which exist in doped superconducting compounds above T_c.

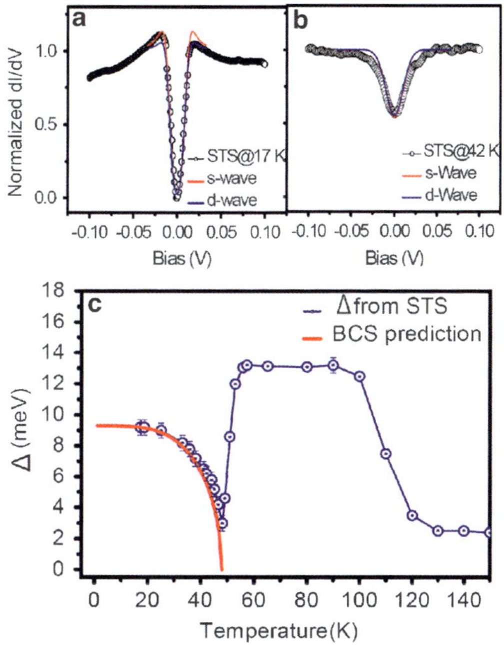

Fig. 2.45 STS-study of $NdO_{0.86}F_{0.14}FeAs$ [110]. (**a**) Tunnel spectrum at $T = 17$ K; (**b**) tunnel spectrum at $T = 42$ K; (**c**) temperature dependence of the superconducting gap Δ and the pseudogap Δ_{PG}s

2.4.4 *Tunnel and Photoemission Spectroscopies (STS, PES, ARPES)*

As an example, we discuss the results obtained for the $NdO_{0.86}F_{0.14}FeAs$ compound with $T_c = 48$ K [110] by the scanning tunnel spectroscopy method. The tunnel spectra obtained at 17 K show a suppression of the electron density within ± 10 meV, whereby two gaps of the widths 9 and 18 meV are revealed. They both close at T_c, but only one of them follows in its behaviour the Dynes formula [119] for the tunnel current. Remarkably, at $T > T_c$ another gap, of not superconducting nature, is detected; it emerges drastically at $T = T_c$ and remains unchanged throughout a broad temperature range up to 120 K (Fig. 2.45).

The temperature dependence of the pseudogap is completely different from that in cuprates, where the pseudogap appears due to the interaction of electrons with spin fluctuations. The same interaction is responsible for the Cooper pairing; hence, the superconducting gap and the pseudogap are driven by the same mechanism and do overlap as the temperature varies. In the sample studied of the above FeAs-type compound, the pseudogap disappears as the superconductor gap opens, so that they seem to be quasi in competition. The nature of this phenomenon is not so far clarified, although it may be suggested that the pseudogap is also related to spin fluctuations.

In another STS study, that of the $SmO_{0.85}FeAs$ compound with $T_c = 52$ K [108], the V-shaped gaps in the spectrum were well mapped on the theory curves

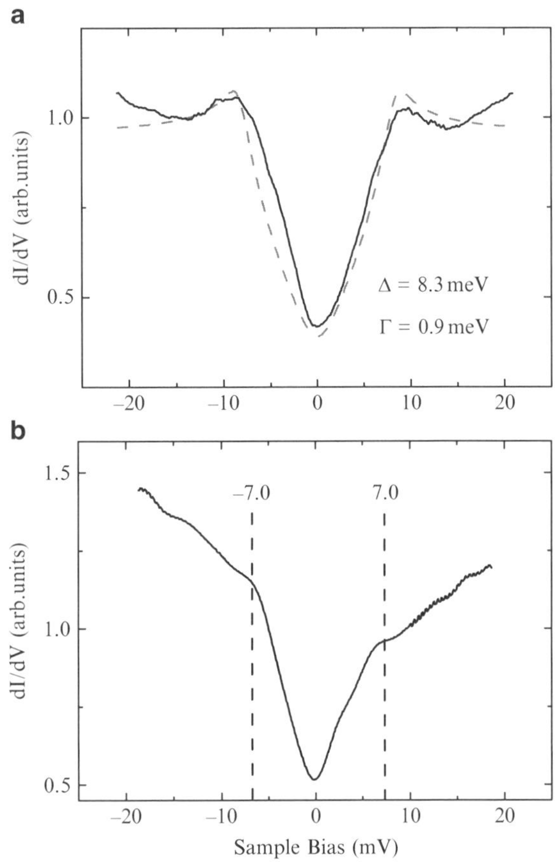

Fig. 2.46 Tunneling spectra of the $SmO_{0.85}FeAs$ at 4.2 K taken (**a**) in a region where coherence peaks were observed and (**b**) in a region where only sharp gap edges were found with a peak at $V = 0$, revealing a presence of zeros in the superconducting gap [108]

corresponding to the *d* symmetry of the order parameter. In some cases, ZBP peaks at $V = 0$ have been detected, which are compatible with the *d* symmetry of the order parameter (Fig. 2.46).

The superconducting compound $NdO_{0.86}F_{0.14}FeAs$, hence with the composition very close to that discussed above and $T_c = 53$ K, was studied by ARPES [111]. The only gap of about 13–18 meV has been detected on the hole sheet of the Fermi surface around the Γ point. The measurements at different angles to the crystallographic axes of the FeAs plane revealed a certain anisotropy (Fig. 2.47).

An inspection of Table 2.3 brings us to a conclusion that the data concerning the symmetry of the superconductor order parameter are, as of now, not conclusive. The most studied so far is the $LaO_{1-x}F_xFeAs$ system. All NMR measurements give a power-law dependence $1/T_1 \sim T^3$ which cannot be unambiguously interpreted: it may signify either the presence of gap zeros on the Fermi surface, or, in the presence of nonmagnetic impurities, the $s^{\pm}$ type of symmetry. PCAR investigations of this system give contradictory results as well. Aiura et al. [115] argues towards the *s* type

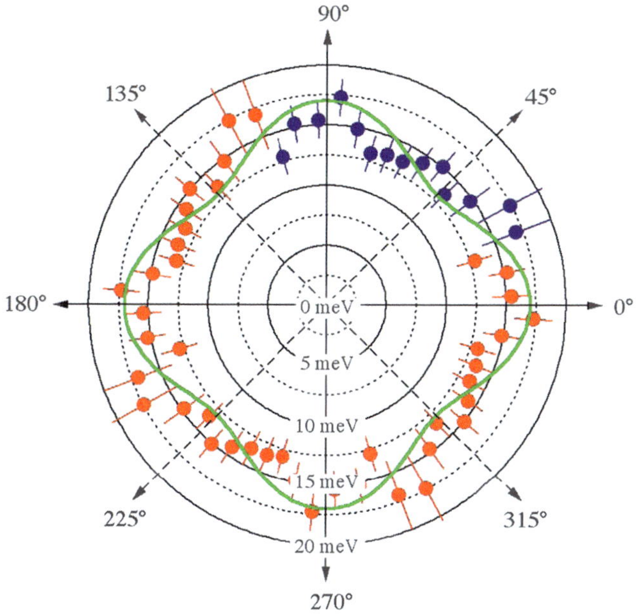

Fig. 2.47 Superconducting gap on a hole sheet of the Fermi level, centred at Γ, after the ARPES study of a $NdO_{0.9}F_{0.1}FeAs$ single crystal at $T = 20\,K$ [111]

symmetry, whereas [128] reports ZBC peaks, corresponding to the presence of gap zeros. In Nd-, Sm- and Pr-containing systems, the situation is equally ambiguous.

At the same time, none of the results gained by different methods seem to contradict the suggestion about the $s^{\pm}$ symmetry of the order parameter. An ultimate conclusion could have been done following the analysis of the data obtained for other FeAs-type systems, considered in the Chaps. 2 and 3. We note a single important fact following from Table 2.3: in many cases, the pseudogap is detected at temperatures substantially superior to the T_c. Previously, such phenomenon has been detected in cuprates, and now it rests to verify whether its nature in FeAs-related systems is the same, i.e. whether it is due to interactions of electrons (or holes) with spin fluctuations.

Chapter 3
Compounds of the AFe_2As_2 (A = Ba,Sr,Ca) Type

3.1 Crystal and Electronic Structure

3.1.1 Crystal Structure

Following the LaOFeAs-type compounds, which served as a starting point in the study of FeAs-based high-T_c superconductors, $BaFe_2As_2$ [13], $SrFe_2As_2$ [129] and other AFe_2As_2 (A = K, Cs, Sr) compounds [130] have been synthesized, which under doping turned superconducting. A discovery of superconductivity with $T_c = 38$ K in $Ba_{1-x}K_xFe_2As_2$ led to a new rise of research activity concerning the systems built on the basis of FeAs motives.

The crystal structure of $BaFe_2As_2$ is shown in Fig. 3.1. It is tetragonal with the $I4/mmm$ space group, built of FeAs planes (the same as in LaOFeAs), separated by Ba layers. There is only one FeAs-unit per unit cell of LaOFeAs, whereas in $BaFe_2As_2$ there are two. In the unit cell of $BaFe_2As_2$, the Fe–As distance is smaller than in LaOFeAs; consequently, one can expect larger Fed – Asp hybridization for $BaFe_2As_2$ and hence a broader d band in the electron spectrum. The distance between the neighbouring Fe atoms within the FeAs layers is also smaller in $BaFe_2As_2$.

According to the data of [13], the lattice parameters in this compound are: $a = 3.9090$ Å, $c = 13.2121$ Å. Therefore, the unit cell size in the basal plane is about the same as in *Re*OFeAs compounds, whereas along the c axis the size is substantially larger.

At 140 K, $BaFe_2As_2$ undergoes a structural phase transition from tetragonal into orthorhombic phase, of the space group $Fmmm$, like in *Re*OFeAs-compounds, whereby the four equal Fe–Fe distances split into two pairs of 2.808 and 2.877 Å. Structural transition is accompanied by magnetic ordering on Fe atoms, again like in *Re*OFeAs. The behaviour outlined for $BaFe_2As_2$ is also typical for other AFe_2As_2 compounds.

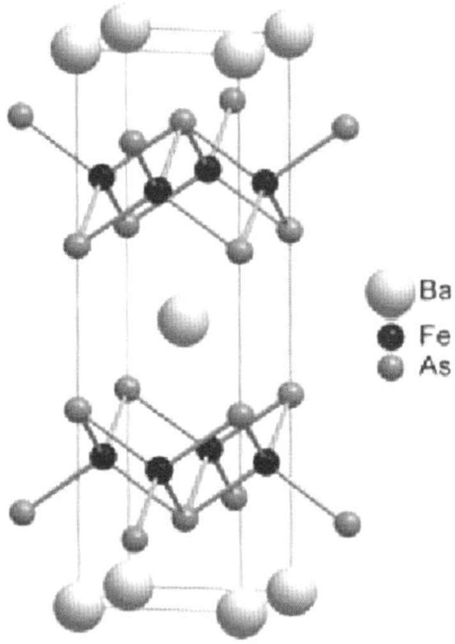

Fig. 3.1 Crystal structure of the $BaFe_2As_2$ compound

3.1.2 LDA Calculations of the Electronic Structure

Electronic structure of $BaFe_2As_2$ has been calculated within the LDA in a number of works [86, 131–134]. Their results being very close, we outline the calculations of total and partial densities of states after [133], where they are given in comparison with the density of states for LaOFeAs, Fig. 3.2. We see big similarity in both total and Fed-partial densities of states (DOS) between the both compounds, in particular within the energy range around the Fermi level. It might have been expected, because in both compounds the Fe atoms are situated in the same environment in the centres of As tetrahedra.

Energy dispersion curves in the vicinity of the Fermi level are also similar in both compounds, because they are primarily shaped by the Fed states. In $BaFe_2As_2$, there are three hole pockets near the Γ and two electron ones near the X point, in analogy with LaOFeAs where the electron pockets are centred at M. We note that the Brillouin zones for the compounds compared are not identical, so that the X point for $BaFe_2As_2$ has to be compared with M for LaOFeAs.

The Fermi surfaces as calculated within the LDA are very close for two compounds. In both cases, there are five sheets of approximately cylindrical shape: the three (hole) ones pass through the centre of the Brillouin zone, having the Γ–Z line as their axis, and two other (electron) cylinders are situated at the corners of the Brillouin zone, along the M–A lines (Fig. 3.3).

In addition to $BaFe_2As_2$, the electronic properties have been calculated in two other compounds, $BaNi_2As_2$ [135] and $BaRh_2As_2$ [136], in which Fe is replaced by other transition elements. $BaNi_2As_2$ is superconductor with a low $T_c = 0.7$ K, its band structure resembles that of $BaFe_2As_2$; however, the Fermi level is shifted upwards due to the fact that the Ni^{2+} ion has two d electrons more than the Fe^{2+}. Consequently, the Fermi surface of $BaNi_2As_2$ is larger, and electronic properties

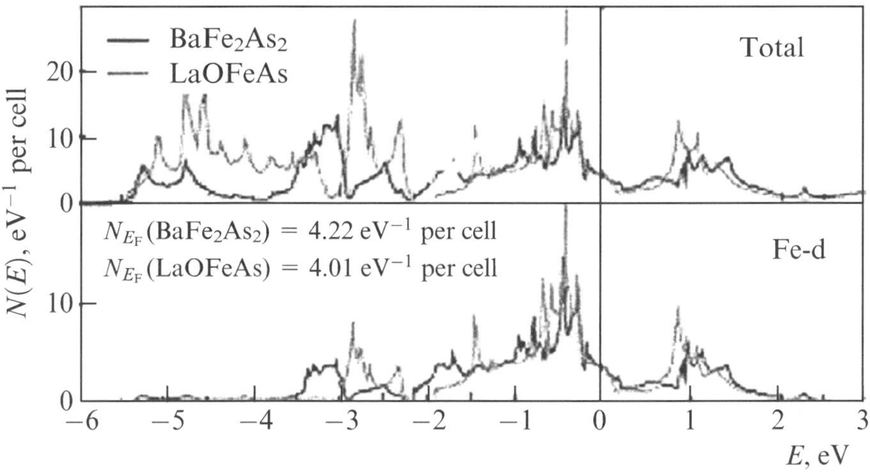

Fig. 3.2 Total and partial densities of states of LaOFeAs and $BaFe_2As_2$ compounds as calculated in the LDA [133]

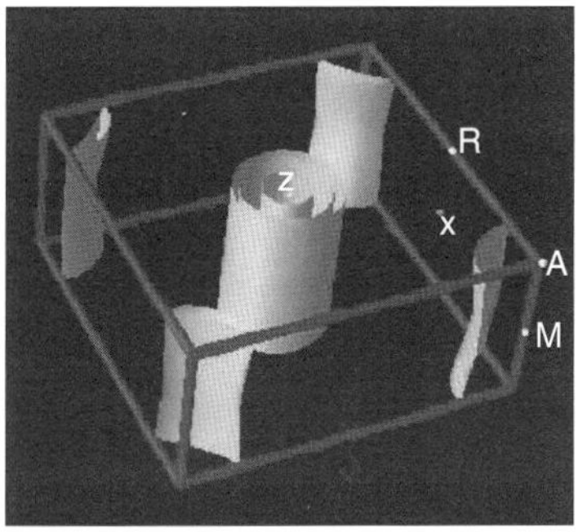

Fig. 3.3 Fermi surface of $BaFe_2As_2$ [133]

of this compound are much different from those of $BaFe_2As_2$. A similar enlargement of the Fermi surface has been found in LaONiP, with respect to LaOFeP. The electron–phonon coupling constant in $BaNi_2As_2$, $\lambda = 0.76$, is enhanced in comparison with $\lambda = 0.21$ for LaOFeAs. The density of states at the Fermi level is $N(E_F) = 3.57\,(eV)^{-1}$ per two spin projections and one formula unit [135]. This compound belongs to superconductors with conventional electron–phonon coupling.

The compounds AM_2As_2 (M = Fe, Co, Ni) exhibit many similarities in their electronic structure: the shape of the density of states is similar, different is just the placement of the Fermi level, due to varying number of d electrons per atom of M. For all these compounds, typical is a moderate hybridization of M3d and As4p orbitals, which makes ~10–20%. No wonder that physical properties of these materials are also similar: they are itinerant magnetics with SDW structure and metallic conductivity.

Differently from the above compounds, the compounds of the same crystal structure with Mn as a 3d element exhibit a semiconductor-type conductivity with a gap of about 0.2 eV at the Fermi level, due to antiferromagnetic ordering. For example, in $BaMn_2As_2$ the electrical resistivity does sharply fall with temperature till ~150 K after which its metallic-type increase starts.

The electronic structure of $BaMn_2As_2$, along with that of $BaMn_2Sb_2$, was calculated in [137, 138]. In [138], calculations have been done under different assumptions about the magnetic ordering, trying the ferromagnetic, SDW and AFM structures. The energies of all magnetically ordered phases were found lower than that of the nonmagnetic phase, whereby the lowest energy was obtained for the G-type AFM structure. From spin-polarized densities of states, a strong hybridization of *d* and *p* states is seen, which is different for the majority-spin and or minority-spin components. The compounds with Mn exhibit more localized-type of magnetism than the AFe_2As_2 compounds. Differently from $BaMn_2As_2$, $BaCr_2As_2$ is metal with a strong hybridization of Cr*d* and As*p* states. Its Fermi surface contains two large pockets, centred near Γ [139].

Even more important differences in the electronic structure of AM_2As_2 (M = Mn, Co, Ni, ...) compounds from Fe-containing ones occur when the transition metal substituent is copper. In [140], electronic structure of two compounds, $BaCu_2As_2$ and $SrCu_2As_2$, has been calculated. As experimental studies show, the differences from the Fe-based compounds are strong. These materials are neither antiferromagnets nor superconductors. The LDA calculation shows that the orbitals of Cu do form a narrow band, situated by 3 eV below the Fermi level, so that all states of Cu ions are occupied, and they are chemically inert in compounds. Fermi surfaces in these compounds are large and of pronounced three-dimensional character.

The larger part of theoretical and experimental studies of the electronic structure of FeAs-compounds relates to nonmagnetic state. Of special interest are those – so far not numerous – studies in which first-principles calculations and experimental studies of the Fermi surface have been performed for magnetically ordered SDW state. Among such, [141] can be named, where LDA calculations of the Fermi surface, along with ARPES measurements, have been done for stoichiometric $BaFe_2As_2$.

Partial densities of Fe*d* states in the nonmagnetic and the SDW phases are shown in Fig. 3.4 within a narrow energy interval around the Fermi level. In Fig. 3.4b, the densities of states for majority- and minority-spin components are depicted separately. It is seen that in the magnetically ordered phase, the states of one orbital only, d_{xy}, are not negligible at the Fermi level. It means that in the real compound, a lowering of temperature through the T_N point and an onset of antiferromagnetic ordering should be accompanied by an orbital ordering.

As AMF odering sets on, a gap should appear in the electron spectrum, and a reconstruction of the Fermi surface occurs. In place of a quasi-two-dimensional surface comprising two cylinder-shaped hole sheets in the Brillouin zone centre and two electron sheets at its corners, an essentially three-dimensional surface is formed (Fig. 3.5).

The thus predicted reconstruction of the Fermi surface agrees with ARPES measurements, done with linearly polarised photons. The technique used did also permit to settle the orbital ordering in the magnetically ordered phase. The resulting

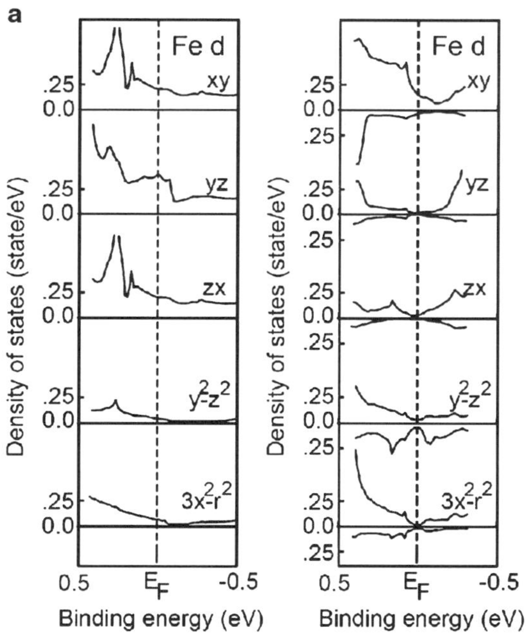

Fig. 3.4 Densities of states for five d-orbitals of Fe in the $BaFe_2As_2$ compound, calculated for nonmagnetic phase (**a**) and the SDW-phase (**b**) [141]

ARPES data agree well with recent measurements of the Fermi surface in magnetically ordered $BaFe_2As_2$ done with the help of quantum oscillations [142, 143].

We point out some further works in which the Fermi surface of other compounds has been studied. Thus, in [144] the electronic structure has been calculated for KCo_2As_2 and KFe_2As_2 compounds, which are neither superconducting nor magnetic, as well as for their intermediate binary alloy, $KFe_yCo_{2-y}As_2$. The calculated Fermi surfaces are shown in Fig. 3.6. In the limiting cases $y = 2$ and $y = 0$, we deal with stoichiometric compounds whose Fermi surface properties are maximally different, the one corresponding to the hole and the other to the electron conductivity, according to different valences of the transition elements, Fe^{2+} and Co^{3+}. On varying y from 0 to 2, the system changes from one limit towards the other. As $y = 1$, the $KFeCoAs_2$ compound has properties equivalent to those of $BaFe_2As_2$, after counting the valences of the elements from which these compounds are formed. We see that in the limiting cases KFe_2As_2 and KCo_2As_2, the nesting between electron and hole sheets is missing, that is apparently what explains why these compounds are neither antiferromagnetic nor superconducting (on their doping). On the other hand, the $KFeCoAs_2$ compound might happen to be antiferromagnetic and lay foundation of a new line of superconductors. The question remains, how to synthesise such compound.

In another work [145], the Fermi surface of the $SrFe_2P_2$ compound was studied, which is neither antiferromagnetic nor superconducting under doping. The

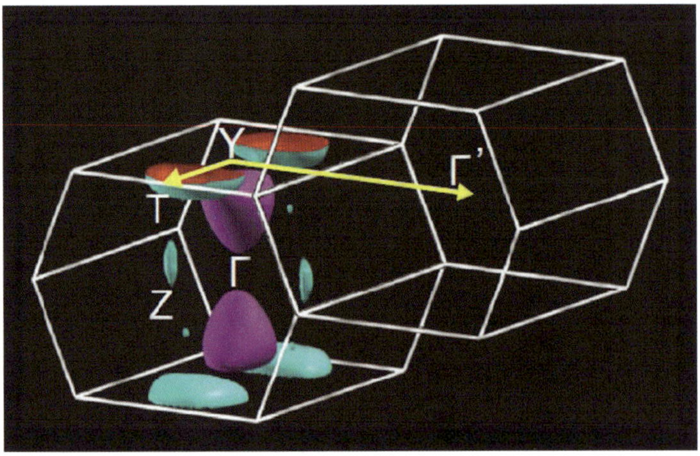

Fig. 3.5 Fermi surface of the $BaFe_2As_2$ compound, calculated for orthorhombic magnetically ordered SDW phase [141]

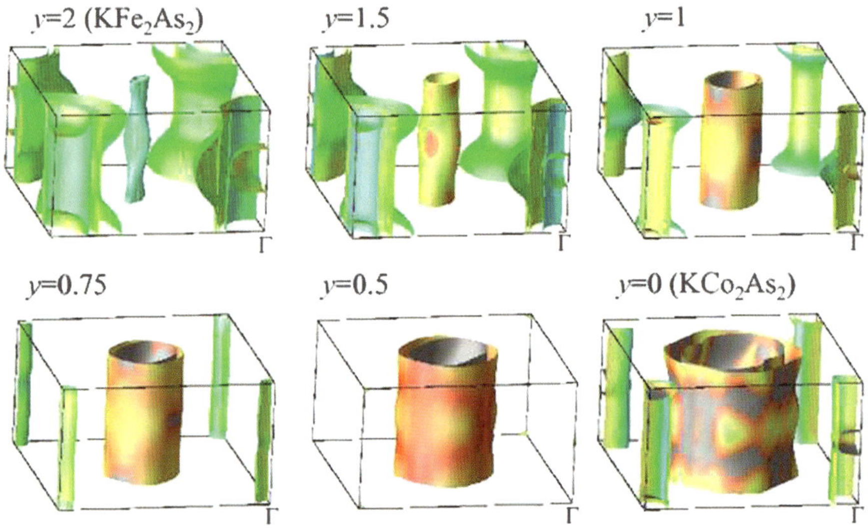

Fig. 3.6 Calculated Fermi surfaces of the $KFe_yCo_{2-y}As_2$ compounds, for six different values of y [144]

measurements of quantum oscillations in the dHvA effect have shown that the Fermi surface is three-dimensional: the hole sheets around the Γ point and the electron ones near X form cylinders which are strongly distorted along the c axis, consistently with numerical calculations. The Fermi surface topology does not show any nesting between the hole and electron sheets, typical for many FeAs-compounds, which are superconducting. This fact, similarly to how it was discussed in the previous case, explains why $SrFe_2P_2$ is not an antiferromagnet. The measurement of quantum fluctuations permitted to determine the effective electron masses in this compound. They change within the interval from 1.3 m_e on the smaller hole sheet to 2.1 m_e for the inner electron sheet. As is other cases, this confirms the conclusion about weak electron correlations in the compounds on the basis of iron and pnictides (As, P).

A dispersion of the electron bands along the k_z direction is typical for a number of compounds, which are superconductors. In ARPES studies, it is difficult to extract the variations of spectra with the k_z component of the wave vector. The recent data for the $Ba(Fe_{1-x}Co_x)_2As_2$ superconductors revealed a noticeable three-dimensional character of electron sheets, see [146].

3.1.3 Experimental Studies of the Fermi Surface

An experimental verification of the above theory conclusions was done, for $BaFe_2As_2$ single crystals and potassium-doped superconducting compound $Ba_{1-x}K_xSrFe_2As_2$, with the help of ARPES [147, 148]. According to the results obtained in [147], the Fermi surface of undoped $BaFe_2As_2$ consists of two small round pockets (hole ones), centred in Γ, and a much larger (electron) pocket centred in X (Fig. 3.7). The Fermi surfaces calculated for $BaFe_2As_2$ and shown in Figs. 3.3 [133] and 3.7 [147] are in good agreement.

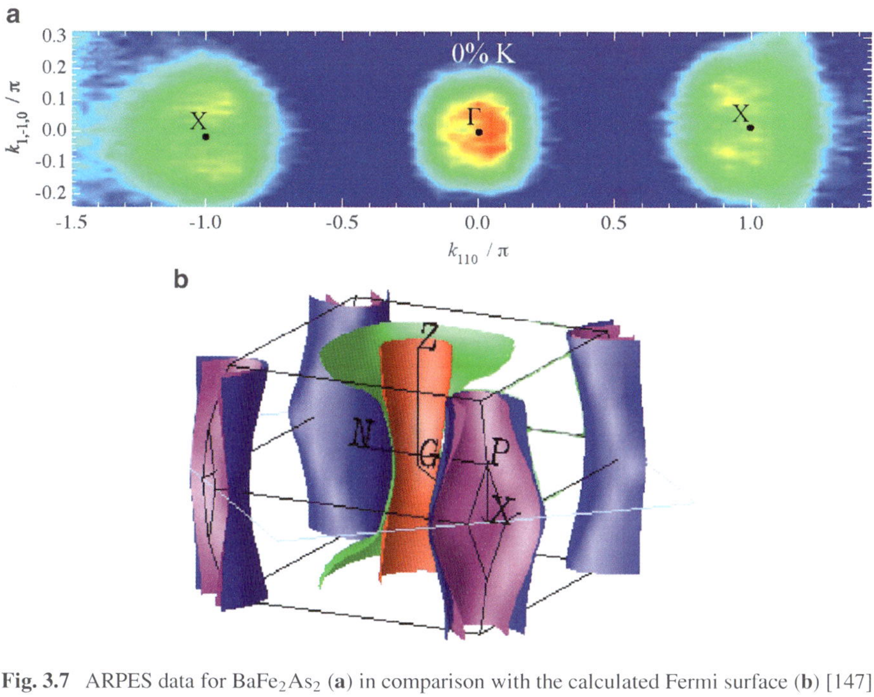

Fig. 3.7 ARPES data for $BaFe_2As_2$ (**a**) in comparison with the calculated Fermi surface (**b**) [147]

In another ARPES study [148], the thus outlined picture got refined. In the vicinity of Γ in $BaFe_2As_2$, three sheets of the Fermi surface have been clearly detected, in agreement with the most LDA calculations – however, quantitative disagreements with the calculated results were visible. It was noted meanwhile that a displacement of the Fermi level downwards by 0.2 eV recovers an agreement between the LDA calculation results and the ARPES data. An suggestion was raised [148] that thus necessary energy shift witnesses a certain role of correlations, neglected in LDA.

An other compound, $BaRh_2As_2$, as follows from measurements of its transport and thermodynamical properties on single crystals [136], is not at all superconducting; moreover, within the temperature range from 2 to 300 K it does not exhibit any magnetic ordering nor structure phase transition. The density of states is $N(E_F) = 3.49\,(\text{eV})^{-1}$ per both spin projection and one formula unit, with the dominating contribution coming from the Rh4*d* states.

Let us provide the calculation data for synthesized $TlFe_2Se_2$ compound [149]. It has large magnetic ordering temperature $T_N = 450$ K, as revealed from Mössbauer effect measurements. LDA calculations [149] show that the magnetic ground state is an AFM structure with antiparallel spin setting on neighbouring Fe atoms. The electronic structure has many similarities with that of other FeAs-based compounds. A difference comes from the fact that Tl^+ is an electron donor, providing 0.5 electrons per Fe atom. This results in a considerable increase of the electron pockets at the corners of the Brillouin zone and, correspondingly, shrinking of the hole pockets at zone centre, so that nesting and the onset of the SDW structure get detuned. This is the reason why in stoichiometric $TlFe_2Se_2$, a purely AFM structure emerges. A deficiency in thallium would reduce the size of electron pockets, therefore it cannot be ruled out that $Tl_xFe_2Se_2$ would turn superconducting.

A detailed ARPES study of stoichiometric $BaFe_2As_2$ was done in [150] at two temperatures, $T = 20$ K and 300 K. No substantial difference was found between the electron spectra in paramagnetic tetragonal and antiferromagnetic orthorhombic phases. Two hole and one electron pockets were found, in agreement with LDA calculations, which also have demonstrated a quasi-two-dimensional character of spectrum.

An unexpected refinement of Fermi surface topology was done in [151] for $Ba_{1-x}K_xSrFe_2As_2$. Along with two hole surfaces concentric around the Γ point, electron pockets around X and Y were found, each one decorated by four hole segments protruding outwards towards the nearest Γ points. As a whole, this feature looks like a propeller centred at X (Y), see Fig. 3.8. Therefore, the structure of the Fermi surface in the AFe_2As_2 compounds turns out more complicated than that found so far in *Re*OFeAs.

Preceding calculations of the electronic structure of AFe_2As_2 have been done for stoichiometric composition. In [152], ARPES measurements of band structures for two non-stoichiometric $Ba_{1-x}K_xFe_2As_2$ compounds, the ultimately doped one ($x = 1$, $T_c = 3$ K) and the other optimally doped ($x = 0.4$, $T_c = 37$ K), have been reported. The Fermi surfaces for both compounds, extracted from the ARPES data, are shown in Fig. 3.9.

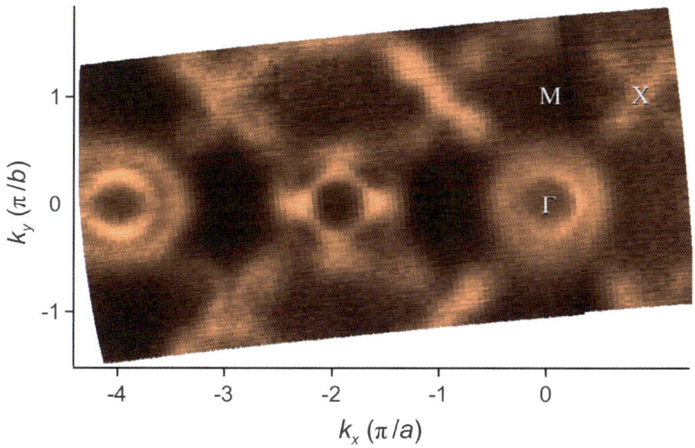

Fig. 3.8 Fermi surface topology for $Ba_{1-x}K_xFe_2As_2$ [151]

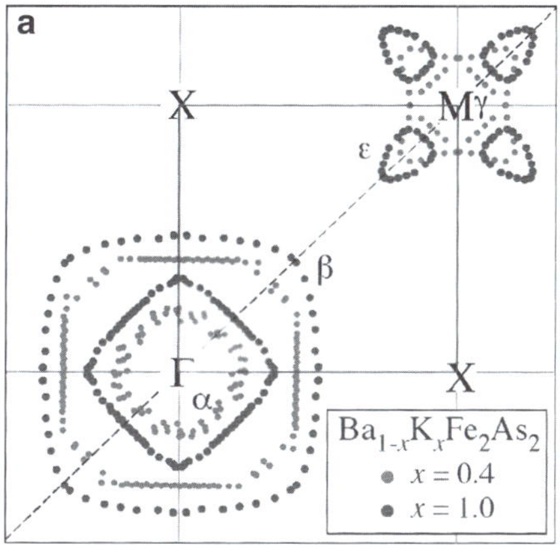

Fig. 3.9 Fermi surface of the $Ba_{1-x}K_xFe_2As_2$ compound at $x = 1$ and $x = 0.4$, recovered from the ARPES data [152]

In the optimally doped sample, two hole surfaces α and β are well seen, along with the ε-sheets around the M point, and the electron pocket γ. In the overdoped sample, due to an increased potassium content which provides the hole carriers, the electron pocket near M does fully disappear, whereas the hole pockets grow. In case of the optimal doping, we have to do with a good nesting on the $\mathbf{Q} = (\pi, 0)$ vector connecting the hole pocket α and the electron pocket γ, in the overdoped sample the nesting is absent because of the disappearance of the electron pocket. This difference in electronic structures of the optimally doped and overdoped compounds does convincingly demonstrate a role of electron scattering processes, which overthrow the electrons from α to Γ surface due to spin fluctuations, in the formation of a superconducting state.

In another work [153], the differences in the Fermi surface structure between the underdoped ($x = 0.25$; $T_c = 26$ K) and the optimally doped ($x = 0.4$; $T_c = 37$ K) samples have been confirmed. Besides this, in underdoped samples the ARPES

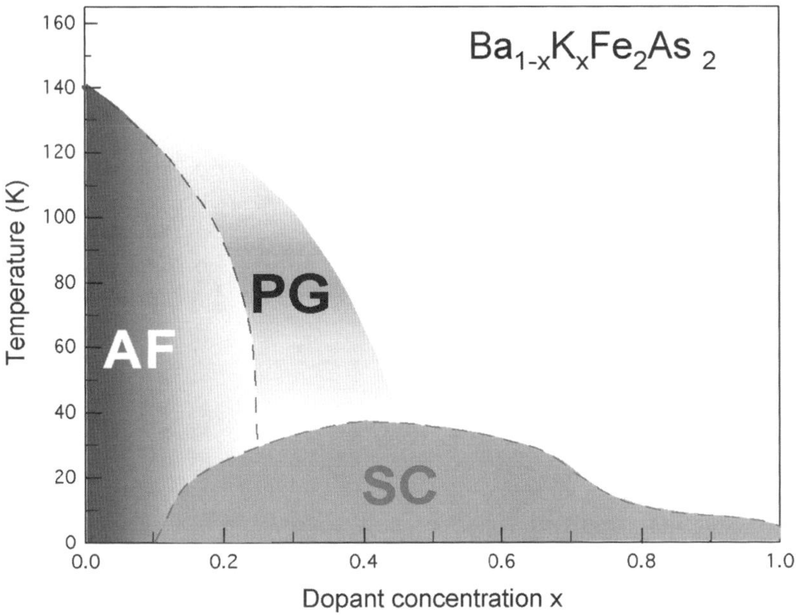

Fig. 3.10 Phase diagram in the (T, x)-plane, with the pseudogap (PG) state indicated after the ARPES data of [153]

measurements [153] have indicated the presence of a pseudogap, which exists up to temperatures of about T_N. A schematic phase diagram of the $Ba_{1-x}K_xFe_2As_2$ compound is shown in Fig. 3.10. The pseudogap state is formed due to the interaction of electrons with spin fluctuations and presents, apparently, an universal property of metallic systems which are close to magnetic phase transition. The phase diagram shown in Fig. 3.10 is similar to those established in cuprates.

The potassium-doped $Ba_{1-x}K_xFe_2As_2$ compound has been studied earlier by the ARPES method [154, 155]. Two hole pockets around Γ have been revealed, along with a rather complicated propeller-like picture in the vicinity of X and Y points of the Brillouin zone. In a recent work [156], the $Ba_{1-x}K_xFe_2As_2$ compound with $T_c = 32$ K has been thoroughly studied by high-resolution ARPES method. The central part of the propeller-shaped structure was found to be situated exactly in the X (Y) points, and the propeller blades directed along the X–Γ and Y–Γ lines. The measurements have been done at two temperatures, 10 and 75 K, and in both cases this structure was preserved, even if at 75 K the intensity of the spectrum was weaker. The central part of the observed complex structure corresponded to electron pockets, and the peripheric parts (blades) – to hole pockets. By changing the vector of photon polarization, it was easy to separately change the intensity of hole and electron pockets, from which it follows that these sheets of the Fermi surface belong to different bands. LDA calculations do not reproduce such shape of the Fermi surface, and its explanations demand for further studies.

ARPES investigations of $BaFe_2As_2$ and $CaFe_2As_2$ have been also done in another work [157], where it was shown that under magnetic ordering below T_N the Fermi surface gets reconstructed. In particular, in the ab plane the parallel segments have been found, which are related by the nesting vector which is much smaller than the wave vector of the SDW structure, $(\pi\pi)$. It was suggested that this nesting can be related to the charge density wave (CDW) structure. A complicated character of the Fermi surface reconstructed by the magnetic ordering needs to be studied further.

High-resolution ARPES measurements have been done on a $BaFe_2As_2$ single crystal, and moreover on an electron-doped $Ba(Co_{0.06}Fe_{0.94})As_2$ and hole-doped $Ba_{0.6}K_{0.4}Fe_2As_2$ compounds [158]. In all three cases, in the vicinity of the X point two electron and two hole bands have been found; however, there are electron bands only which cross the Fermi level whereas the hole bands are situated below it. The electron bonds hybridize with the hole ones giving rise to a gap somewhere along the symmetric Γ–X direction. The authors of [158] argue that their interpretation of the ARPES data for $BaFe_2As_2$ does agree well with LDA calculations and is more convincing than that of [151], where a propeller-like structure of the Fermi surface in the vicinity of the X point has been reported. Both works [151, 158] are consistent in the sense that the electronic structure of spectrum in the vicinity of X in the $BaFe_2As_2$ compound is more complex than the earlier LDA calculations [131, 133] had suggested.

LDA calculations of compounds of the $BaFe_2As_2$-type result in Fermi surfaces, which might be considered as quasi-two-dimensional ones in the first approximation only. In fact, the cylinders do vary somehow as we probe them along the k_z direction. How strong the three-dimensional features are pronounced can be judged on the basis of ARPES experiments. Thus, the measurements on $BaFe_2As_2$ and on an electron-doped compound $Ba(Fe_{1-x}Co_x)_2As_2$ did reveal a three-dimensional character of the corresponding sheets of the Fermi surface [146, 159].

Interesting results have been obtained for $CaFe_2As_2$ [160]. It turned out that on structural magnetic transition, a change of effective dimension of the electron spectrum takes place. At $T > T_s$, the electron spectrum is quasi-two-dimensional, but at $T < T_s$ it turns three-dimensional. Specifically, for $T > T_s$ the electron and hole pockets remain in a reasonably good approximation, cylindrical whereas for $T < T_s$ they become ellipsoidal. This indicates that for a superconducting state to develop in FeAs layers, the dimensionality of the electron system may be not that important as it is broadly assumed to be.

Now we discuss the results of Fermi surface recovery from the data on quantum oscillations in magnetic field, which method permits to detect extremal cross-sections of the Fermi surface by planes normal to the magnetic field direction. Measurements of quantum oscillators in a $BaFe_2As_2$ single crystal at low temperatures [143] provided cross-sections, which do not agree with LDA calculations of the Fermi surface in a nonmagnetic state [134]. In the presence of a long-range magnetic order (SDW structure), the Fermi surface gets reconstructed, however, as it was shown in [161], the gap is formed not over the whole Fermi surface, therefore the compound in question is a metallic antiferromagnet. In [143], spin-polarized LDA

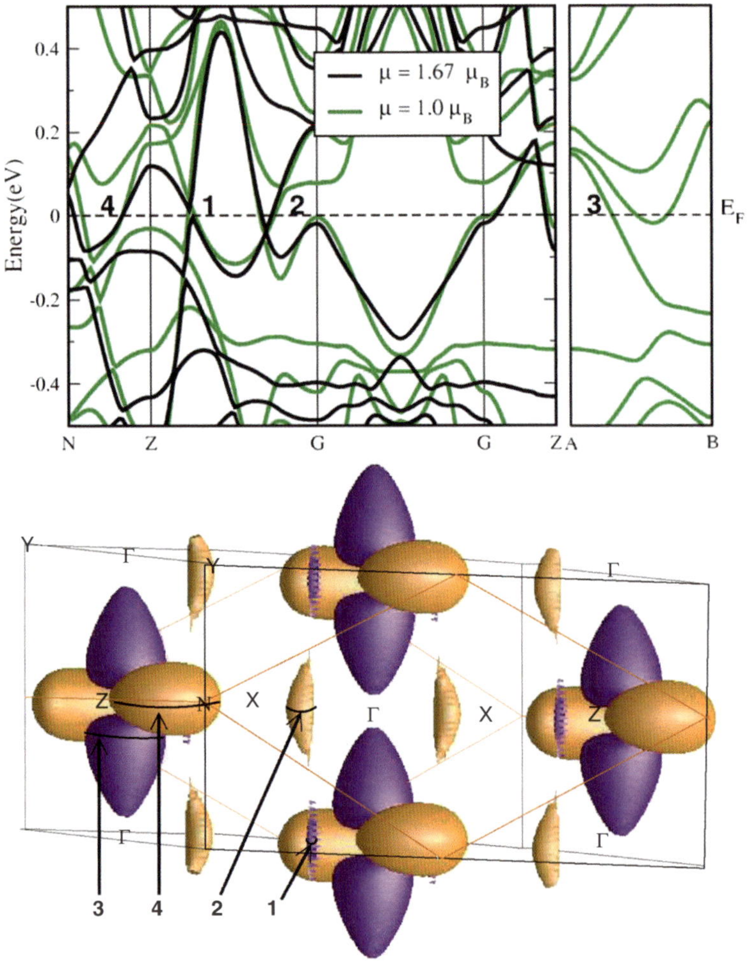

Fig. 3.11 Band structure and the Fermi surface as calculated for $BaFe_2As_2$ taking into account the SDW structure [143]

calculations have been done, and the resulting Fermi surface of antiferromagnetic $BaFe_2As_2$ is shown in Fig. 3.11.

The calculation of the band spectrum has been done for two values of mean magnetic moment, $\mu = 1.67\,\mu_B$ and $\mu = 1\,\mu_B$. In the bottom panel of Fig. 3.11, the extremal cross-sections (numbered and indicated by arrows) of the sheets of the Fermi surface are shown, for magnetic field directed along the c axis. From the observed dHvA effect, three pockets have been determined, which occupy 1.7%, 0.7% and 0.3% of the paramagnetic Brillouin zone. The obtained small pockets agree with the shape of the Fermi surface shown in Fig. 3.11. The effective mass on each sheet is of the order of 1.4 m_e (m_e being the free electron mass).

Quantum oscillations have also been measured for the $CaFe_2As_2$ compound [162]. The largest pocket occupies less than 0.05% of the volume of paramagnetic Brillouin zone, which agrees with the magnetically reconstructed Brillouin zone, resulting from the magnetic order. In another compound, $BaNi_2P_2$, which turns superconducting with $T_c \sim 3$ K under doping, the measurement of quantum oscillations revealed six extremal cross-sections of the Fermi surface, which belong to four bands. The Fermi surface turned out to be large, comprising about one electron and one hole per formula unit. The effective mass is $m^* \approx 2\,m_e$. The identified extremal cross-sections and effective masses are in good agreement with the calculated reconstructed Fermi surface [163].

In [164], quantum oscillations have been measured in the $CaFe_2P_2$ compound, close to the state of collapsed tetragonal phase in $CaFe_2As_2$. The clarified topology of the Fermi surface, which is a three-dimensional one, is markedly different from that for $CaFe_2As_2$. In the corners of the Brillouin zone, electron sheets, strongly modulated along the c direction, are situated. Around the Brillouin zone centre, instead of a hole cylinder, a surface is found in the shape of two pillows, which are placed at the upper and lower faces of the Brillouin zone symmetrically with respect to Γ. No nesting is possible that apparently explains why no superconductivity is observed in this system. An enhancement of electron mass on hole and electron sheets has a factor of 1.5 – seemingly, due to electron–phonon interaction.

3.1.4 $(Sr_3Sc_2O_5)Fe_2As_2$ and Other Similar Compounds

In [165], a new class of layer FeAs-systems, whose crystal structure is more complicated than that of AFe_2As_2, has been synthesized and characterized. In the $(Sr_3Sc_2O_5)Fe_2As_2$ compound, instead of an atom A, a group $Sr_3Sc_2O_5$ is inserted between the FeAs layers. Consequently, the distance L between consecutive FeAs-layers gets substantially enlarged. This met a big interest because of an observation that the maximal T_c values in different classes of compounds do correlate with L.

Indeed, in the LaOFeAs systems the maximum $T_c \sim 55$ K and $L \sim 8.7$ Å. In the systems of the AFe_2As_2 type, the maximum $T_c \sim 38$ K, with $L \sim 6.4$ Å. Hence, on increasing L in this row of compounds the maximum T_c does also increase. From extrapolating this tendency, one could expect in the new class of compounds, where $L \sim 13.4$ Å, the maximum T_c to be even higher. In the stoichiometric $(Sr_3Sc_2O_5)Fe_2As_2$ compound, however, neither lattice nor magnetic instability has been discovered, therefore it is not yet clear whether this compound might turn superconducting under doping or external pressure. Nevertheless, an elucidation of its electronic structure is of interest.

The crystal structure is shown in Fig. 3.12 (space group $I4/mmm$). Calculated values of the lattice parameters, $a_0 = 4.0952$ Å, $c = 26.3935$ Å are in good agreement with experimental data $a_0 = 4.0678$ Å, $c = 26.8473$ Å [165]. Electronic structure of the new compound has been calculated by the standard full-potential linearized augmented plane wave (FLAPW) method [166]. The results obtained fall

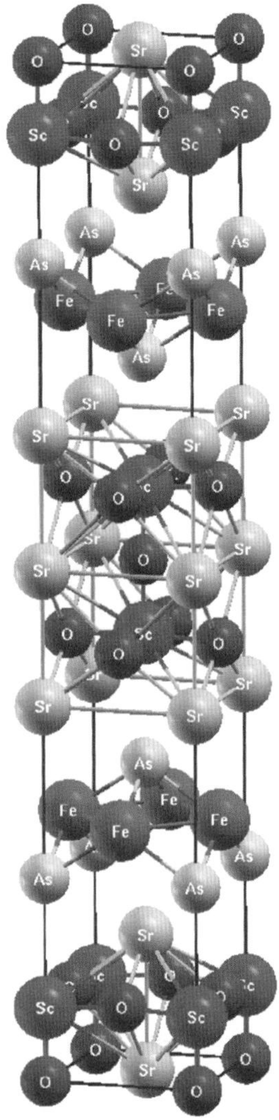

Fig. 3.12 Crystal structure of the $(Sr_3Sc_2O_5)Fe_2As_2$ compound [165, 166]

into the common scheme for all types of the FeAs-systems, namely: the states at the Fermi level are primarily formed by the Fe3d orbitals; the Fermi surface is multi-sheet one and contains three hole pockets in the Brillouin zone centre, and two electron pockets in the corners. The density of states at the Fermi level is $N(E_F) = 1.456\,(\text{eV})^{-1}$ per Fe atom; for comparison $N(E_F) = 1.860\,(\text{eV})^{-1}$ in $BaFe_2As_2$. This conclusion leaves hope that the new compound $(Sr_3Sc_2O_5)Fe_2As_2$ might become a prototype of a new row of superconductors. However it remains to clarify why it does not exhibit a structural nor magnetic instability.

In [167], a synthesis of a new member of the FeAs-family, with $T_c = 17\,\text{K}$, has been reported. The chemical formula $(Sr_4Sc_2O_6)Fe_2P_2$ reveals a crystallographic structure composed of the FeP layers (analogous to the FeAs ones), interlaced with layers made of $Sr_4Sc_2O_6$ complexes. The compound has tetragonal space group $P4/nmm$ with the following experimentally determined lattice parameters: $a = 4.016$ Å; $c = 15.343$ Å. Its electronic structure was calculated in [168], whereby the optimized lattice parameters $a_{calc} = 4.008$ Å and $c_{calc} = 15.444$ Å were found in reasonable agreement with the experimental data. Standard LDA calculations yielded an electronic structure quite typical for all FeAs-compounds. Thus, the DOS near the Fermi level is formed by the d states of Fe atoms. The Fermi surface consists of four sheets: two hole ones, concentric around the Γ–Z direction and two electron ones along the M–A line. All sheets are parallel to the k_z direction. Parameters of the sheets of the Fermi surface and the density of states at the Fermi level have values typical for FeAs-systems. Also, the electronic structure of a hypothetical compound $(Sr_4S_2O_6)Fe_2As_2$, which differs only slightly from the true one, constructed from the FeP planes, has been calculated. If such compound would ever be synthesized, one could expect an advent of a new row of FeAs-based superconductors.

A synthesis of new complex materials $(Sr_3S_2O_5)Fe_2As_2$ and $(Sr_4Sc_2O_6)Fe_2P_2$ gives a chance to attend an arrival of further compounds, where the FeAs layers are separated by motifs built of complex atomic components. In [169–171], a synthesis of new compounds with complex insertions into the AFe_2As_2 structure is reported: $(Sr_4Sc_2O_6)M_2As_2$ (M = Fe, Co), $(Sr_4M_2O_6)Fe_2As_2$ (M = Sc, Cr). Of special interest is the $(Sr_4Sc_{2-x}Ti_xO_6)Fe_2As_2$ compound, in which a substitution of Sc by Ti increases the carrier concentration and leads to superconductivity. It is reported in [171] that under doping with 30% Ti, the onset of superconductivity, as estimated by resistivity measurements, occurs at 45 K.

Motivated by this result, the authors of [172] performed an LDA calculation of the electronic structure of (Sr_4ScTiO_6) Fe_2As_2. It turned out that the states of the Ti ion come close to the Fermi level, therefore the (Sr_4ScTiO_6) complexes yield a contribution to conductivity, differently from other FeAs-systems, in which the atomic interlayers between the FeAs-layers do not give any immediate effect on conductivity. This result clarifies in part why the titanium-doped compound $(Sr_4Sc_2O_6)$ F_2As_2 is superconducting.

It is obvious that the Fe_2As_2 class of compounds with interlaced complexes will get a further extension and might become a germ of a new family of FeAs-type superconductors. Indeed, Zhu et al. [173] reports a synthesis of the $(Sr_4Ti_{1.2}Cr_{0.8}O_6)Fe_2As_2$ compound which turned out to be a superconductor with $T_c = 6$ K. In [174], new Ni-based compounds with the composition $(Sr_4Sc_2O_6)(Ni_2Pn_2)$, Pn = P, As, have been synthesized. At stoichiometry, they are superconductors, albeit with not high T_c values. Thus, the phosphorus-containing compound has $T_c = 3.3$ K, and arsenic-containing one – $T_c = 2.7$ K. Moreover, two further compounds of a similar type have been synthesized, $(Sr_2CrO_3)Fe_2As_2$ and $(Ba_2ScO_3)Fe_2As_2$ [175], which, however, are not superconducting. The whole class of AFe_2As_2 compounds, where A stands for a multiatomic complex, possesses the same tetragonal space group $P4/nmm$ with the cell elongated along the c direction, because of a large size of complexes. This is undoubtedly a new class of superconducting compounds on the basis of Fe, providing great opportunities of chemical variation among the complexes. The electronic structure of such compounds, as the already performed calculations [168, 176] have demonstrated, is similar to the electronic structure of other classes of the FeAs-based compounds.

The superconducting transition temperature in these new FeAs-systems may strongly vary with pressure. Thus, [177] reports that the pressure applied to superconducting compounds $(Sr_4Se_2O_6)Fe_2As_2$ and $(Sr_4Se_2O_6)Fe_2P_2$ changes their T_c in different senses. In the former compound, T_c drifts from 36.4 K at ambient pressure to 46.0 K at $P \sim 4$ GPa. Contrary, in the second compound T_c drops from 16 K to 5 K at $P \sim 4$ GPa. This difference is apparently explained by a difference in the positioning of pnictide atoms relative to the plane formed by the Fe atoms. Thus, in the phosphorus-containing system, the "height" $h_P = 1.20$ Å, whereas in the As-based compound, $h_{As} = 1.42$ Å. On the other hand, in the NdOFeAs, $h_{As} = 1.38$ Å, i.e. in between the h_P and h_{As} values for the complex systems discussed above. This possible correlation has yet to be checked for other compounds with complex atom motives inserted between the FeAs-layers.

A remarkable discovery of the recent time is the finding of high-temperature superconductivity (with $T_c = 37.5$ K) in stoichiometric compound $Sr_4V_2O_6Fe_2P_2$ [178]. Similarly as the $Sr_4Sc_2O_6Fe_2P_2$ compound in which superconductivity with $T_c = 17$ K under stoichiometry has been detected earlier, the present system is not an antiferromagnet [179]. The Hall effect measurements indicate that the dominating conductivity is of the electron type. This compound synthesized in [178] is the first one among the whole row of the FeAs-systems in which superconductivity exists under stoichiometry, and moreover with quite high T_c. This fact demands a thorough analysis, since getting it explained would be important for understanding the mechanism of superconductivity in the FeAs-systems.

3.2 Superconductivity

3.2.1 Doping

Record-setting values of $T_c = 38$ K were obtained by doping $BaFe_2As_2$ with potassium. The charge carriers in $Ba_{1-x}K_xFe_2As_2$ are holes, which is confirmed by Hall effect measurements and expected from the crystallochemistry of the compound. Due to closeness of atomic radii of Ba and K, a complete substitution of the one by the other is possible, so that the $(Ba_{1-x}K_x)Fe_2As_2$ compound can be obtained in the whole range $0 < x < 1$. The measurements have shown that superconductivity exists throughout broad x range but the maximum value $T_c = 38$ K occurs for $x = 0.4$ (Fig. 3.13).

A discovery of superconductivity in $BaFe_2As_2$ under electron doping, induced by a substitution of Fe atoms by Co, stimulated big activity among the researchers. In [182], based on systematic measurements of electrical resistivity, heat capacity, magnetic susceptibility and the Hall coefficient, the (x, T) phase diagram of

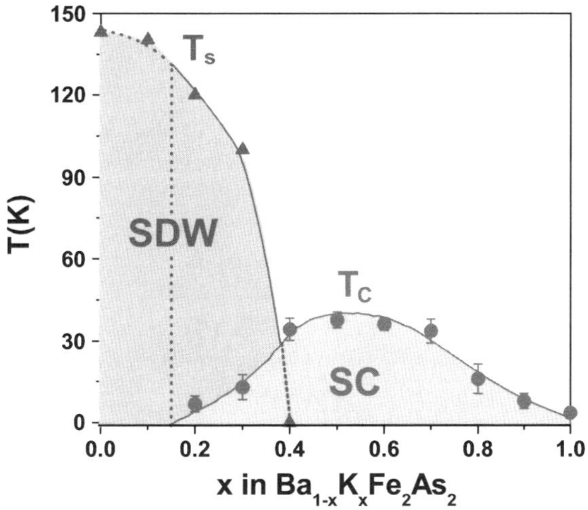

Fig. 3.13 Temperatures of magnetic and superconducting transition in $Ba_{1-x}K_xFe_2As_2$ as functions of the level of doping [181]

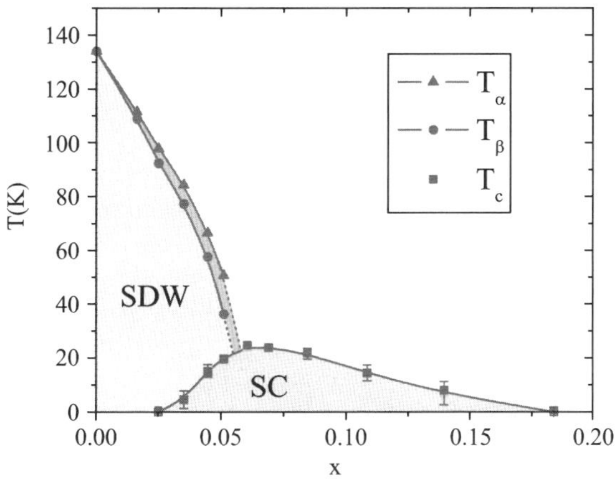

Fig. 3.14 Phase diagram of the $Ba_{1-x}Co_xFe_2As_2$ with the electron doping [182]

$Ba(Fe_{1-x}Co_x)_2As_2$ has been constructed (Fig. 3.14). At low Co concentrations, as the temperature decreases, a phase transition from tetragonal into orthorhombic phase occurs, accompanied by an onset of SDW magnetic ordering. Two closely lying lines of a phase transition have been found. As in the other FeAs-based compounds, the phase transition temperature does rapidly decrease as the dopant concentration grows. The superconductivity appears at $x \approx 0.025$ and persists till concentrations $x \approx 0.16$ – into the range where the magnetic ordering is already suppressed; however, there is an interval $x = 0.025$–0.06, where the superconductivity and the SDW phase do coexist for $x = 0.025$–0.06. As the authors of [182] carefully note, this region has yet to be thoroughly studied, since a different explanation can be given for the "coexistence" of magnetic and superconducting phases.

Similar observations have been done in two other publications, which appeared almost simultaneously [183, 184]. In the former, an analogous phase diagram as above, with a splitting of the structural (magnetic) phase transition, has been found. In the latter publication, NMR measurements on undoped and overdoped samples have been done and, at low temperatures, the quantum critical point found, which separates the magnetic and the superconducting phases. In [185], combined studies on a single crystal of $Ba(Fe_{0.9}Co_{0.1})_2As_2$ revealed $T_c = 25$ K and a high anisotropy of the upper critical field, $\Gamma = H_{c2}^{ab}/H_{c2}^{c} = 3.5$.

According to [186], the pristine compound $BaFe_2As_2$ doped with cobalt is a superconductor with $T_c = 22$ K. The Hall effect data indicate that the charge carriers are electrons, that is, again, expected from the crystallochemistry of the compound. These facts emphasize a big difference of FeAs-systems from cuprates, in which a substitution of Cu in the CuO_2 planes suppresses superconductivity. In the sample studied, $Ba_2Fe_{1.8}Co_{0.2}As_2$, a substitution of Fe by Co results in a substantial disorder in the FeAs-planes, not only does not suppress superconductivity, but in fact helps it, as it destroys the AFM ordering in the initial system. These studies have been continued by the authors with the use of NMR, whereby it was demonstrated that $BaFe_{2-x}Co_xAs_2$ with $T_c = 22$ K reveals a pseudogap state with the pseudogap magnitude $\Delta_{PG} \approx 560$ K [187]. Note that a similar situation has been earlier detected in the LaO(Fe,Co)As system [38]. Moreover there are reports about a substitution of a rare-earth element by sodium, resulting in a high T_c. Thus, in $Eu_{0.7}Na_{0.3}Fe_2As_2$ the superconductivity with $T_c = 34.7$ K was detected [188].

Similar results have been obtained under doping by cobalt of the $SrFe_2As_2$ compound. In the doped $SrFe_{2-x}Co_xAs_2$, within the x range $0.2 < x < 0.4$, superconductivity with $T_c = 20$ K was detected [189]. An existence of superconductivity under the conditions of such strong disorder makes an assumption about the p- or d-symmetry of the order parameter rather problematic, since it is known that such superconductivity is suppressed already by quite low degree of disorder. LDA calculations done for $x \geq 0.3$ in [189] show that the AFM ordering must have been completely suppressed, and consequently the conditions for superconductivity to appear may be created.

In [190] it was reported that in $CaFe_{2-x}Co_xAs_2$, the superconductivity appears already at $x = 0.06$ and has $T_c = 34.7$ K. Detailed discussion of these questions is presented in review [191].

A substitution of Fe in the AFe_2As_2 compounds by other elements (Ni, Pd, Ru, Rh) leads to similar effects as the substitution by Co. Thus, in the $BaFe_{2-x}Ru_xAs_2$ compound [192], the doping with Ru does suppress the long-range magnetic order and results in a superconducting state with the maximum $T_c = 20.8$ K at $x = 0.75$. In $SrFe_{2-x}Ru_xAs_2$ [193], the suppression of antiferromagnetism and onset of superconductivity does also happen, with $T_c = 13.5$ K at $x = 0.7$. In $SrFe_{2-x}Rh_xAs_2$ [194], the superconductivity sets on at $x > 0.1$, and the T_c reaches its maximum of 22 K at $x = 0.25$. Under substitution of Fe by palladium in the $SrFe_{2-x}Pd_xAs_2$ compound [195], the maximum $T_c = 8.7$ K is for $x = 0.15$. In all these systems, the (T, x) phase diagrams are similar to those for the case of Co substituting Fe. Differently from the doping of the AFe_2As_2 compounds with Pd, Ru, or Rh, their doping with Cr does suppress the long-range magnetic order without resulting in superconductivity [196]. In the $SrFe_2As_2$ compound, a substitution of Fe by Ir adds electron carriers, and in $SrFe_{2-x}Ir_xAs_2$ with $x = 0.5$ the superconductivity appears with $T_c = 22.3$ K and critical field $H_{c2}(0) \approx 58$ T [197].

A study of (T, x)-phase diagrams for $SrFe_{2-x}M_xAs_2$ compounds (M = Rh, Ir, Pd) was continued in [198]. In total, the phase diagrams turned to be similar to those of the compounds with Co and Ni substituting Fe. The maximum T_c in the compounds with Co, Rh and Ir have close values, although the masses of these ions are markedly different. This provides an additional argument in favour of non-phonon mechanism of pairing.

Under full replacement of Fe by, for example, Ni, a substantial change of the compound's properties takes place. Thus, it was detected for a single crystal $BaNi_2As_2$ [199] that the structural and magnetic phase transitions at $T_0 = 130\,K$ are of the first order, the Hall coefficient is negative and the superconducting transition temperature is low: $T_c = 0.7\,K$. The upper critical field is anisotropic and has an initial slope $dH_{c2}^c/2dT = -0.19\,T/K$, $dH_{c2}^{ab}/dT = -0.40\,T/K$.

Superconductivity with $T_c \approx 34$–$36\,K$ was achieved in $K_{1-x}Ln_xFe_2As_2$ (Ln = Sm, Nd, La) [200]. The pristine compound KFe_2As_2 ($x = 0$) turned out to be superconducting with $T_c \approx 3\,K$, showing neither a structural nor an SDW phase transition. The doped compounds with high T_c have hole carriers, similar to the case of the earlier studied AFe_2As_2 (A = Ba,Sr,Ca) systems doped by potassium. This confirms a dominating role of the FeAs-layers in the formation of superconductivity in $Z_{1-x}K_xFe_2As_2$ compounds with different Z elements.

In the $BaFe_2As_2$ compound, not only Fe but also As can be substituted. The effects of phosphorus doping in the $BaFe_2(As_{1-x}P_x)_2$ compound have been studied in [201] for $0 < x < 0.54$. Notwithstanding the isovalent type of doping, the SDW state of the pristine $BaFe_2As_2$ compound was suppressed by it. At $x \approx 0.3$, a linear dependence of magnetic susceptibility on T was observed, and a superconducting state was formed with $T_c \approx 31\,K$. On further doping, non-Fermi-liquid anomalies did gradually disappear, and the system entered the Fermi liquid regime. LDA calculations have shown that the Fermi surfaces of two limiting stoichiometric compounds, $BaFe_2As_2$ and $BaFe_2P_2$, are substantially different, notably the hole pockets in the latter case are larger. Therefore on doping with P, the hole pockets do grow whereas the electron ones remain unchanged, so that the nesting is detuned, leading to a suppression of the SDW ordering.

A large part of recent works on superconductivity in FeAs-systems deals with compounds of the $BaFe_2As_2$ type, notably those doped with Co [202, 203]. For a dopant, Ni can be taken instead of Co. In the $BaFe_{2-x}Ni_xAs_2$ compound, the maximum T_c is 21 K. Recently, a substitution of Fe by Ni has been studied in the $SrFe_{2-x}Ni_xAs_2$ system, and $T_c \approx 10\,K$ obtained. It is remarkable that the maximum of T_c in this compound is achieved at the concentration of Ni which is two times smaller than when taking Co as a dopant. This is related to the fact that Ni delivers two times more excess electrons (compared to Fe) than Co does. The evolution of T_c with concentration x for different dopants is shown in Fig. 3.15 for the $SrFe_{2-x}M_xAs_2$ compounds [206]. In all cases, on doping the initial stoichiometric $SrFe_2As_2$, the temperature T_N of the SDW magnetic ordering decreases as x grows, and superconductivity sets on at the full suppression of the magnetic order. In [207], the (H, T) phase diagram has been studied in detail, and the anisotropy of the upper critical field with its components $H_{c2\parallel}$ and $H_{c2\perp}$, parallel and perpendicular to the

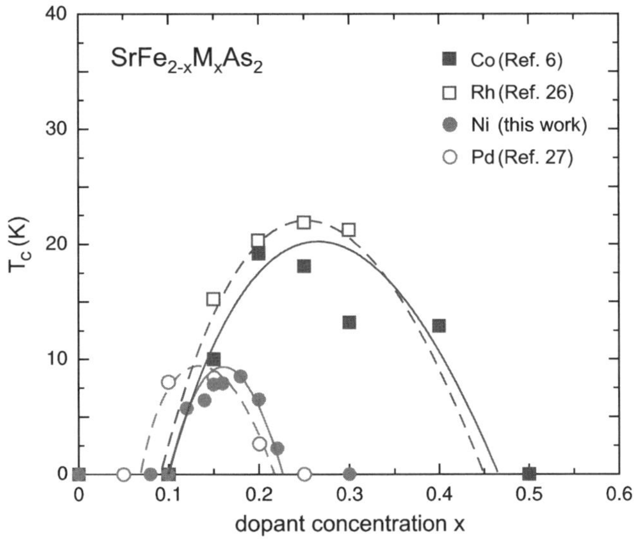

Fig. 3.15 Comparison of $T_c(x)$ dependencies for some $SrFe_{2-x}M_xAs_2$ compounds at different dopants M substituting Fe; M = Co [204], Rh [205], Ni [206], Pd [194]

ab plane of the crystal, has been determined. The measurements of the critical field are usually done near T_c, and from the dH_{c2}/dT value at $T = T_c$, using a known extrapolation, the field magnitude $H_{c2}(0)$ at $T = 0$ is obtained. This method results in an overestimated value of $H_{c2}(0)$. Thus, in [208], in the $(Fe,Co)_2As_2$ superconductor the H_{c2} was measured up to the fields of 45 T, and by extrapolation the value $H_{c2\|}(0) \approx 70$ T has been obtained.

In [207], the (H, T) phase diagram for the $BaFe_{1.84}Co_{0.16}As_2$ single crystal has been constructed from immediate measurements of electric resistivity in the field, at low temperatures, without applying any extrapolations (Fig. 3.16). It is seen from the Figure that $H_{c2\perp} = 50$ T and $H_{c2\|} = 55$ T, i.e. they are substantially smaller than the earlier reported result $H_{c2\|}(0) \approx 70$ T of [208]. These values exceed the paramagnetic limit $\mu_B H_p = 1.84\,T_c$, which reveals a non-standard superconductivity of this compound. The anisotropy parameter $\Gamma = H_{c_{2\|}}(0)/H_{c_{2\perp}}(0)$ shown in the inset of Fig. 3.16 exhibits its maximum value of 3.4 for $T_c = 21$ K and decreases at lower temperatures.

From the values obtained of $H_{c2\|}(0)$ and $H_{c2\perp}(0)$, the coherence length can be estimated along the formulae:

$$\xi_\| = (\Phi_0/2\pi H_{c2\perp})^{1/2}, \quad \xi_\perp = \Phi_0/2\pi\xi_\| H_{c2\|},$$

whence it follows that at $T = 0.7$ K, $\xi_\| = 25.8$ Å and $\xi_\perp = 24.7$ Å, and at $T = 22$ K both ξ values equal 31.9 Å. The coherence length in the perpendicular direction, $\xi_\perp$, is much larger than the thickness of the superconducting layer $d = c/2 = 6.49$ Å, which indicates that superconductivity in this compound

Fig. 3.16 $(H-T)$ phase diagram of $BaFe_{1.84}Co_{0.16}As_2$ in the fields perpendicular to the *ab*-plane and directed along the *c* axis [207]. Inset: temperature dependence of the anisotropy parameter γ

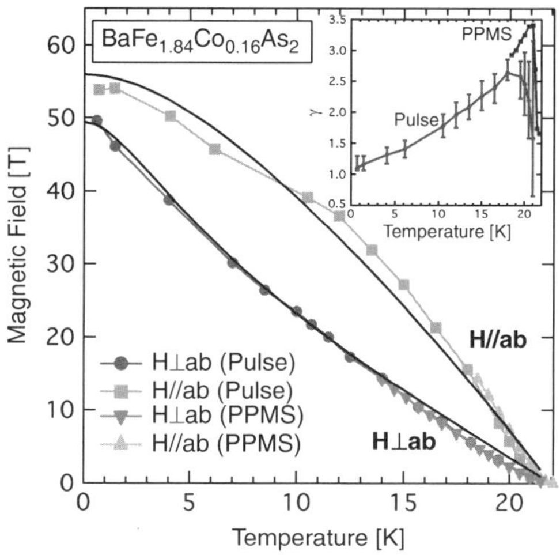

does not split into superconductivity of individual FeAs-layers and hence possess a three-dimensional and not two-dimensional character.

3.2.2 *Coexistence of Superconductivity and Magnetism*

A common rule for different FeAs-compounds is the following: under doping of a pristine stoichiometric compound, first a long-range magnetic order appears, and then the superconducting state sets on. Hereby whatever goes on at the boundary between the magnetic and the superconducting phases needs a detailed investigation. In some compounds, e.g. $CeO_{1-x}F_xFeAs$ [12], the both phases exclude one another, but in other examples such as $Ba_{1-x}K_xFe_2As_2$ [209], $Ba(Fe_{1-x}Co_x)_2As_2$ [183,210] and $SmO_{1-x}F_xFeAs$ [75], the regions of the magnetic and the superconducting phases do overlap, and the question poses whether they cohabitate in the same volume, or coexist side by side as dispersive fractions.

In Fig. 3.17, the phase diagram with such overlapping regions for an electron-doped compound $Ba(Fe_{1-x}Co_x)_2As_2$ is shown, as obtained from neutron diffraction in [211]. From the entirety of experimental data the authors draw a conclusion that superconductivity appears within the orthorhombic phase of this compound, and the region of phase coexistence is microscopically homogeneous.

A similar phase diagram has been obtained by authors of [212], in which work the coexistence region spreads up to the Co concentration $x = 0.06$. In [213], NMR spectra and spin–lattice relaxation rate $1/T_1$ on ^{75}As nuclei have been studied. An existence of static magnetic moment on each Fe atom has been shown. Below $T_c = 21.8$ K, superconductivity in the whole volume of the sample was detected. A comparison of these two observations allows to unambiguously state that in the $Ba(Fe_{1-x}Co_x)_2As_2$ compound, the coexistence of magnetic ordering

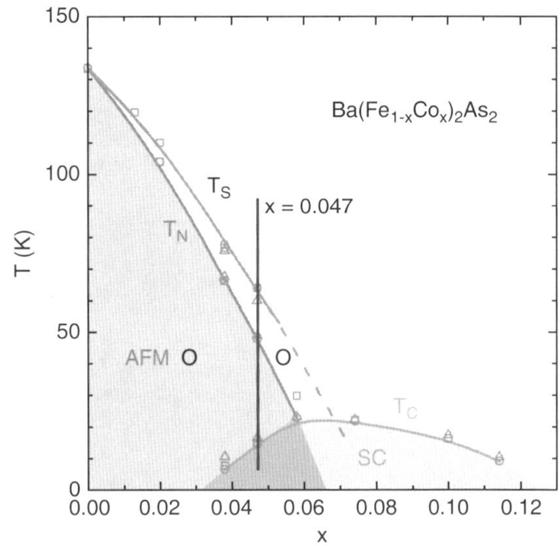

Fig. 3.17 Phase diagram of $Ba(Fe_{1-x}Co_x)_2As_2$ [211]

and superconductivity occurs at the atomic level. This experimental result is well described by the theory outlined in Sect. 5.6.3.

In [214], the μSR spectroscopy was applied for a study of two compounds: an electron-doped $BaFe_{2-x}Co_xAs_2$ and hole-doped $Pr_{1-x}Sr_xFe_2As_2$. In the first case, an existence of magnetism within the superconducting phase has been established. On the contrary, in the hole-doped compound a mesoscopic decomposition into two phases, – the almost unchanged antiferromagnetic normal one, and nonmagnetic and superconducting one, – has been found. With the help of magnetic and transport measurements, it was moreover demonstrated that in the $Sr_{1-x}K_xFe_2As_2$ compound, at low level of doping, a coexistence of magnetism and superconductivity takes place [215].

Based on measurements of transport properties and the upper critical field [89], the phase diagram of $Ba(Fe_{1-x}Co_x)_2As_2$ was constructed, which turned out to be of the $Sr_{1-x}K_xFe_2As_2$ type, as is seen from Fig. 3.17. The authors do not rule out an existence at $T = 0$ of the quantum critical point separating the antiferromagnetic and the superconducting phases, as a possible alternative to a hypothesis of the latter's coexistence.

Indications of coexistence of magnetism and superconductivity have been found in doped AFe_2As_2 compounds, in which electron inhomogeneity, related to chemical inhomogeneity, is inevitable. In [216], for the first time a coexistence of these two order parameters in stoichiometric $SrFe_2As_2$, under application of pressure, has been detected. In the (T, P) plane, the coexistence range near the pressure values of 5 GPa has been found. The superconductivity appears in a narrow range of pressures. NMR spectra at 5.4 GPa indicate a simultaneous existence of magnetically order and superconducting phase; however, they remain spatially separated. The magnetism exists over ranges of nanosize volume in the orthorhombic phase, whereas the superconductivity – within the regions of tetragonal phase. This makes a beautiful example of self-organization in a homogeneous structure of a chemically pure system.

3.2.3 *Effect of Pressure*

Besides being dependent on doping, the physical properties of the AFe_2As_2 compounds are subject to substantial changes under external pressure. Thus, [217] shows that the undoped $SrFe_2As_2$ undergoes structural and magnetic phase transitions in the volume at $T_0 = 205$ K. Under applied pressure, this temperature decreases and, according to estimates, T_0 may drop to zero at the pressure of 4–5 GPa. Under pressure of slightly above 2.5 GPa, the electrical resistivity sharply falls down, indicating a tendency towards superconductivity.

Among unique properties of the systems with double FeAs-layers, we point out the induction of superconductivity by applying pressure to stoichiometric AFe_2As_2 (A = Ba,Sr) compounds, discovered in [218]. In $SrFe_2As_2$ under a pressure of 28 kBar, the superconductivity with $T_c = 27$ K is induced, and in $BaFe_2As_2$ under 35 kBar – $T_c = 29$ K (Fig. 3.18). A similar effect occurs in $CaFe_2As_2$ [219], where superconductivity appears under the pressure of 0.35 GPa.

In [220], a thorough investigation of this compound has been done using neutron diffraction methods. An astonishing result has been obtained: at a fixed temperature ($T = 50$ K) and while applying pressure, a "collapse" of the initial tetragonal structure occurs, which means that the unit cell volume drops down drastically by about 5%, without changing the cell symmetry. On subsequent increase of pressure,

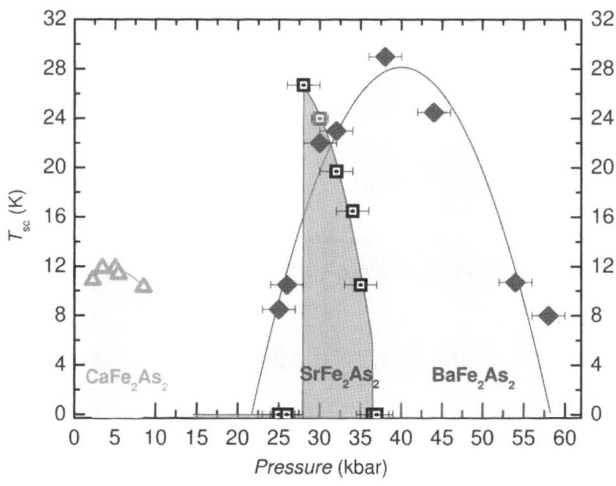

Fig. 3.18 Superconducting transition temperature as function of pressure applied to the AFe_2As_2 (A = Ca,Sr,Ba) compounds [218]

a collapsed tetragonal structure remains stable, maintaining its cell parameters, up to maximal values of the applied pressure of about 0.6 GPa. Hereby, the structural transition from the tetragonal into orthorhombic phase and the associated magnetic transition remain blocked once the superconducting phase emerges. Remarkably, numerical LSDA calculations indicate that the magnetic moments disappear on the transition into the collapsed phase; neutron diffraction studies do not detect them either. The collapse can be otherwise achieved while the pressure is kept fixed and the temperature gradually lowered.

Summarizing, for this compound the following phase diagram on the (T, P) plane is realized. At $P \leq 0.35$ GPa and high T, the substance is in the initial tetragonal phase, which can transform into the collapsed phase, corresponding to a normal metal. Within the region of existence of the tetragonal phase, along a certain interval of P and at low enough T, the superconductivity takes place. A transition into the superconducting phase occurs from the collapsed tetragonal phase.

The results of theory analysis of the situation outlined on the basis of LDA calculations is depicted in Fig. 3.19. In Fig. 3.19a, b, calculated total energy is presented

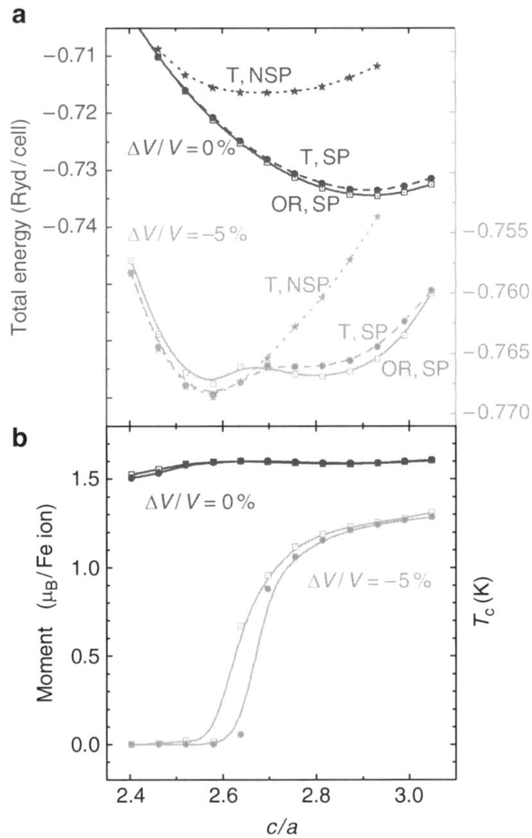

Fig. 3.19 Results of neutron diffraction studies and total energy calculations for $CaFe_2As$ [220]. (**a**) Spin-polarized (SP) and non-polarized calculations of total energies ($\Delta V/V = 0\%$ and $\Delta V/V = -5\%$) for the tetragonal (T) and orthorhombic (OR) phases. (**b**) For the "collapsed" phase, the moment on Fe is frozen at the minimal total energy in the spin-polarized calculation

Fig. 3.20 Phase diagram for $CaFe_2As_2$, obtained from neutron diffraction and X-ray diffraction studies under hydrostatic pressure. O: orthorhombic phase; T: tetragonal phase; cT: collapsed tetragonal phase [221]

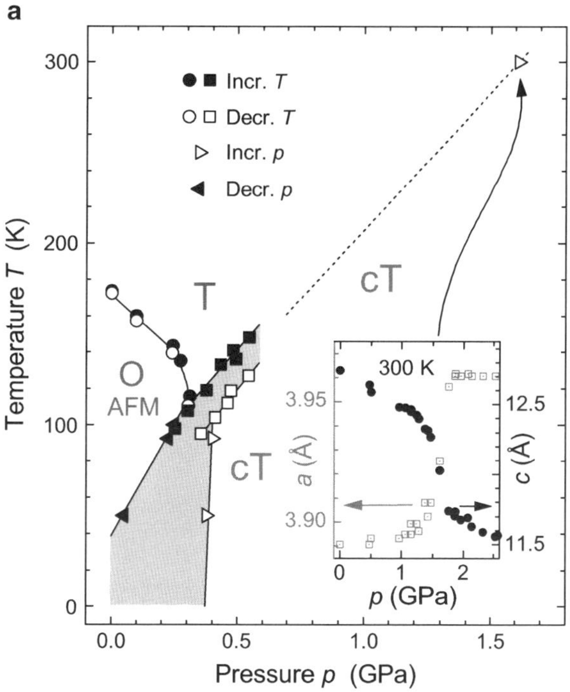

as function of the c/a parameter, for two values of $\Delta V/V = 0$ and -0.05, i.e. corresponding to zero pressure and to some elevated applied one. The calculations have been done for the tetragonal (T) and orthorhombioc (OR) phases in two variants: spin-polarized (SP) one and non-spin-polarized (NSP). As is seen, in both cases the energy of the orthorhombic phase at zero pressure is lower than that of the tetragonal phase that is consistent with experiment. At the pressure corresponding to a reduced volume, $\Delta V/V = -0.05$, the minimum of energy is realized in the tetragonal phase, which confirms an existence of the collapsed tetragonal phase. At the same conditions, as is seen from Fig. 3.19b, magnetic moments in the collapsed phase disappear. In Fig. 3.20, the phase diagram of the $CaFe_2As_2$ compound is shown.

Therefore, the most favourable situation for an onset of superconductivity in $CaFe_2As_2$ occurs within the collapsed tetragonal phase, in which the magnetic ordering disappears, as follows from neutron diffraction studies and calculations, reported in [185, 220].

In [222], by measuring electrical resistivity and magnetic susceptibility of a $CaFe_2As_2$ single crystal under pressure, the phase diagram was constructed which confirms that shown in Fig. 3.20. In Fig. 3.21, the black squares relate to the temperature T_0 of the structural transition, as determined from the anomaly of $\rho(T)$ under increase of pressure; light squares – the same, for decreasing pressure. The rhombi represent a temperature anomaly of the magnetic susceptibility.

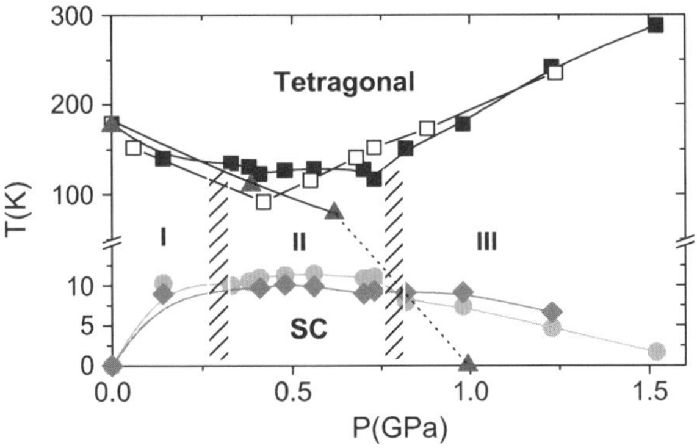

Fig. 3.21 (T, P) phase diagram for $CaFe_2As_2$ at three pressures, corresponding to the I, II and III regions. The region I for $P < 0.3$ GPa corresponds to the orthorhombic phase which is formed from the tetragonal phase; the II region for $0.3 \lessapprox P < 0.8$ GPa is that of collapsed tetragonal phase; the region III is for $P > 0.8$ GPa [222]

As is seen, in the compound under investigation the structural and magnetic phase transitions in the collapsed phase are separated. A reason for this might be the variation of the ratio of exchange interactions J_1/J_2. Calculations within the Heisenberg model show that $J_1/J_2 = 1/2$ for $CaFe_2As_2$ at the ambient pressure [223]. Apparently, this ratio is lower in the collapsed phase, that means an enhancement of spin frustration and an increased role of spin fluctuations, which suppress the magnetic ordering, and are favourable for superconducting pairing. Therefore, in $CaFe_2As_2$ the three effects – structural phase transition, magnetic ordering and superconductivity – are closely related within the collapsed phase.

To elucidate the reasons for different behaviour under pressure of AFe_2As_2 (A = Ca,Ba,Sr) compounds, detailed calculations of electronic structure and its evolution with pressure for these systems have been done in [224]. The calculations were performed within the DFT, making use of molecular dynamics for maintaining optimal values of the lattice parameters at any given pressure [225]. The results obtained are in good agreement with those of DFT calculations which used experimental values of all lattice parameters. In Fig. 3.22, the calculated variations of the unit cell volume and lattice parameters a, b are shown in their dependence on the reduced pressure P/P_c, where P_c is the critical pressure at which the transition from tetragonal into the orthorhombic phase occurs. We see a clearly pronounced structural phase transition of the first kind, accompanied by a magnetic phase transition from SDW into the non-magnetic phase.

Band structure calculations for $CaFe_2As_2$ under pressures $P/P_c = 0$ and $P/P_c = 0.87$ show changes of the electron spectrum close to the Fermi surface. Calculations at $P/P_c = 1.04$ (within the collapsed phase) exhibit substantial modifications of the Fermi surface, which destroy the nesting. In Fig. 3.23, the calculated Fermi surfaces of $CaFe_2As_2$ for three domains in the (P, T) phase diagram are shown, after [226]. In the collapsed tetragonal phase, the Fermi surface

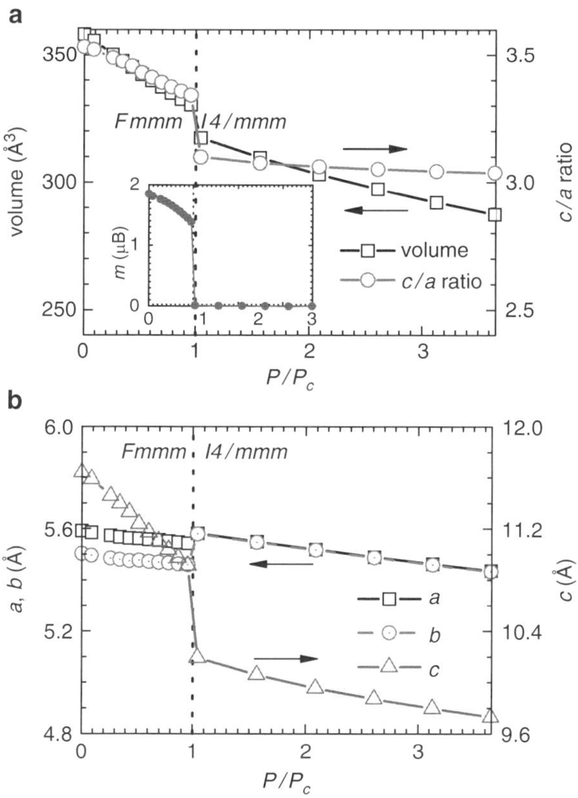

Fig. 3.22 Calculated values of volume (**a**) and lattice parameters a and b (**b**) as functions of the applied pressure for the $CaFe_2As_2$ compound [224]. In the inset, the variation with pressure of the mean magnetic moment per Fe atom is shown

is clearly three-dimensional. It contains horizontal hole "pillows" and vertical electron pockets in the Brillouin zone centre. Therefore, the very phase transition and the suppression of the magnetic moment in the tetragonal phase are consequences of the reconstruction of the electron spectrum and of the violation of nesting at the critical pressure P_c.

This conclusion correlates with the electron structure results for two other compounds, for A = Ba and Sr. The calculation of the cell volume and the a, b parameters of $SrFe_2As_2$ shows their small jump at $P = P_c$, revealing a weak phase transition of the first kind. In $BaFe_2As_2$, no jump is detected, and the phase transition from orthorhombic into tetragonal phase is a second-kind phase transition. In this process, the Fermi surface of the latter compound does not fully lose its nesting properties.

Therefore, the analysis of the evolution of electronic structure in AFe_2As_2 compounds reveals that the variations of their properties over A = Ca, Ba, Sr is related to corresponding differences of T_c. The calculations performed indicate that in the

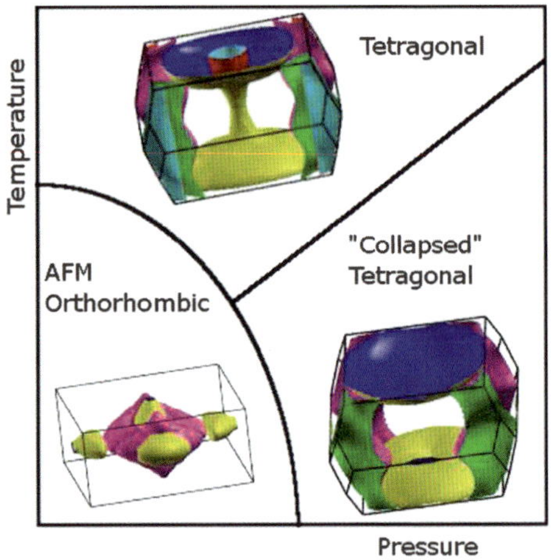

Fig. 3.23 Schematic phase diagram of the $CaFe_2As_2$ compound with the drawing of the Fermi surface corresponding to each region of the (P, T)-plane [226]

vicinity of the Fermi level, the d_{xy}, d_{xz}, d_{yz} orbitals of the Fe3d shell are dominating, therefore the minimal model for discussing superconductivity in AFe_2As_2 must be a three-band model.

We note in passing that a different scenario of structural and magnetic phase transitions in $CaFe_2As_2$ exists, based on the localized approach and using the well-known Anderson model. In this scenario, the driving force behind the observed structural/magnetic phase transition is believed to be a selective-orbital Mott transition [227].

Noteworthy, the tetragonal collapsed phase has been detected in yet another compound, $SrNi_2P_2$ [228]. This compound is not an antiferromagnet, but at 325 K it exhibits a structural phase transition into the orthorhombic phase. Under pressure of 4 kBar at room temperature, it transforms into a collapsed phase.

Isostructural phase transitions under pressure were observed in some strongly correlated systems, for instance in cerium, where such transitions are of electronic nature, being induced by strong Coulomb correlations. It is possible that the same applies to $CaFe_2As_2$. On the other side, in [220] the further references can be found in the works on the *Re*OFeAs system, in which similar anomalies in the pressure dependence of lattice constants had been reported, notably, the correlations between T_c and the unit cell size in $ReO_{1-\delta}FeAs$. In view of the above, one can presume that oxygen vacancies play a double role in triggering the superconductivity: they modify the number of carriers and apply a "chemical pressure".

In Fig. 3.24, the phase diagram (T, P) for $EuFe_2As_2$ [229], as obtained from measurements of $\rho(T)$ under hydrostatic pressure, is shown. The temperature of magnetic ordering T_0 in the Fe sublattice is gradually decreasing towards 90 K at $P = 2.3$ GPa. The superconductivity sets on at the pressure $P = 2.0$ GPa and has $T_c = 29.5$ K within the antiferromagnetic phase, whereby T_c does not depend on the applied pressure till $P = 2.66$ GPa. As the temperature drops below T_c, the superconductivity is preserved up to the antiferromagnetic ordering temperature T_N on the

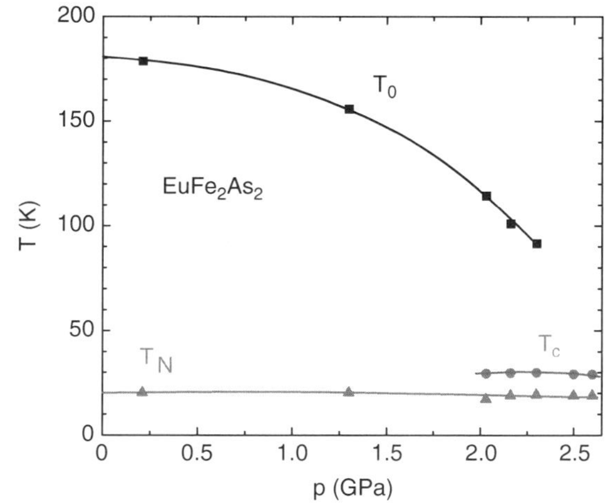

Fig. 3.24 (T, P) phase diagram of $EuFe_2As_2$. T_0 is the temperature of antiferromagnetic ordering in the FeAs layers, T_N is the ordering temperature in the Eu sublattice, T_c is the superconducting transition temperature [229]

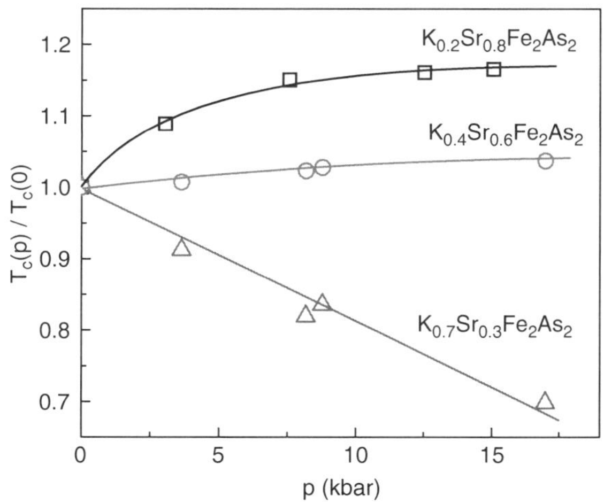

Fig. 3.25 Relative variation of the superconducting transition temperature with pressure, for the $K_{1-x}Sr_xFe_2As_2$ compounds with $x = 0.3$; 0.6; 0.8 [232]

Eu sublattice, and disappears for $T < T_N$. A similar situation was earlier observed in the Chevrel phase $GdMo_6S_8$ [230] and got a name of reentrant superconductivity.

An indication of magnetic ordering of Eu in the $EuFe_2As_2$ system within the superconducting phase was also obtained on doping. Thus, Mössbauer effect measurements on ^{57}Fe and ^{151}Eu nuclei of an $Eu_{0.5}K_{0.5}Fe_2As_2$ sample with $T_c = 33$ K have shown [231] that superconductivity coexists with short-range magnetic order, which sets on in the Eu sublattice at temperatures inferior to 15 K. At the same time, a magnetic ordering in the Fe sublattice, existing in the undoped compound with $T_N = 190$ K, gets completely suppressed on 50% substitution of europium by potassium.

A systematic study of the pressure effect on the hole doping in AFe_2As_2 compounds has been done in [232, 233]. In Fig. 3.25, it is shown that in the $K_{1-x}Sr_xFe_2As_2$ compounds, the pressure can increase or decrease T_c, depending on the dopant

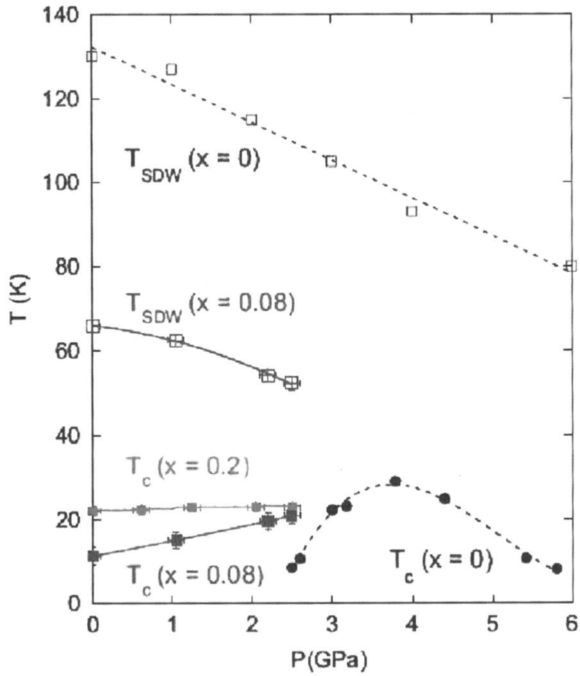

Fig. 3.26 (T, P) phase diagram of the $BaFe_{2-x}Co_xAs_2$ compound at $x = 0$; 0.08 and 0.2 [234]. The results for $x = 0$ are taken from [186]

concentration x [232]. Comparing the effect of pressure with the hole doping in the stoichiometric KFe_2As_2, we conclude that the pressure induces the transfer of electrons from the FeAs planes into the KSr plane. A similar mechanism of the T_c variation with pressure is observed in cuprates.

An effect of pressure in the potassium-doped $BaFe_2As_2$ compound was studied in [233]. It turned out that $Ba_{0.4}K_{0.5}Fe_2As_2$ remains tetragonal down to low temperatures, and superconductivity appears at temperature close to 30 K. The effect of pressure is in lowering the T_c at the rate −0.21 K/kBar.

The influence of pressure (of up to 2.5 GPa) on transport properties of AFe_2As_2-type compounds was studied taking $BaFe_{2-x}Co_xAs_2$ [234] as an example; the results are shown in Fig. 3.26. The undoped compound is not a superconductor at the ambient pressure; superconductivity sets on at $P > 2$ GPa (to be compared with the phase diagram in Fig. 3.14). Under doping with cobalt, the spin ordering temperature gets lowered, and superconducting state appears already at zero pressure. For $x = 0.08$, T_c increases by the factor of two, from 11 to 22 K. At the same time, for $x = 0.2$, the T_c does not depend on the pressure applied, whereas the spin ordering turns out to be completely suppressed. The comparison of results for undoped ($x = 0.08$) and optimally doped ($x = 0.2$) compounds show the crucial role of the systems' closeness to the magnetic ordering in the mechanism of formation of superconducting state. Indeed, in the optimally doped sample, with the magnetic

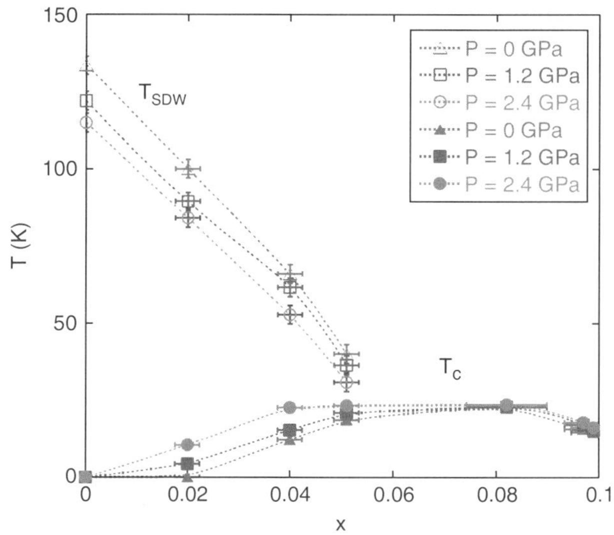

Fig. 3.27 $(T-x)$ phase diagram of the $Ba(Fe_{1-x}Co_x)_2As_2$ compound under pressure $P = 0$; 1.2; 2.4 GPa. *Open* and *shaded symbols* indicate T_{SDW} and T_c, correspondingly [236]

order fully suppressed, an application of pressure does not affect T_c, whereas in the undoped sample, which caries the rests of magnetic ordering, T_c is increased under pressure by two times. A common conclusion from the study undertaken is that pressure, similarly as doping, promotes superconductivity.

In [235], an unusual dependence of T_c on pressure, derived from high-resolution dielectric measurements and an analysis of the measured heat capacity in terms of the Ehrenfest relation, has been detected for $Ba(Fe_{0.92}Co_{0.08})_2As_2$. T_c turned out to be strongly anisotropic with respect to the pressure being applied along the c or a axes. Namely, the relations $dT_c/dP_a = 3.1$ K/GPa and $dT_c/dP_c = -7.0$ K/GPa have been found, which hints for a strong coupling between T_c and the c/a ratio.

In [236], the (T, x)-phase diagram of $(Fe_{1-x}Co_x)_2As_2$ under pressure of up to 2.75 GPa has been obtained (Fig. 3.27). As is seen, the pressure reduces the magnetic ordering temperature at all levels of doping. In what regards the superconducting state, the situation is more complicated. The superconductivity is not induced by pressure in the underdoped regime ($x < 0.051$). In the range of doping $0.02 \lesssim x \lesssim 0.051$, which corresponds to underdoped regime, T_c does drastically increase under pressure; however, in the optimally doped regime $0.082 \lesssim x \lesssim 0.099$, the pressure has a weak effect on T_c. In [236], different scenarios of this situation are discussed, including an existence of a quantum critical point between the magnetic ordering and the superconducting state.

An influence of pressure on the T_c has been also studied in a number of stoichiomeric 122-compounds: $SrFe_2As_2$ [237, 238] and $EuFe_2As_2$ [239]. In the $SrFe_2As_2$ single crystal, superconductivity appears under application of uniaxial pressure, but the critical pressure depends on the transmitting medium used in the

Fig. 3.28 $(T - P)$ phase diagram for stoichiometric $SrFe_2As_2$. *Dark triangles* and *circles* are determined from the data on electrical resistivity and magnetic susceptibility, correspondingly [240]

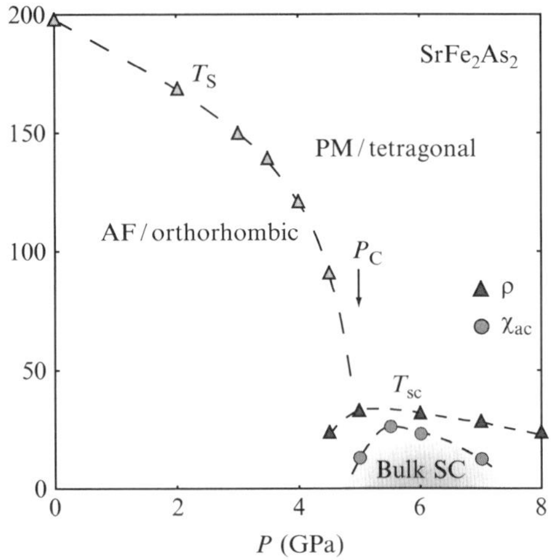

anvils. In the almost perfectly hydrostatic medium, P_c turned out to be 4.4 GPa, but when using a medium which transferred the stress uniaxially, $P_c \approx 3.4$–3.7 GPa. Terashima et al. [239] reported inducing the superconductivity with $T_c \approx 30$ K by pressure $P = 28$ kbar. The measurements of magnetic susceptibility indicated an antiferromagnetic ordering of the Eu atoms with $T_c \approx 20$ K, which coexists with superconductivity.

In another work [240], the $BaFe_2As_2$ compound was studied over a broad region of pressures up to 8 GPa. It was unexpected that the antiferromagnetic order persisted up to the maximum applied pressure, and superconductivity did not emerge. In contrast, for the $SrFe_2As_2$ compound the structural and magnetic phase transitions were suppressed at $P \approx 5$ GPa, and then a superconducting state was formed with the T_c achieving its maximum near to 6 GPa (Fig. 3.28). The fact that the SDW phase and superconductivity coexist in a very narrow region hints that these two order parameters are in competition. The importance of nonhydrostatic pressure in inducing of superconductivity in CaFe2As2 is shown in [241].

3.2.4 *Symmetry of the Superconducting Order Parameter*

Figure 3.29 depicts the intensities of the quasiparticle peaks reconstructed from the primary ARPES data. Two circles, concentric around Γ, are seen, revealing two hole sheets of the Fermi surface. Around the *M* point, light spots emerge, similar to those found in [151].

On the two hole sheets of the Fermi surface around Γ, superconducting gaps of different magnitude have been detected. The gap on the inner sheet turned out to be somehow anisotropic, varying with the asimutal angle between 10 and 12 meV, whereas the gap on the outer sheet is isotropic and equals roughly 8 meV. In the

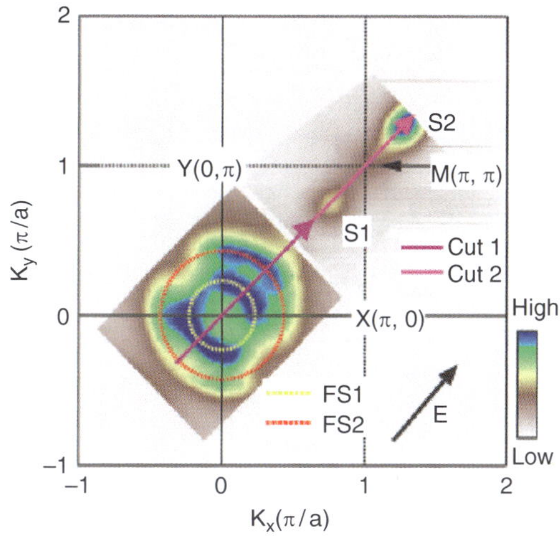

Fig. 3.29 Sheets of the Fermi surface of (Ba,K)Fe_2As_2, after the data of [154]

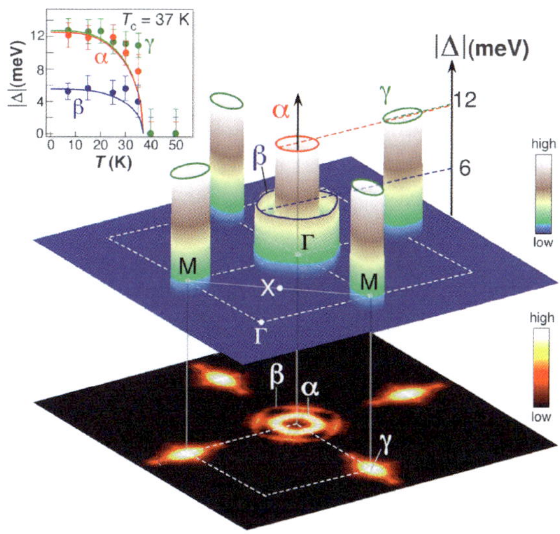

Fig. 3.30 ARPES data of the study of the Fermi surface and superconducting gaps in the $Ba_{0.6}K_{0.4}Fe_2As_2$ compound [155]

vicinity of the M point, on the electron sheet of the Fermi surface a gap is found which equals 10 meV at $T = 25$ K; the gap disappears at $T = T_c$. No zeros of the superconducting gaps have been found over the Fermi surface. The authors of [154] draw a conclusion that in the compound under study, a superconducting state with several gaps and s-symmetry of the order parameter is realized.

On the other single crystal with $x = 0.40$ and $T_c = 37$ K, a thorough ARPES study of the Fermi surface and superconducting gap has been done [155]. The results obtained are depicted in Fig. 3.30. On the bottom plane, the intensity of quasiparticle peaks at the Fermi surface, as recovered from the measurements of photoelectron spectra, is shown by bright spots. Two bright circles in the centre correspond to two hole sheets of the Fermi surface around Γ, which are indicated as α and β in the

Table 3.1 ARPES data concerning the gap in diferent FeAs-compounds. α_1 and α_2 are inner and outer hole sheets of the Fermi surface centred at Γ; β_1 – the electron sheet centred at M

Reference	[93]	[155]	[154]	[245]	[244]	[246]
T_c	53 K	37 K	35 K	53 K	37 K	32 K
$\Gamma{:}\alpha_1$	20	12.5	12	15	12	9.2
$\Gamma{:}\alpha_2$	–	5.5	8	–	6	< 4
$M{:}\beta_1$	–	12.5	10	–	11	9±2

Table 3.2 Values of $2\Delta/kT_c$ for superconducting compounds, in which the gap is measured by ARPES [93, 154, 155, 244–246], by PCAR[105, 247], and from heat capacity [248]

Reference	[93]	[155]	[154]	[245]	[244]	[246]	[105]	[247]	[248]
Large Δ	9	8.1	8.2	6.8	7.5	6.8	3.7	9.6	4
Small Δ	–	3.6	5.5	–	3.9	< 3	–	3.4	–

figure. Superconducting gaps over these sheets are isotropic and, at low temperatures, equal to $\Delta_\alpha \approx 12$ meV and $\Delta_\beta \approx 6$ meV. As temperature grows, the both gaps disappear simultaneously at $T = T_c$. On the electron sheet γ in the vicinity of the M point, an isotropic superconducting gap of $\Delta \approx 12$ meV is found. In the upper plane of Fig. 3.22, a sketch of superconducting gaps is assembled, in which the cylinders' height represents the relative gap magnitude at different sheets of the Fermi surface. These data are consistent with the results of the gap measurement in $Ba_{0.6}K_{0.4}Fe_2As_2$ (having $T_c = 37$ K) by infrared spectroscopy [242], where $\Delta \approx 12$ meV has been measured. They also agree with the ARPES results of [154], discussed above.

Further ARPES studies on a good-quality single crystal of $Ba_{0.6}K_{0.4}Fe_2As_2$ [243] have confirmed the measured values of gap on the hole sheets of the Fermi surface and on the electron sheet Γ. Moreover, the gap on the second electron sheet δ of the Fermi surface has been measured and turned out to be smaller (~6 meV) than the three others (~10–12 meV). Further on, detailed measurements of the **k**-dependence of the superconducting order parameter have shown that the latter can be well approximated by the function $\Delta(\mathbf{k}) = \Delta_0 \cos k_x \cos k_y$, that is consistent with the $s^{\pm}$-symmetry of the order parameter.

In [244], systematic ARPES studies are outlined for two compounds: $Ba_{1-x}K_x Fe_2As_2$ ($T_c = 37$ K) and $Sr_{1-x}K_xFe_2As_2$ ($T_c = 26$ K). Two concentric Fermi surfaces around the Γ were detected and the further one – in the vicinity of the M point. The superconducting gap was found to possess a small anisotropy. The largest magnitude of $\Delta = 12 \pm 2$ meV is at the inner Fermi surface near Γ; then it reduces on the outer central Fermi surface and further on towards M, but then it increases again near the corner of the Brillouin zone. Such behaviour, in combination with other observed facts, agrees with an assumption that in these superconductors, the symmetry of the order parameter is of the $s_{x^2y^2}$ or $s^{\pm}$ type, like in some *Re*OFeAs compounds. Experimental results of the study of Fermi surface in different compounds are listed in Tables 3.1 and 3.2.

The measured velocity at the Fermi surface turned out to be small, of merely 0.7 eV·Å. Using the relation $\xi = \hbar V_F/\Delta$, we arrive at an estimate for the correlation

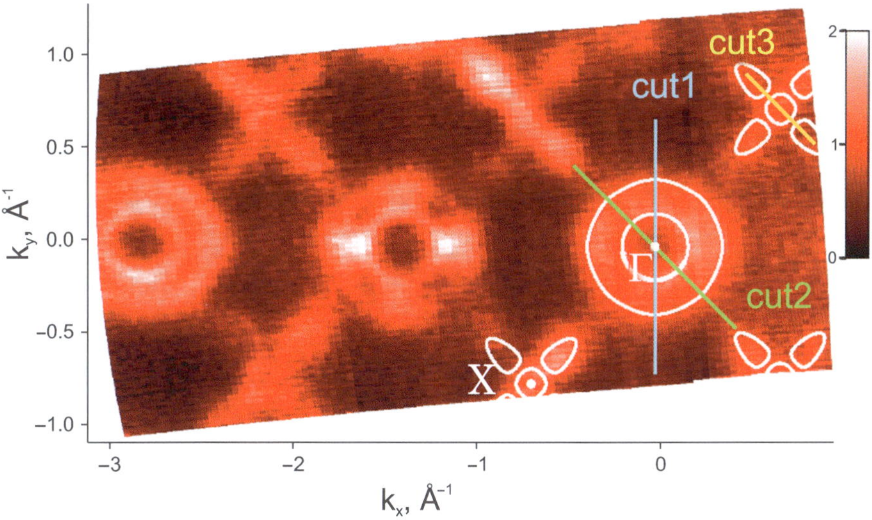

Fig. 3.31 Distribution of the photoemission intensity at the Fermi level from ARPES measurements on the $Ba_{1-x}K_xFe_2As_2$ superconductor with $T_c = 32$ K [246]. *White lines* pass through the intensity maxima, and represent the sheets of the Fermi surface. *Straight lines* mark three cuts in the **k**-space, along which the photoemission spectra have been measured

length: $\xi \lesssim 20$ Å, that is about $4a$ (a being the lattice parameter). Such small magnitude of ξ is unknown for superconductors with the phonon pairing mechanism.

In the same work, the quasiparticles dispersion curves were measured, in which the kinks in the energy ranges of 40 ± 10 and 18 ± 5 meV have been detected. The magnitudes of magnetic interactions J_1 and J_2 determining the magnetic structure of undoped compounds of the given type are approximately 20–50 meV that corresponds well to the high-energy kink. The lower kink may plausibly be due to interaction between electrons and phonons.

The authors of [246] performed ARPES measurements on a good-quality single crystal of $Ba_{1-x}K_xFe_2As_2$. In Fig. 3.31, the intensity distribution over the Fermi surface ($\omega = 0$) is shown. Straight lines mark cross-sections in the **k**-space, along which the spectra have been measured. Evaluation of intensities along the cuts across the sheets of the Fermi surface is shown in Figs. 3.32 and 3.33.

It follows from these figures that on the inner Γ sheet, the gap equals 9.1 ± 0.7 meV along the Γ–M direction (cut1 in Fig. 3.31) and 0.7 ± 0.1 meV along Γ–X (cut2). The gap at the outer Γ-surface is less than 4 meV, and at the X pocket – 9.3 ± 2 meV. The gap at the narrow sheets (cut3) is estimated to be 9 meV.

The $2\Delta/kT_c$ parameter, which in the BCS theory equals 3.52, makes 6.8 for the inner Γ sheet and the X pocket of the blades, but less than 3 for the outer Γ sheet.

Two gaps have been found in $Ba_{1-x}K_xFe_2As_2$ ($T_c = 32$ K) while exploring the penetration depth λ [249]. At low T, their magnitudes are 9 and 1.5 meV, that is in agreement with the results of [246].

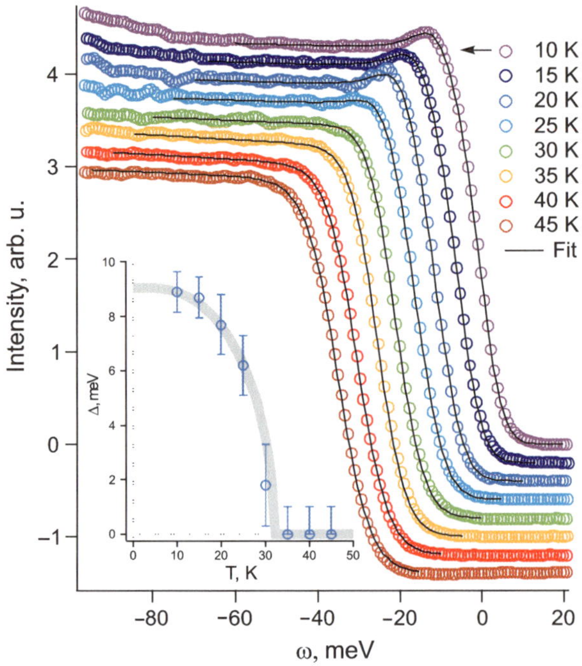

Fig. 3.32 Evolution of the emission spectra along the *cut1* of the inner sheet of the Fermi surface (see Fig. 3.31) near Γ with temperature. In the inset: temperature dependence of the superconducting gap extracted from these data with the use of the BKFA fitting [246]

A recent investigation of the $Ba_{1-x}K_xFe_2As_2$ compound by ARPES and μSR methods [250] has confirmed an existence of two gaps, $2\Delta_1/kT_c = 1$ and $2\Delta_2/kT_c = 6.8$, that agrees with the above results. NMR studies indicated $1/T_1 \sim T^3$, thus apparently confirming the $s^{\pm}$-symmetry of the superconducting order parameter [251].

The first study of $Ba_{1-x}K_xFe_2As_2$ by PCAR has been done on a single crystal with $x = 0.45$. The measurement results for conductance, as function of the applied voltage on one of the point contacts, are shown in Fig. 3.34. The spectrum reveals beyond doubt an existence of two gaps, the larger one $\Delta_L = 9$–11 meV and the smaller one, $\Delta_S = 2$–5 meV, whose temperature dependence is depicted in the inset. Above T_c, a pseudogap in the electron density of states at the Fermi surface, apparently stemming from fluctuations of the SDW order parameter, is detected.

In some spectra for the *ab* plane, the ZBC peaks at zero voltage have been found. They may point towards a presence of zeros in the superconducting order parameter. Earlier, the ZBC peaks were registered in a number of superconductors of the FeAs group [105, 109, 113, 252].

In the experiments on infrared spectroscopy [242] and measurements of the penetration depth of electromagnetic radiation, the superconductor $Ba_{1-x}K_xFe_2As_2$ [253] does as well reveal two gaps at the Fermi surface. In μsR experiments on the

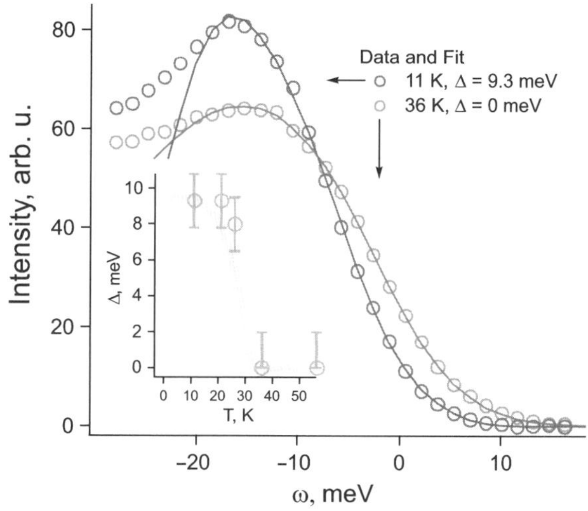

Fig. 3.33 Evolution of ARPES spectra along the *cut3* (see Fig. 3.31). In the inset: temperature dependence of the gap at the sheet of the Fermi surface near the $X(M)$ point [246]

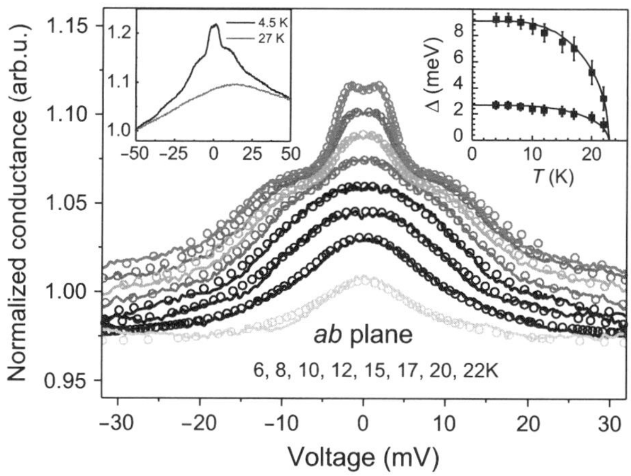

Fig. 3.34 Normalized PCAR spectrum of the $Ba_{0.55}K_{0.45}Fe_2As_2$ superconductor in the *ab*-plane of a single crystal [247]. The *left* inset: raw data for two temperatures. The *right* inset: temperature dependence of two gaps; the *solid line* shows the BCS-behaviour of the energy gap

same system with optimum doping, also two gaps are found with the magnitudes of the $2\Delta/kT_c$ parameter which exceed by far the value 3.52 of the BCS theory [254]. The above described properties of the $Ba_{1-x}K_xFe_2As_2$ system find confirmation in the measurements of the temperature dependence of heat capacity [255]. It follows from transport and thermodynamic measurements in the normal phase of this system [256] that, on variation of x, a crossover takes place from the behaviour of electrical resistivity as $\rho \sim T^2$ towards the linear one $\rho \sim T$. Such change indicates a turn from the Fermi-liquid type of system behaviour towards that dominated by a strong interaction of quasiparticles with bosons.

We point out moreover an ARPES study of a more complex system $(Sr,Ba)_{1-x}(K,Na)_xFe_2As_2$ [257], which reveals the properties of the basic $Ba_{1-x}K_xFe_2As_2$ system: a presence of hole sheets of the Fermi surface along with electron ones. In the dispersion curves at energies 40±15 meV, the kinks were detected which reveal the scattering of quasiparticles on quantum fluctuations and phonons.

A piece of information about the symmetry of superconducting order parameter can be also obtained by measuring temperature and field dependencies of the electron heat capacity. In [248], such measurements have been done on a single crystal $Ba_{0.6}K_{0.4}Fe_2As_2$ with $T_c = 36.5$ K. It was found that the $\gamma(T)$ coefficient does not depend on temperature and, at low temperatures, grows almost linearly with the magnetic field H. These data indicate that in the compound under investigation, the superconducting order parameter does not have zeros. The experimental data fall onto a curve corresponding to s symmetry of the order parameter with the gap value $\Delta = 6$ meV. This number agrees with the smallest gap found in the ARPES experiment.

Therefore, it can be stated that in superconducting compound $Ba_{1-x}K_xFe_2As_2$, a multi-gap superconductivity occurs, apparently with the s symmetry of the order parameter. The magnitude of $2\Delta/kT_c$ is approximately 7, thus pointing towards a case of strong coupling in the theory of superconductivity.

However, this conclusion is not confirmed for all doped $BaFe_2As_2$ compounds. Thus, in [258] via measuring the magnetic field penetration depth λ and heat capacity κ it was shown that in the phosphorus-doped $BaFe_2(As_{1-x}P_x)_2$ compound ($T_c = 30$ K), a line of zeros exists in the superconducting order parameter. Astonishing is the fact that the electronic structure of this compound, like that of the other, potassium-doped, compound with the same magnitude of T_c, are very close, to an extent that the sizes of their hole and electron pockets are practically identical. Nevertheless, some tiny differences lead to different shape of gaps on the Fermi surface. This reveals that in doped $BaFe_2As_2$, two kinds of superconducting state exist which are nearly degenerate in energy. It is not clear so far which differences in the Fermi surface do lift this degeneracy.

3.2.5 Measurements on the Josephson Contacts

The measurement of the Josephson current as function of the applied magnetic field may yield a direct information concerning the symmetry of the superconducting order parameter. Such measurements have been done in cuprates (see references in [259]) using a corner contact (Fig. 3.35). For a conventional superconductor [259] with s symmetry of the order parameter, the critical current is given by the formula [260]:

$$I_c = I_0 \left[\frac{\sin(\pi\Phi/\Phi_0)}{(\pi\Phi/\Phi_0)} \right],$$

where Φ is magnetic flux through the contact, Φ_0 – the flux quantum. This dependence of I_c on Φ is shown in Fig. 3.35a. For a superconductor which has the phase

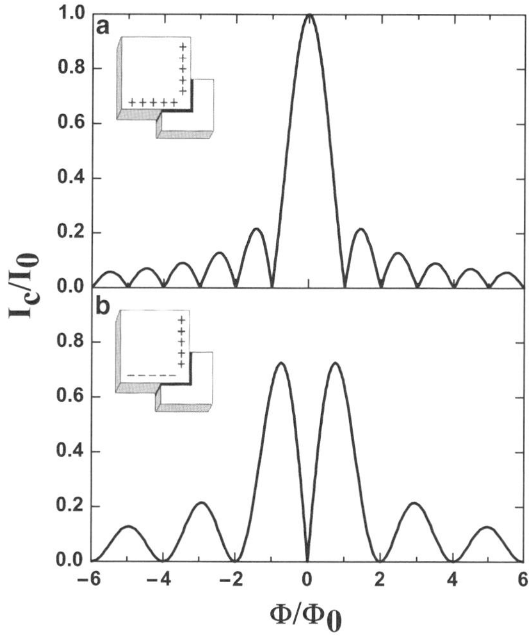

Fig. 3.35 Josephson current at a corner contact for a superconductor with zero phase difference (**a**) and with phase difference equal to π (**b**) [259]

difference π between its perpendicular contact faces, I_c in terms of Φ is described by the formula

$$I_c = I_0 \left[\frac{\sin^2(\pi\Phi/2\Phi_0)}{(\pi\Phi/2\Phi_0)} \right],$$

which dependence is shown in Fig. 3.35b.

For cuprates, the diffraction pattern as in Fig. 3.35b was obtained and hereby the $d_{x^2-y^2}$ symmetry of the order parameter established. In [259] on a sample $Ba(Fe_{0.9}Co_{0.1})_2As_2$ with $T_c = 22$ K, the pattern as in Fig. 3.36 has been obtained [259], from which it follows, via comparison with Fig. 3.35, that there is no phase shift between the a–c and b–c faces of a crystalline sample. This means that the symmetry of the order parameter in the FeAs-systems is not the same as in cuprates. In another work [261], the current in planar and point contacts of the $(Ba,K)Fe_2As_2$ superconductor was measured. The measurements have shown the p and $d_{x^2-y^2}$ symmetries of the order parameter and are also consistent with the existence of the $s^{\pm}$-symmetry of the superconducting order parameter.

We bring attention to [262], in which Josephson contact was studied between two superconductors, $SrFe_{1.74}Co_{0.26}As_2$ and $Ba_{0.23}K_{0.77}Fe_2As_2$, with electron and hole conductivity, respectively. The contact made of two single crystals oriented along the c direction, on applying the magnetic field in the ab plane, gives us a symmetric Fraunhofer picture of the Josephson current of the same type as shown in Fig. 3.36.

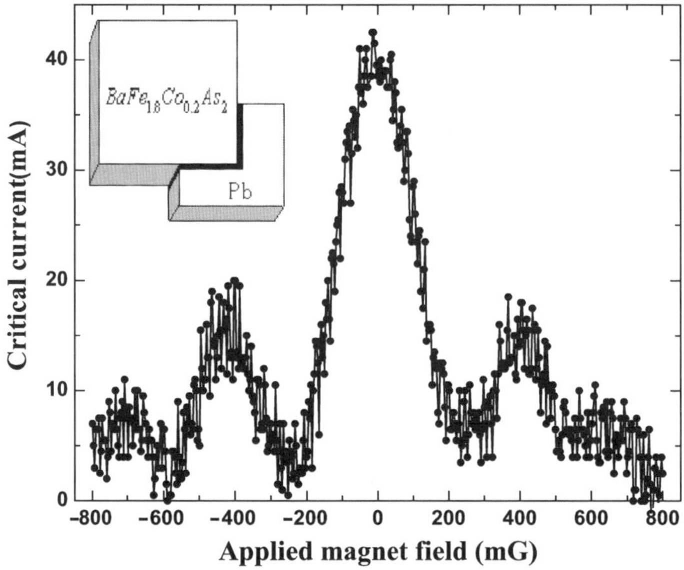

Fig. 3.36 Josephson current measured at a corner contact of the $Ba(Fe_{0.9}Co_{0.1})_2As_2$ superconductor [259]

3.2.6 *Critical Fields*

Similar to the *Re*OFeAs compounds, the bilayer AFe_2As_2 ones are characterized by high critical fields. Studies on a single crystal of $(Ba_{0.55}K_{0.45})Fe_2As_2$ with $T_c = 45\,K$ have shown [263] a moderate anisotropy of the upper critical field H_{c2}: H_{c2}^{ab}/H_{c2}^{c} is around 3.5 in the temperature range close to T_c and decreases at lower T, reaching 1.5 at $T \approx 20\,K$. An estimate of H_{c2} for this compound is $H_{c2}(T=0) \approx 75\,T$ [263]. At lower level of potassium doping ($x = 0.23$–0.4), the anisotropy of the upper critical field decreases, and the critical fields get higher: $H_{c2}^{ab}/H_{c2}^{c} \approx 2$, $H_{c2}^{ab}(0) = 300\,T$, $H_{c2}^{c}(0) = 210\,T$ [264].

The electron concentration in the AFe_2As_2 compounds is by an order of magnitude higher than in LaOFeAs, therefore the density of the superconducting component in them must be higher. For superconductors with low concentration of carriers n, the Uemura plot is known, the relation [265] connecting the density ρ_s with the penetration depth $\lambda(0)$. For quasi-two-dimensional systems, $\rho_s \sim 1/\lambda_{ab}^2(0)$, which relation holds well in cuprates and in doped compounds of the LaOFeAs type. A test of Uemura plot was done for $Ba_{0.6}K_{0.4}Fe_2As_2$ [266], on a single crystal with $T_c = 36.2\,K$. From measured magnetization curves, the lower critical field $H_{c2}(T)$ as function of temperature has been extracted, and the density of superconducting carriers determined making use of the relation

$$\rho_s(T) = \frac{\lambda_{ab}^2(0)}{\lambda_{ab}^2(T)} = \frac{H_{c1}(T)}{H_{c1}(0)}\,.$$

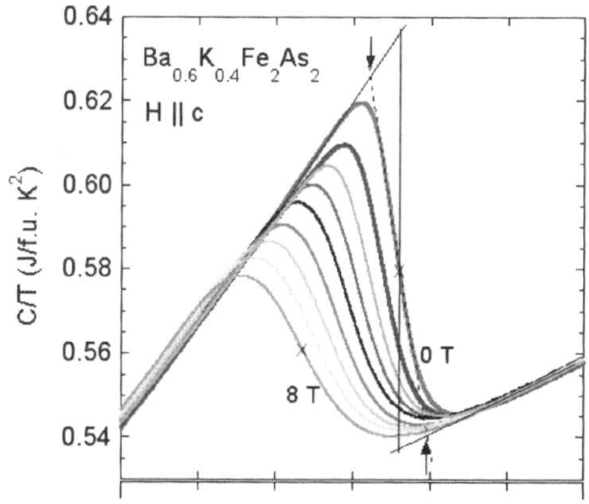

Fig. 3.37 Temperature dependence of heat capacity in the $Ba_{0.6}K_{0.4}Fe_2As_2$ superconductor in the field oriented along the crystalline c axis [270]

It turned out that the Uemura's relation does not hold for AFe_2As_2 compounds because of high density of superconducting carriers in the latter.

We mentioned already that in the compounds of the LaOFeAs type, μSR measurements revealed a possibility for static magnetism and superconductivity [267] to coexist within a narrow concentration range in the vicinity of the antiferromagnet/superconductor phase boundary. Such state has also been detected in μSR studies of a $(Ba_{0.55}K_{0.45})Fe_2As_2$ single crystal [268]. These studies have been also done on single crystals of a number of compounds – $(Ba_{0.5}K_{0.5})Fe_2As_2$ (T_c = 37 K), $(Sr_{0.5}Na_{0.5})Fe_2As_2$ ($T_c = 35$ K) and $CaFe_2As_2$ under applied pressure of up to 10 kBar [269].

The behaviour of superconducting $Ba_{1-x}K_xFe_2As_2$ in magnetic fields was thoroughly studied in [270] on the basis of heat capacity measurements. Figure 3.37 depicts the temperature dependence of specific heat in different fields for a single crystalline sample with $T_c = 34.6$ K and $x = 0.4$. From these data, by the method of entropy-conserving construction, the temperature derivatives of H_{c2} in the vicinity of T_c have been derived. For the fields **H** || **c**, it followed $\partial H^c_{c2}/\partial T = -6.5$ T/K, whereas for the fields oriented in the basal plane – $\partial H^{ab}_{c2}/\partial T = -17.4$ T/K. Using the Werthamer–Helfand–Hohenberg formula of [43], see (2.1) one can estimate the upper critical field $H_{c2}(0)$ at $T = 0$. The results are $H^c_{c2}(0) = 155$ T and $H^{ab}_{c2}(0) = 415$ T. These values, however, must be strongly overestimated, because the paramagnetic limit H_p, determined by the relation $\mu_0 H_p = 1.84\,kT_c$, makes in this case 64 T.

Previously performed measurements of magnetoresistivity [271] and penetration depth [263] on the samples with $T_c = 28$ K and 29.5 K, respectively, and moreover the study of magnetoresistivity [208] on a sample with $T_c \approx 22$ K have shown that the upper critical field in these samples is inferior to 100 T.

The coherence length calculated from the above cited data for the $Ba_{0.6}K_{0.4}Fe_2As_2$ compound turned out to be: $\xi_{ab} \approx 1.2$ nm and $\xi_c \approx 0.45$ nm, whereby the Ginsburg–Landau parameter $K = \lambda/\xi$ assumes values $K_c \approx 100$ and $K_{ab} \approx 260$. This implies that the superconductor in question is that of the second kind.

To conclude, we discuss the results of study concerning the lattice of fluxoides in the AFe_2As_2 compounds. In [272], the lattice of fluxoides has been studied by small-angle neutron diffusion for an electron-doped compound $Ba(Fe_{0.93}Co_{0.07})_2As_2$ with $T_c = 21$ K. The measurements have been done at the temperature of 2 K in the fields applied along the c axis. It turned out that this system does not exhibit any long-range order, but a strong short-range one. Over the regions of about 5 distances between the fluxoides, a hexagonal symmetry with the axis of the sixth order was detected. A disordering of the lattice is, most probably, related to a strong pinning. This disordering is of orientational character in the sense that the distance between fluxoides tends to be preserved everywhere, and the distortion affects, primarily, lattice directions.

We outline further on the measurement results of small-angle neutron diffraction on the lattice of fluxoides. For this study, the superconductor $Ba(Fe_{0.93}Co_{0.07})_2As_2$ with $T_c = 21$ K has been chosen. The measurements have been done at $T = 2$ K in the fields of 450–810 mT. In the case of an ideal lattice of fluxoides, the Bragg peaks would have been detectable. Small-angle diffraction in all fields has shown that the intensity of the magnetic scattering of neutrons forms circles, indicating that the fluxoides are disordered. The radii of the circles in the $\mathbf{q}$-space reveal the size of domains over which the short-range order is maintained. An average distance between the fluxoides is $a \sim (\Phi_0/B)^{1/2}$, where Φ_0 is the flux quantum, and B – the applied field. The radius of a circle q in the low-angle distribution in the $\mathbf{q}$-space determines the mean domain size $l{\sim}1/q$, corresponding to several a distances. Within a single domain, a short-range order exists, characterized by a sixth-order hexagonal axis. The Bitter decoration does visually reveal a large disordering of fluxoides. This finds a natural explanation by a strong pinning, since a doped compound has defects due to deviation from crystallization ordering.

3.3 Magnetism

3.3.1 Stoichiometric Compounds

Magnetic ordering of Fe atoms in stoichiometric AFe_2As_2 (A = Ba, Ca, Sr) compounds was probed by neutron diffraction in [273–277]. In all compounds studied, the same magnetic structure, shown in Fig. 3.38, has been found. Within the basal plane, the magnetic moments of Fe atoms build ferromagnetic chains along one of the orthorhombic axes, whereas along the other axis of the basal plane and also along the c axis the orientations of spins are alternating. Thus, a three-dimensional AFM structure is formed with the same motif as in the LaOFeAs compounds. The difference is only in the arrangement of antiferromagnetically coupled planes. In $BaFe_2As_2$, the adjacent planes belong to the same unit cell, whereas in LaOFeAs – to two adjacent cells. With all the atoms but Fe removed from the crystal structures of both groups of compounds, the remaining magnetic lattices would be identical.

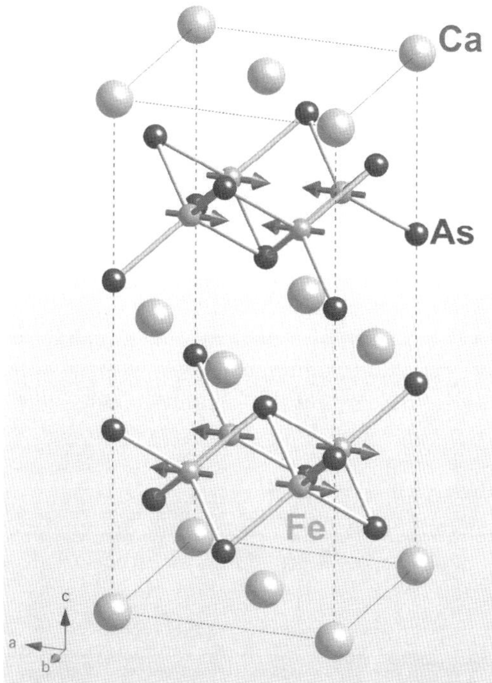

Fig. 3.38 Magnetic structure of the $BaFe_2As_2$ compound [275]

As in the LaOFeAs-type compounds, the magnetic ordering appears simultaneously with, or close to, the structural phase transition from the tetragonal into the orthorhombic phase.

Let us turn now to some details of magnetic and structural phase transitions in different AFe_2As_2 compounds. The first neutron diffraction study of $BaFe_2As_2$ was done on a polycrystalline sample [274]. The temperature variation of the (101) magnetic reflection has shown that a second-kind magnetic phase transition occurs at $T_N = 143$ K; at the same temperature, the structural transition into the orthorhombic phase takes place. A subsequent study done on a single crystal [273] led to different conclusions in what regards the phase transitions (the magnetic structure as such has been confirmed). It turned out that the magnetic ordering disappears at the same temperature T_N; however, the structural transition into tetragonal phase is retarded, and was not yet detected at room temperature. Meanwhile, the magnetic phase transition remains that of the second kind. The magnetic moment per Fe site in the ordered phase is $\sim 0.8\,\mu_B$, whereby the magnetic moments align along the shortest orthorhombic axis. The latter neutron diffraction study of $BaFe_2As_2$ done on a good-quality single crystal [278] confirmed the results earlier obtained on a polycrystal [274].

Similar results have been obtained in a neutron diffraction study of a $CaFe_2As_2$ single crystal [275]. A structural phase transition on cooling was detected at $T = 172.5$ K, and on heating – at $T = 173.5$ K, so that the temperature hysteresis of

$\sim$1 K turned out to be much smaller than for $BaFe_2As_2$. A magnetic transition into the above discussed magnetic structure (Fig. 3.38) occurs simultaneously with the structural transition. The magnetic moments at Fe atoms are $\sim 0.81\,\mu_B$, as in $BaFe_2As_2$.

A neutron diffraction study on a single crystal of $SrFe_2As_2$ [276] revealed a structural transition at $T_0 = 220$ K, and magnetic ordering proceeds at the same temperature into the magnetic structure shown in Fig. 3.38. These results agree with those from neutron diffraction study on a polycrystalline sample [277], according to which the temperature of both phase transitions is $T_0 = 205$ K. The magnetic moments at Fe atoms turn out to be $\sim 1\,\mu_B$.

The character of magnetic and structural phase transition was studied in detail by neutron diffraction methods on a large good-quality single crystal of $BaFe_2As_2$ [279]. It was found that a joint second-kind phase transition occurs, with simultaneous appearance of magnetic and structural order parameters at $T_N = 136.0$ K. In any case, if this transition happens to be a weak first-kind phase transition, the difference between T_N and T_c should not exceed 0.5 K. An analysis of the temperature behaviour of both order parameters, done by measuring the intensities of magnetic and structural Bragg peaks, allowed to determine the critical index of the order parameter component $\phi^2 \sim (1 - T/T_N)^{2\beta}$, which turned out to be $\beta = 0.103 \pm 0.028$. The magnitude of this index is close to $\beta = 1/8 = 0.125$ for the two-dimensional Ising model, to the universality class of which do the FeAs-systems belong.

Therefore in all AFe_2As_2 compounds studied, the same magnetic structure, appearing along with the structural phase transition, has been detected. The similar situation exists for the LaOFeAs-type compounds. This means that a unique mechanism of magnetic ordering exists in all compounds built from the FeAs-planes, independently of what interlacing planes, formed of which atoms, make their separation. In all these compounds, a strong coupling occurs between the magnetic moments on Fe atoms and the structural distortions.

The observed antiferromagnetic structures can be described by the Heisenberg model with effective exchange interactions, the antiferromagnetic ones J_1 and J_2 between nearest and next-nearest neighbours in the basal plane, and the J_z between the planes [65,280,281]. For $J_1 < 2\,J_2$, the ground state of this model is a collinear antiferromagnet, namely the SDW structure, in which the ferromagnetic chains are alternating in their spin directions. The ground state is doubly degenerate, because ferromagnetic chains can be directed either along the x or along the y axis.

With an interaction between the spin system and the lattice taken into account, this degeneracy can be lifted; then along with the magnetic ordering, a structural phase transition from tetragonal into the orthorhombic phase may occur. The simplest model accounting for the spin–lattice interaction in a frustrated spin system can be represented by the following Hamiltonian [282]:

$$H = \sum_{\langle ij \rangle} \left[J_1 (1 - \alpha_1 y_{ij}) \mathbf{S}_i \mathbf{S}_j + 1/2 K_1 y_{ij}^2 \right] + \sum_{\langle\langle ij \rangle\rangle} \left[J_2 (1 - \alpha_2 y_{ij}) \mathbf{S}_i \mathbf{S}_j + 1/2 K_2 y_{ij}^2 \right]. \tag{3.1}$$

Here, $y_{ij} = \left|\mathbf{u}_j - \mathbf{u}_i\right|$, $\mathbf{u}_i$ being a displacement of the atom i in the ab plane, and α_1, α_2 – the electron–phonon interaction constants. Thus, the mechanism of the spin–lattice coupling in the model of (2.2) is the modulation of exchange interactions.

In the spin wave approximation, the energy of the above model has been calculated under a condition of strong frustration, $J_2/J_1 \geqslant 0.5$, and minimized over the displacements y_{ij}. As a result, the minimum energy was that of an SDW structure with the arrangement of magnetically ordered spins along the shortest axis of the orthorhombic crystal; along the longer axis, the spins tend to arrange themselves antiferromagnetically. This is exactly the spin and crystal structure as observed in undoped FeAs-compounds. The thus outlined magneto-structural phase transition is a variety of the spin-Peierls transition. In [283], such magnetostructural transition has been justified within microscopic theory of the AFe_2As_2 compounds, using an LSDA calculation of electronic structure in structurally distorted crystal, in which phonon modes with $\mathbf{q} = 0$ were calculated. A strong softening of exactly those phonon modes has been demonstrated, which condense to form an orthorhombic phase with an SDW magnetic structure linked to it.

An instability of stoichiometric compounds with other compositions has been revealed in anomalies of thermodynamical and transport properties, as well as via Mössbauer effect measurements on ^{57}Fe. Thus, in $EuFe_2As_2$, structural and magnetic phase transitions have been found at $T_0 = 190\,\mathrm{K}$ [284]. Under doping with potassium, in the $Eu_{1-x}K_xFe_2As_2$ system, the magnetic transition was suppressed, and superconductivity with quite high $T_c = 32\,\mathrm{K}$ resulted [285]. Magnetic transitions detectable by neutron diffraction and by other methods, and the relation between their suppression by doping and the onset of superconductivity, are apparently a common feature of the AFe_2As_2 systems.

In some AFe_2As_2 compounds, where rare-earth elements appear as the A atoms, at low enough temperatures the effects related to their magnetic ordering can be observed. They are the most pronounced in $EuFe_2As_2$, because of a large magnetic moment of Eu atoms [286]. Under stoichiometric composition, $EuFe_2As_2$ has SDW structure in the FeAs-sublattice. Near $T_N \approx 20\,\mathrm{K}$, an AFM ordering appears in the Eu sublattice, whereby, due to a strong coupling to the Fe sublattice, a mutual influence develops of the two magnetic orderings.

The $EuFe_2As_2$ compound and the doped ones exhibit metamagnetism, i.e. phase transitions from AFM into the FM state driven by magnetic field. In [287], phase diagrams of the pristine and doped compounds in the (T, H) plane are constructed. It is in the fields of about 1 T that the above transition may occur (Fig. 3.39).

In the $EuFe_{2-x}Co_xAs_2$ compound, with an increase of Co concentration x, the SDW ordering in the Fe layers is suppressed, and superconducting state sets on; simultaneously, at 17 K, a magnetic ordering appears in the Eu layers. In the magnetic field, antiferromagnetic ordering of Eu atoms is replaced by a feromagnetic one, i.e. a metamagnetic transition takes place. It is remarkable that superconductivity coexists with this ferromagnetic order [288].

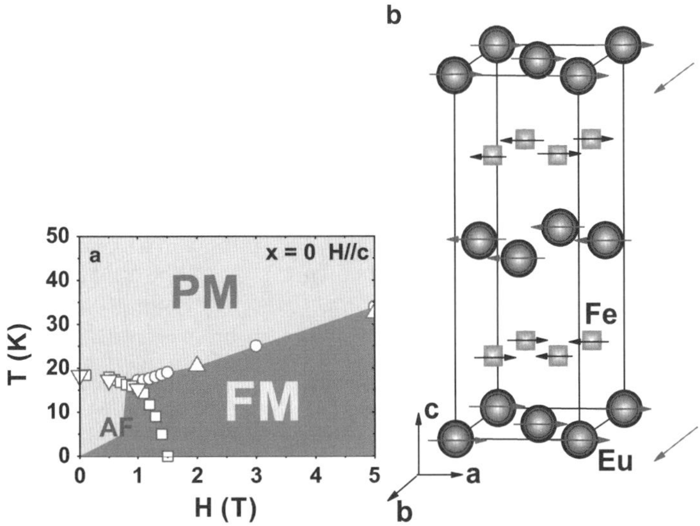

Fig. 3.39 (**a**) Magnetic phase diagram in the (T, H)-plane for $EuFe_2As_2$, obtained from the data on electrical resistivity, magnetic susceptibility and specific heat in magnetic field; (**b**) a possible magnetic structure [287]

3.3.2 Doped Compounds

The effect of doping of $EuFe_2As_2$ by substituting Fe atoms by Ni was analyzed in [289, 290] by measuring the temperature dependence of electrical resistivity, magnetic susceptibility, and thermo electromotive force. The resulting phase diagram is shown in Fig. 3.40. With Ni doping, the magnetic ordering within the FeAs-layers is gradually suppressed, whereas the AFM ordering of Eu atoms persists up to the concentration $x = 0.05$. At further increase of x, a transition occurs on the Eu lattice from AFM into the FM phase. Meanwhile, no superconductivity was detected at any concentrations $0 < x \leqq 0.2$. An absence of superconductivity in $EuFe_{2-x}Ni_xAs_2$ is somehow unexpected, because on replacing Fe by Co in $EuFe_2As_2$, an induced superconductivity occurs. Apparently, this is related to an onset of ferromagnetic ordering in the Eu sublattice, even if the Eu$4f$ states lay deeply below the Fermi level, and their overlap with the Fe$3d$ states is negligible.

The doping of the basic compound, $BaFe_2As_2$, with potassium suppresses T_0 of the structural and magnetic transitions, whereas simultaneously the superconductivity appears, as is seen from Fig. 3.13. A more detailed study of the magnetic state close to onset of the superconducting phase revealed a region of coexistence of antiferromagnetic and superconducting states. An analogous coexistence of antiferromagnetism and superconductivity was detected for $BaFe_{2-x}Co_xAs_2$ [291], see Fig. 3.41. This makes a difference of AFe_2As_2 systems from the $LaO_{1-x}F_xFeAs$ ones, in which, as we have seen, the superconductivity sets on when the AFM ordering disappears completely.

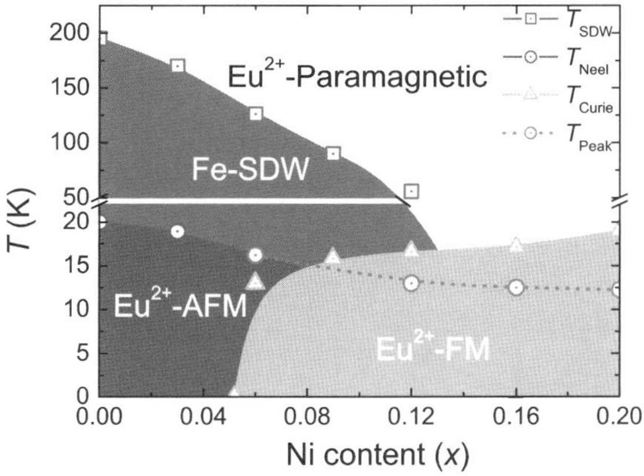

Fig. 3.40 Magnetic phase diagram of $EuFe_{2-x}Ni_xAs_2$ [289]

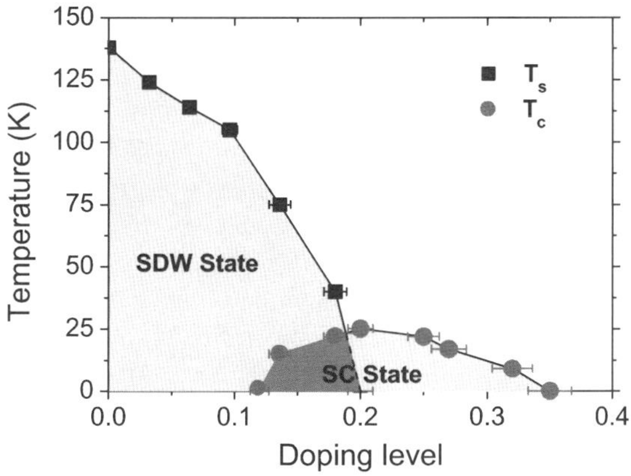

Fig. 3.41 Phase diagram of $BaFe_{2-x}Co_xAs_2$ [291]

An analysis of different compounds out of the AFe_2As_2 class has shown that superconductivity appears under doping of those initial compounds which exhibit an SDW instability. The suppression of the latter by dopants creates conditions for a superconducting state to set on. In this relation, it is interesting to address the results of study of a new compound $BaRh_2As_2$ [292] with the same crystal structure as AFe_2As_2: it contains RhAs-layers, where the Rh atoms are surrounded by tetrahedra made of As. The electronic properties of this compound resemble those of all AFe_2As_2. Nevertheless, in $RhFe_2As_2$ no traces of any structural or AFM ordering have been found in the temperature range from 2 to 300 K. It would have been utterly interesting to study doped compounds under a view of a possible superconductivity in them. This would shed additional light onto the role of antiferromagnetic instability in the FeAs-systems for the formation of superconducting state.

In [293], the (T, x) phase diagrams of the $Ba(Fe_{1-x}M_x)_2As_2$ compounds, M = Rh, Pd, were constructed and compared to diagrams for the compounds with M = Co, Ni. In spite of an essential difference between the 3d and 4d elements, there are similarities between these two cases, in what the effects of the Rh doping resemble those of a doping with Co, and the effects of Pd are similar to those of Ni. At the same time, there are differences on doping with 3d and 4d elements.

Let us further discuss a problem of coexistence of superconductivity and magnetism in AFe_2As_2 compounds. Indications towards such possibility are given by experimental studies of underdoped $(Ba_{1-x}K_x)Fe_2As_2$ ($x < 0.4$), see [209]. Polycrystalline samples of the $0 < x < 0.3$ composition were studied by ^{57}Fe Mössbauer spectroscopy. Structural phase transition from orthorhombic into tetragonal phase has been registered throughout up to $x \approx 0.3$. It is remarkable that the superconducting state was detected already at $x > 0.1$, by measuring electrical resistivity and magnetic susceptibility. A splitting of Mössbauer lines was seen in all samples with $0.1 < x < 0.2$, but not at $x = 0.3$. The authors of [209] claim a coexistence of superconductivity and magnetic ordering in underdoped system $(Ba_{1-x}K_x)Fe_2As_2$, where no mesoscopic phase separation was detected. The picture outlined applies to the vicinity of optimum potassium concentration, $x \approx 0.4$. It should be noted, however, that these results are at variance with the measurements done on single crystals [268, 269].

3.3.3 Magnetic Excitations

AFe_2As_2, as all FeAs-systems, reveals quasi-two-dimensional character of electronic properties, as a consequence of their layered structure. It appears interesting to find out whether quasi-two-dimensionality is also pronounced in their magnetic properties. A direct information on magnetic interactions can be gained from measurements of spin waves by inelastic neutron scattering. Such measurements on single crystals of stoichiometric compounds $SrFe_2As_2$ [294], $BaFe_2As_2$ [295] and $CaFe_2As_2$ [223] indicated a substantial anisotropy of magnetic interactions, albeit not that large as to justify a conclusion in favour of quasi-two-dimensionality of their magnetic properties.

The doping enhances quasi-two-dimensionality of electronic properties, favouring superconductivity, that is what has motivated a study of magnetic excitations in $Ba_{0.6}K_{0.4}Fe_2As_2$ [296]. However, because of dealing with a polycrystal, no unambiguous answer to the question posed above could have been obtained. In [297], a study of magnetic excitations in $Ba(Fe_{0.92}Co_{0.08})_2As_2$ with the use of inelastic neutron scattering has been undertaken. Magnetic excitations in this material turned out to be more quasi-two-dimensional than in stoichiometric $BaFe_2As_2$. In the normal phase ($T > T_c$), the excitations in the vicinity of the antiferromagnetism vector $\mathbf{Q}$ were gapless, but they turned into those with a gap on cooling to $T < T_c$. The observed energy gap was 9.6 meV that corresponds to $5\,kT_c$. It is noteworthy that the energy gap in magnetic excitations in cuprates is of about the same magnitude.

Spin fluctuations in different FeAs-compounds were studied by inelastic neutron scattering in [298, 299]. Previous researches determined the spin wave spectrum in the SDW phase of some FeAs-compounds, and revealed the resonance mode in the superconducting phase of some 122 systems.

Among the latest publications, we mark out [298], in which the spin fluctuations were studied in the $BaFe_{1.96}Ni_{0.04}As_2$ system. They revealed a different behaviour from that in a Co-doped $BaFe_2As_2$, which under stoichiometry has $T_N = T_S = 149$ K. Upon substitution of Fe by Co, T_S and T_N rapidly fall down as the dopant concentration grows, and a superconducting state is formed, within which a resonance mode is detected [211, 300].

In contrast to results for Co-doped $BaFe_2As_2$, the Ni-doped compound $BaFe_{1.96}Ni_{0.04}As_2$ exhibits other features of spin fluctuations. At the given level of doping, the compound remains antiferromagnetic and tetragonal at high temperatures. At $T_S = 97$ K, the phase transition into the orthorhombic phase takes place, and at $T_N = 15$ K the superconductivity occurs, so that we deal with coexistence of superconductivity with magnetic ordering. Neutron studies indicate that spin fluctuations in this compound are not related to superconductivity, as is the case in the Co-doped $BaFe_2As_2$. The Ni doping leads to a different effect: the three-dimensional spectrum of spin fluctuations in stoichiometric $BaFe_2As_2$, on Ni doping, becomes essentially two-dimensional, as a result of a weakening of exchange interaction J_c between the FeAs-layers. This two-dimensional structure, when subject to electron doping, may become an important factor in the appearance of superconductivity – at least, in the 122-type compounds. This fact demands a further theoretical elaboration.

Another question which demands clarifying is which model has to be adopted for a description of experimentally measured spin excitations (spin waves) in a AFe_2As_2 compound with SDW-ordering, the localized Heisenberg model or an itinerant one? Let us consider, as an example, a study of $CaFe_2As_2$ by methods of neutron spectroscopy [301, 302].

In Fig. 3.42, the spin wave spectrum is depicted along three symmetric directions of the three-dimensional Brillouin zone [301]. The solid line corresponds to the

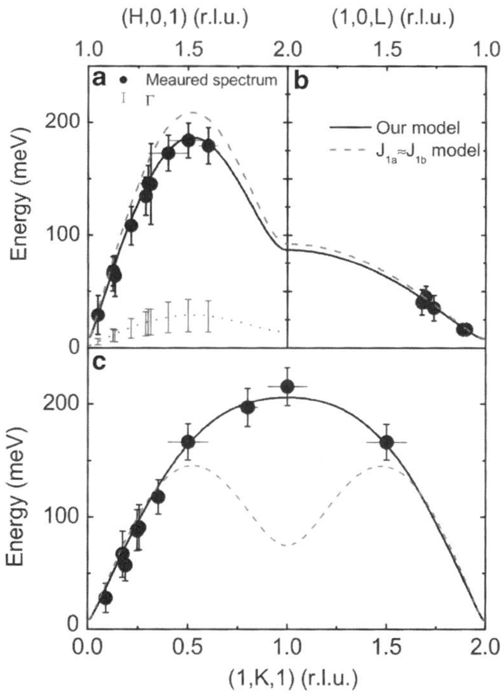

Fig. 3.42 Spin wave spectrum in $CaFe_2As_2$, after the neutron measurements of [301]

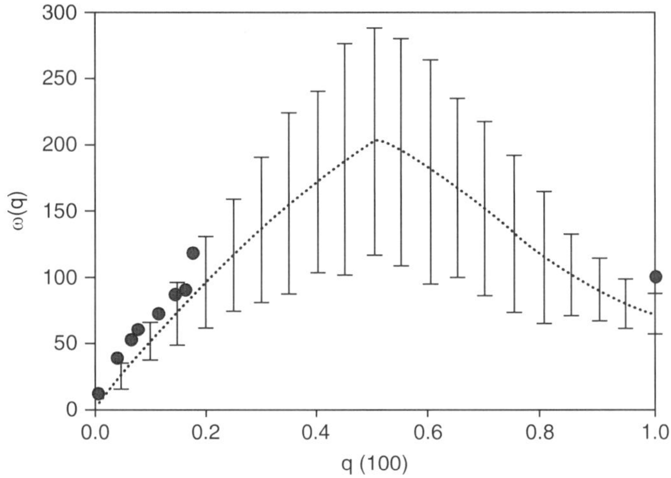

Fig. 3.43 Spin wave spectrum of $CaFe_2As_2$ (*dotted line*) in comparison with the decay of spin excitations in the itinerant model [302]

spin wave spectrum as calculated in the Heisenberg model, with exchange integrals between nearest and next-nearest neighbours (Fe atoms) fit to the experimental data. It follows from the figure that the spin wave spectrum has three-dimensional nature; the whole bandwidth of spin excitations makes 200 meV, and even near the top of this band the spin waves are well defined.

In another work [302], spin waves have been measured with energies up to 100 meV. In Fig. 3.43, the experimental points fall onto the dashed curve which describes the spectrum of collective excitations in an itinerant model. This dashed line passes through the maximum of the spectral distribution of spin fluctuations $\mathrm{Im}\chi(\mathbf{q}, \omega)$, whereas vertical bars indicate the widths of corresponding peaks. It is seen that the peak width increases sharply for excitation energies of around 100 meV. It is namely in this region where well-defined spin waves have been measured. Beyond this range (ω >100 meV), one observes a damping of collective excitations via their decay into electron-hole pairs in the Stoner continuum.

It follows from the comparison of both resulting curves that neither localized nor itinerant models do reflect the spectrum of spin excitations in its entirety. In compounds of the $CaFe_2As_2$ type, the adequate description of spin excitations demands the both aspects, the localized and the itinerant one, to be incorporated.

In conclusion, we refer to the results of [303], which indicate the relation between the strength of spin fluctuations and the temperature of superconducting transition, measured as functions of doping in the $Ba(Fe_{1-x}Co_x)_2As_2$ system, see Fig. 3.44. In the insets of Fig. 3.44a, the structure of the Fermi surface is schematically shown for both optimally doped compound ($x < 0.15$) and an overdoped one, in which the superconductivity is suppressed. In the first case, the sizes of the hole and electron pockets are comparable, therefore a nesting takes place on the (π, π) vector connecting the hole and electron pockets. The nesting induces spin fluctuations which

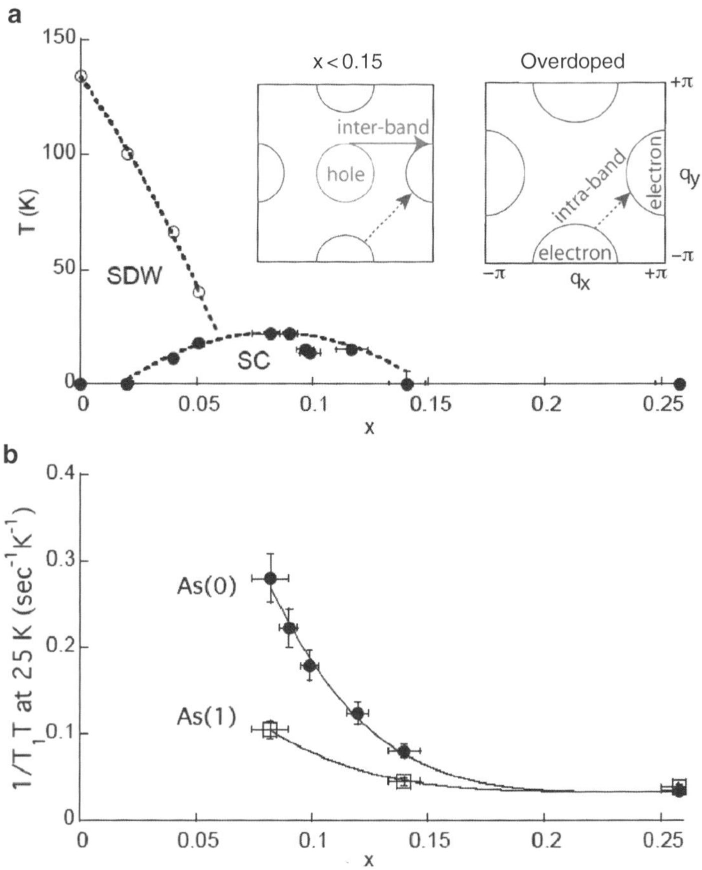

Fig. 3.44 Correlation between T_N, T_c and the magnitude of spin fluctuations, as function of doping in $Ba(Fe_{1-x}Co_x)_2As_2$ compounds [303]. (**a**): Phase diagram in the (T, x) plane; (**b**): $1/T_1T$, measured on As nuclei by NMR

give rise to a superconducting state. In the overdoped regime, due to electron doping, the electron pockets become larger, whereas the hole pocket disappears. As a consequence, the nesting responsible for spin fluctuations, which lead to the Cooper pairing, breaks down. An absence of such fluctuations is seen from Fig. 3.44b, where the $1/T_1T$, measured on the As nuclei, is shown as function of doping. At $x > 0.15$, the value of $1/T_1T$, which determines the strength of spin fluctuations, becomes negligibly small. The As(0) symbol indicates an As atom whose all four neighbours are Fe, and As(1) stands for an As atom for which one of its neighbours is substituted by Co. We see that in the overdoped regime, the strength of spin fluctuations is strongly suppressed, due to a loss of nesting on the (π, π) wave vector, and a superconducting state is not realized.

Chapter 4
Other FeAs-Based Compounds

4.1 Compounds of the FeSe, FeTe Type

4.1.1 Superconducting Properties

A new family of superconductors on the basis of Fe has been found in FeSe, FeTe compounds, whose T_c increased under chemical doping and/or external pressure [16, 304–307]. The crystal structure of these compounds has much in common with the previously discussed ones, where the FeAs layers were an important element. FeSe is also made of stocked FeSe planes; however, differently from the FeAs-type compounds, there is nothing more between the layers. The Fe atoms make a square lattice, so that each Fe atom is within a tetrahedric environment of Se (or Te) atoms. These new materials have tetragonal structure with the $P4/nmm$ space group, exactly like the FeAs-based compounds.

Stoichiometric compounds of this type under ambient pressure are not superconducting. The superconductivity appears either under doping, or in the presence of vacancies, or under pressure. Thus, in $FeSe_{1-x}$ with $x = 0.12$, the superconductivity has been detected with $T_c = 8$ K [304]. T_c soon achieved 27 K in FeSe under external pressure of $P = 1.48$ GPa [16, 305]. The upper critical field was estimated to be $H_{c2} \sim 72$ T. In [305], $T_c = 14$ K has been reported at $x \approx 0.08$ and ambient pressure.

Further investigations lead to even higher T_c values in FeSe-type compounds under pressure [308–311]. The highest value of T_c was 21 K under pressure of 3.5 GPa, when measured from the zero of resistivity, and 37 K at $P = 4.15$ GPa – when measured from the initial fall of resistivity. According to [309], the T_c in $Fe_{1.01}Se$ increases from 8.5 to 36.7 K as pressures rises to 8.9 GPa. Thereby, no static magnetic ordering was detected throughout the whole (P, T) phase diagram. In [308], it was reported that hydrostatic pressure first rapidly enhances T_c up to 37 K at $P \sim 7$ GPa, then pushes it down to 6 K at $P \sim 14$ GPa. Such non-monotonous behaviour of $T_c(P)$ is FeSe follows a sharp anomaly in the minimal interatomic distance as function of pressure.

Detailed studies of structural and superconducting properties of FeSe under pressures of up to 26 GPa were done in [310]. At $P = 12$ GPa, a transition from

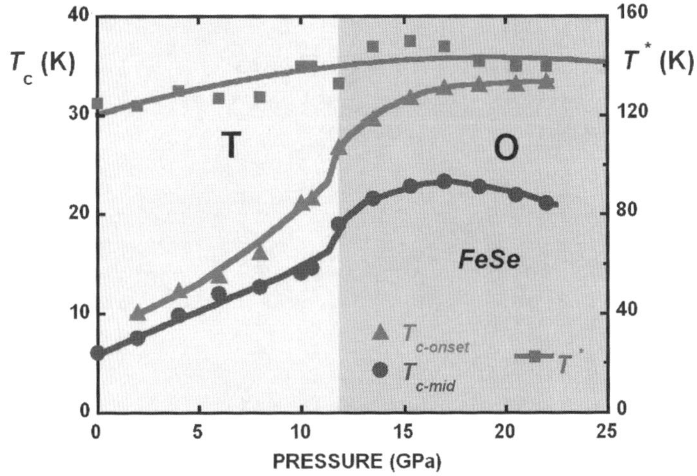

Fig. 4.1 Evolution of the magnetic phase transition temperature T^* and the superconducting transition temperature T_c in FeSe under pressure [310]

tetragonal to orthorhombic structure has been reported. The superior T_c value of 34 K was achieved in the orthorhombic phase under 22 GPa. Discrepancies in qualitative results concerning T_c under pressure as reported in different works is, following the opinion of the authors of [310] related to two reasons: different choice of media transmitting hydrostatic pressure to the sample, and possibly different composition of the FeSe samples which could have contained an excess or deficiency of Fe atoms, i.e. deviations from stoichiometry which are difficult to control on synthesis.

In Fig. 4.1, the phase diagram in the (P, T)-plane, as constructed in [310], is shown. Under pressure of 12.5 GPa, FeSe transforms from tetragonal into orthorhombic phase. In the vicinity of this transition, T_c increases and, already within the orthorhombic phase, passes through maximum. A remarkable fact is an existence of superconducting phase within a magnetically ordered one. This unusual phenomenon has yet to be verified for all FeSe-related compounds.

In $Fe_{1+x}Se$ compounds, T_c does not exceed 8 K, differently from FeAs-compounds where it can be much higher. However, a number of experiments indicate that an application of pressure of the order of several GPa increases T_c to 30 K. In [312], according to magnetic susceptibility measurements of $Fe_{1.02}Se$, a pressure inferior to 1 GPa rises T_c from 7 to 30 K. The authors of [312] suggest that in the interval of pressures 0.5–1.0 GPa, two phases exist, with different superconducting transition temperatures, $T_{c1} \sim 7$ K and $T_{c2} \sim 30$ K. This explains why in some experiments on the $Fe_{1+x}Se$ compounds under pressure, a transition into the superconducting state is strongly smeared.

Further researches have shown that a substitution of selenium in FeSe by sulphur increases T_c by 20%, and substitution by tellurium – by 75%, contrary to the first results of [16]. In isovalent compounds $Fe(Se_{1-y}Te_y)_{0.82}$ [306], the T_c values within 8–14 K have been obtained, with the maximum at $y \approx 0.6$ [313]. At the same time, a substitution of Fe by other transition metals, Co and Ni, suppresses superconductivity.

The symmetry of the superconducting order parameter in FeSe was studied in μSR experiments [314]. For an $FeSe_{0.85}$ sample with $T_c = 8.3$ K, the behaviour of $\lambda_{ab}^{-2}(T)$ has been recovered from the μSR data. The emerging behaviour excluded both the order parameter with gap zeros on the Fermi surface and the isotropic s symmetry. The experimental data seemed consistent either with anisotropic s symmetry, or with two-gap s symmetry. Within the anisotropic s model, the maximal gap value at $T = 0$ was 1.35 meV, that would yield the parameter $2\Delta/kT_c = 3.79$, close to 3.52 of the BCS theory. Within the two-gap model, $\Delta_{01} = 1.63$ meV and $\Delta_{02} = 0.38$ meV, that would result for the $2\Delta/kT_c$ in the values 4.59 and 1.07, correspondingly.

The doping of FeTe with sulphur was studied in [315]. A sample $FeTe_{1-x}S$ with $x = 0.2$ showed $T_c \simeq 10$ K. This superconductor had a high critical field of $H_{c2} \sim 70$ T, whereby the coherence length was ~2.2 nm. The FeSe and FeTe compounds seem to be interesting as prototypes of a new superconductors family, not containing poisonous arsenic, and promising for practical applications.

NMR measurements on the ^{77}Se nuclei have been done for a sample with $T_c = 8$ K, concluding that the resulting spin–lattice relaxation rate $1/T_1$ in the superconducting phase varies with temperature as $1/T_1 \sim T^3$. This taken into account along with the absence of the coherent peak hints for a non-standard superconductivity with zeros of the superconducting order parameter on the Fermi surface and, presumably, d symmetry of the order parameter. Still, it would have make sense to discuss a possibility of explaining this result as a consequence of the $s^{\pm}$ symmetry of the order parameter, which, as was discussed in Chap. 1, would also lead to a power law dependence of $1/T_1$ on T.

The samples $Fe_{1+x}Se$, i.e. those with additional of Fe atoms which enter the interstitials between the FeSe layers were found to exhibit a particular sensitivity towards this excess x [316]. T_c turns out to be extremely sensitive to small variations of the Fe concentration near the stoichiometry. Thus, $T_c \approx 8.5$ K for the most stoichiometric of the achieved compositions, $Fe_{1.01}Se$. At $Fe_{1.02}Se$, T_c drops to 5 K, and slightly beyond this small x value the superconductivity completely disappears. The most intriguing in this is that non-superconducting samples are not antiferromagnets, at variance with the FeAs-type compounds. The measurements of Mössbauer spectra at temperatures 259 K and 5 K in $Fe_{1.01}Se$ and $Fe_{1.03}Se$ samples did not exhibit any difference, even if one of these samples was superconducting with $T_c = 8$ K, and the other did not show any traces of superconductivity down to 0.6 K. At least for the $Fe_{1+x}Se$ system, the situation looks like if superconductivity be not related to the suppression of the antiferromagnetic order, contrary to what is the case in all FeAs-type compounds. This aspect requires a more detailed elucidation. Taking this into account, astonishing is the fact that the electronic structure of $Fe_{1+x}Se$ reveals all the same important features as the FeAs-based compounds do exhibit.

Let us moreover reveal the data on ternary compounds $Fe_{1+y}Te_xSe_{1-x}$ [317] and $Fe_{1+y}Te_{1-x}Se_x$ [318]. For single crystalline sample of $Fe_{1+y}Te_xSe_{1-x}$ with $0 < y < 0.15$, the superconductivity with $T_c = 14\,K$ was only detected near $x = 0.5$. An onset of superconductivity correlates with diminishing the magnetic susceptibility at room temperature and with lower concentration y of excess Fe atoms.

4.1.2 Unusual Magnetic Properties

The $FeSe_x$ compounds in their normal phase exhibit anomalies of magnetic properties at temperatures $T_{s1} \simeq 106\,K$ and $T_{s2} \simeq 78\,K$ [319]. The magnetization M in applied magnetic field and gradually decreasing temperature increases sharply at T_{s1} and then returns to the original $M(T)$ line in T_{s2}. Astonishingly, under pressure both T_{s1} and T_{s2} shift towards higher temperatures (Fig. 4.2). The next anomaly, a diamagnetic one, occurs at $T_c = 8\,K$. These anomalies indicate that superconductivity in the compound given comes about in a situation characterized by strong magnetic fluctuations.

An immediate study of magnetism in these systems have been done by neutron diffraction experiments [320]. In a neutron diffraction pattern, superstructure magnetic reflections were detected which hint for an existence of an incommensurate magnetic ordering. For stoichiometric FeTe, the wave vector of the magnetic structure was found to be $(\delta\pi, \delta\pi)$, differently from the case of FeAs-type compounds, where the wave vector of the SDW structure is $(0, \pi)$. The magnetic structure of

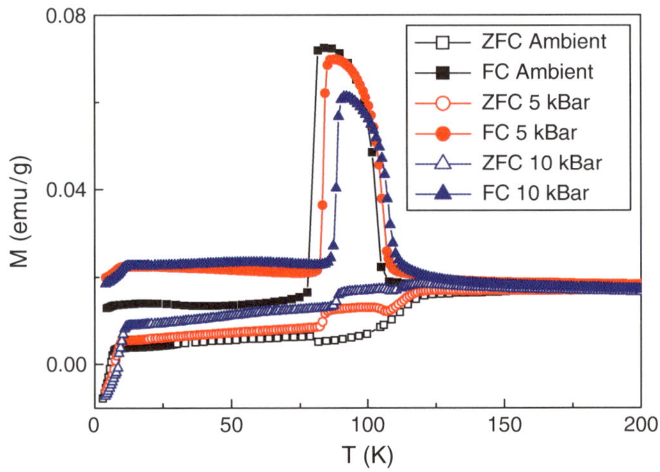

Fig. 4.2 Temperature dependence of magnetization at different pressures for $FeSe_x$, $x = 0.80$. The magnetization was measured in the field of 10 Oe, cooling on field (FC) and in zero field (ZFC) [319]

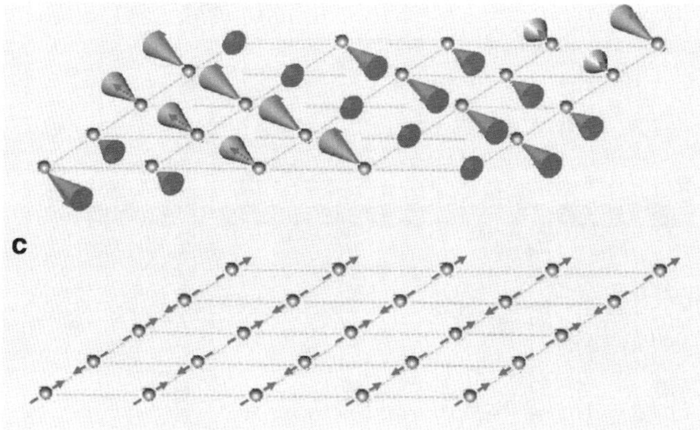

Fig. 4.3 Magnetic structure of the FeTe (**a**), different from that of the FeAs-compounds (**b**) [320]

FeTe is shown in Fig. 4.3a. The modulation occurs along the diagonal of square lattice made of the Fe atoms in the plane. In the plane normal to this direction, all the spins are tilted over the surface of a cone shown at each Fe site. This arrangement is very different from that known to exist in all FeAs-type compounds (Fig. 4.3b).

In the doped $Fe_{1.14}Te$ compound, another wavevector of magnetic structure was determined: $\mathbf{q} = (\pm\delta, 0, 1/2)$, with $\delta = 0.380$. Magnetic moment at a given Fe site situated at $\mathbf{R}$ is

$$\mathbf{M(R)} = M\left[\hat{\mathbf{a}}\cos\mathbf{qR} + \hat{\mathbf{c}}\sin\mathbf{qR} + i\sigma\hat{\mathbf{b}}\cos(\mathbf{qR} + \psi)\right],$$

where $\hat{\mathbf{a}}$, $\hat{\mathbf{b}}$ and $\hat{\mathbf{c}}$ are unit vectors along the tree lattice vectors, and ψ is an arbitrary phase. The above formula defines a spiral structure with rotation of spins in the (a, c) plane and sine-like modulation along the $\hat{b}$ direction.

Magnetic ordering of spins occurs at $T_s = 63\,\mathrm{K}$ in $Fe_{1.14}Te$ and at $T_s = 75\,\mathrm{K}$ in $Fe_{1.076}Te$. Simultaneously with the onset of magnetic ordering a structural phase transition takes place, as in all FeAs-type compounds.

Concerning experimental determination of non-collinear magnetic structure in FeX (X = S, Se, Te) compounds, we note that in [321] an LDA-based study was reported for an analysis of magnetic interactions in the low-temperature phase of FeSe. It was concluded that the ground state must be non-collinear; however, a large continuum of nearly degenerate states was found to be separated by a very small energy from the SDW structure.

Neutron diffraction studies of FeTe have been extended in [322]. A Fe-excessive $Fe_{1.068}Te$ compound undergoes, as the temperature decreases, magnetic and structural transition from the tetragonal phase (space group $P4/nmm$) into monoclinic one ($P2_1/m$). This phase transition is of the first order. The corresponding magnetic structure is shown in Fig. 4.4.

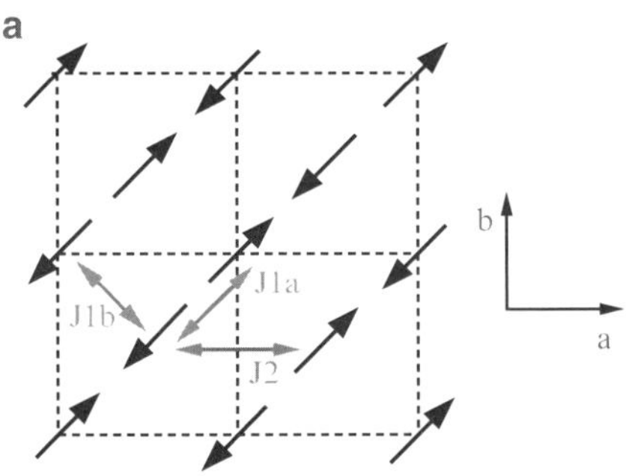

Fig. 4.4 Magnetic structure (**a**) in all stoichiometric FeAs-compounds, and (**b**) in FeTe [322]. The drawing is from [323]. The two-head *arrows* indicate the AFM interactions J_{1a}, J_{2a} and the FM interactions J_{1b}, J_{2b} between the nearest and between the next-nearest neighbouring Fe atoms, correspondingly

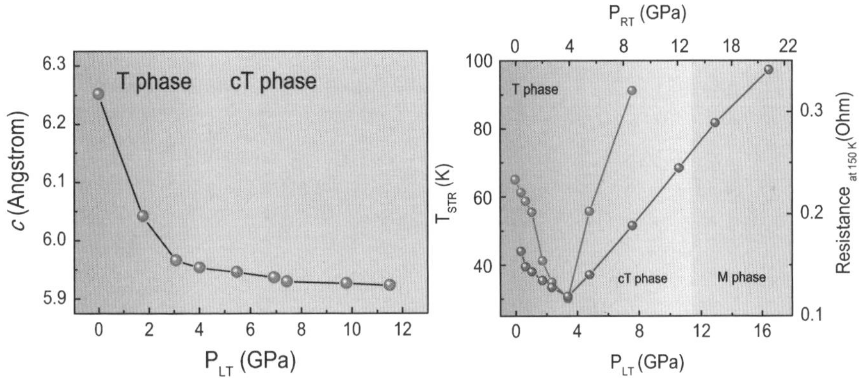

Fig. 4.5 Pressure dependence of the lattice parameter and the structural phase transition temperature in $Fe_{1.05}Te$, after [325]

In FeTe, a collinear antiferromagnetic structure, shown in Fig. 4.4b (in the tetragonal coordinate system), has been found. It differs from the SDW-structure (Fig. 4.4a), which has been found in all FeAs-type systems.

The magnetic structure of FeTe is a double-stripe one, rotated by 45° relative to the standard stripe structure which comes about in the FeAs-type compounds. An evaluation of exchange interactions will be necessary for further discussion of this magnetic structure in the light of subsequent calculations [323] of the electronic structure of FeTe. Therefore, from the comparison of two neutron-diffraction works, [323] and [324], we conclude that magnetic structure of FeTe does substantially depend on stoichiometry (excess of Fe atoms).

Measurements of electrical resistivity and X-ray diffraction on a $Fe_{1.05}Te$ single crystal hinted for an existence of a transition into the collapsed tetragonal phase, analogous to that previously observed in $CaFe_2As_2$ [325]. In Fig. 4.5, this transition is seen from the variation of the lattice parameter and electrical resistivity with pressure. An abrupt decrease of the c parameter at $P = 4\,\mathrm{GPa}$ within the tetragonal T-phase witnesses a transition into a collapsed tetragonal cT-phase. At the same pressure, a sudden drop of the structural transition temperature and electrical resistivity occurs (Fig. 4.5), with their subsequent increase in the collapsed phase.

In the (T, P) phase diagram (Fig. 4.5), a region M is indicated in which, at low temperatures, an increase of electrical resistivity is observed, indicating the semiconductor type of conductivity. Superconductivity was not manifesting itself at any pressures studied.

To fully characterize the cT-phase, it would have been necessary to perform magnetic measurements under applied pressure and to measure the magnitudes of magnetic moments at Fe ions, how it has been done for the $CaFe_2As_2$, where it has been shown that the onset of the collapsed tetragonal phase is accompanied by a disappearance of localized magnetic moments. Magnetic measurements and neutron diffraction studies would have been highly desirable for this compound.

4.1.3 Electronic Structure of Stoichiometric Compounds

LDA calculations for the three compounds FeS, FeSe and FeTe have been performed by the FLAPW method in [326]. Energy dispersion curves and Fermi surfaces in all three compounds turned out to be very close, and similar to those for FeAs-type compounds. In the calculations, experimental values of the lattice parameters have been used, whereas the internal coordinate z_X for a chalcogen atom (X = S, Se, Te) was found from total energy minimization, see Table 4.1.

In Fig. 4.6, calculated density of states of FeSe is given as an example; in two other compounds it is identical, with the only difference that the pseudogap at the Fermi level in FeTe is much smaller. Similarly to how it is in the FeAs-type systems, the partial density of d states is grouped within the range −2 to 2 eV around the Fermi level, whereas the chalcogen p states lie deep below the Fermi level. The DOS at the Fermi level is given in Table 4.1. As is seen from this table, the compounds in question are metals with high states density (but low concentration of carriers), as all FeAs-type compounds also are: for example, $N(E_F)$ of LaOFeAs is $2.62\,\mathrm{eV}^{-1}$. The energy gain for an SDW state is the largest in FeTe (47 meV), and for the same

Table 4.1 Structure parameters and calculated magnetic properties of the FeX compounds, X = S,Se,Te [326]. Density of states at the Fermi level $N(E_F)$ is given in the $(\mathrm{eV})^{-1}$ units per Fe atom and per spin channel; the magnetic moment is given per Fe atom, in Bohr magnetons. The energy gain relative to the nonmagnetic case E_{SDW} is given in meV per Fe atom

Compound	a (Å)	c (Å)	z_X	$N(E_F)$	m_{SDW} (μ_B)	E_{SDW}
FeS	3.6735	5.0328	0.2243	1.35	0.00	0
FeSe	3.765	5.518	0.2343	0.95	0.65	5
FeTe	3.8215	6.2695	0.2496	1.83	1.28	47

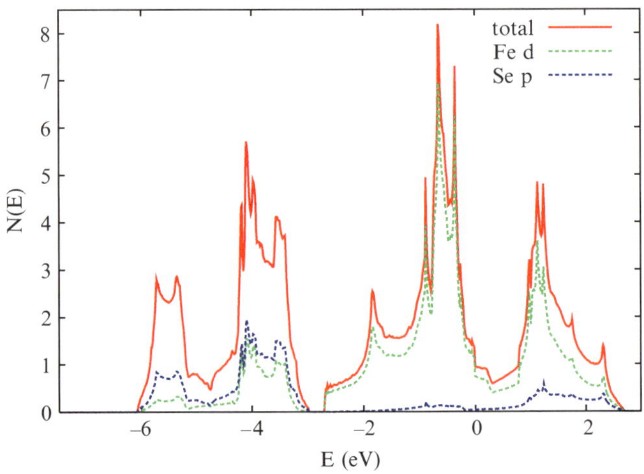

Fig. 4.6 Density of electron states in the FeSe compound, after the LDA calculations [326]

compound the magnitude of the Fe magnetic moment is maximal ($1.3\,\mu_B$). These values can be compared to those for LaOFeAs: the magnetic moment in the SDW is $1.0\,\mu_B$ and the energy gain in the SDW, per Fe atom, is 11 meV.

The Fermi surface in each of three compounds is multi-sheet one: there are two electron cylindrical surfaces around the M point and concentric hole ones around Γ. Moreover, FeS and FeTe possess one more sheet of the Fermi surface within the hole cylinders (see Fig. 4.7).

Diameters of the electron and hole sheets are almost identical, therefore a good nesting occurs with the vector $\mathbf{Q} = (\pi, \pi)$, which connect these sheets. This results in an SDW-type instability of all FeX compounds that has been confirmed experimentally. Therefore, the calculations under discussion show that the compounds of the FeX group are close in their electronic characteristics to the FeAs-type compounds. Such closeness is explained by similarity in crystal structures, notably by the fact that the FeX layers are constructed and assembled in the same way as the FeAs layers in the ferro-arsenic group materials.

In [326], the phonon spectra and electron–phonon interaction parameters have been moreover calculated. The electron–phonon coupling constant λ estimated for FeSe turned out to be very small ($\lambda = 0.17$). Consequently, the observed superconductivity in this compounds cannot be explained by a conventional mechanism of electron–phonon coupling, again similarly to how is the case in the FeAs-type compounds.

Let us make note of works in which, by using inelastic neutron scattering, the density of vibrational states in the phonon spectrum of $FeSe_{1-x}$ has been measured [327], and Raman phonons in FeTe analyzed [328]. They emphasize that no anomalies whatsoever were detected in phonon spectra of these compounds in their both normal and superconducting phases. This implies that the coupling of electrons is mediated by non-phonon mechanisms.

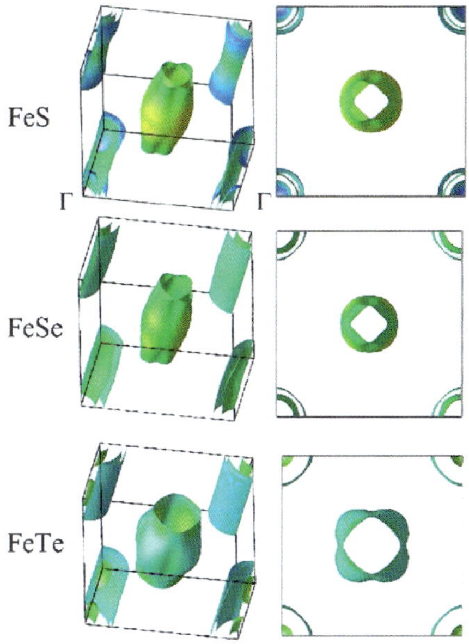

Fig. 4.7 Fermi surfaces of FeS, FeSe and FeTe, obtained in spin-polarized LDA calculations [326]. Note that the Γ point is shown at the corner of the reciprocal unit cell

Table 4.2 Energies of FM, AFM, SDW and bicollinear states, calculated in the LDA for the FeSe and FeTe compounds. The energies are given in eV per Fe atom [329], relative to the nonmagnetic state

Compound	FM	AFM	SDW	*Bi*
FeSe	0.183	−0.101	−0.152	−0.089
FeTe	−0.0897	−0.098	0.156	−0.166

The LDA results of [326] discussed above related to a state of FeX systems without SDW or any other AFM ordering. These results should be revised, taking into account a possible magnetism of the ground state. In [329], LDA calculations for four magnetic states of FeSe and FeTe have been done: ferromagnetic one (FM), double-sublattice antiferromagnetic (AFM), collinear SDW and bicollinear (BC). The two latter states are depicted in Fig. 4.4.

It follows from the total energies of four abovementioned magnetic states, listed in Table 4.2, that in FeSe an SDW-type magnetic structure is realized, similarly to which is known for FeAs-type compounds, whereas in FeTe – a bicollinear structure. The neutron diffraction studies [322] indicate precisely this latter type of magnetic structure.

An attempt to grasp a quite peculiar bicollinear magnetic structure was undertaken in the framework of a localized model, which included, beyond antiferromagnetic interactions between nearest neighbours J_1 and next-nearest neighbours J_2, further interactions between next-next-nearest neighbours J_3, which may be of the

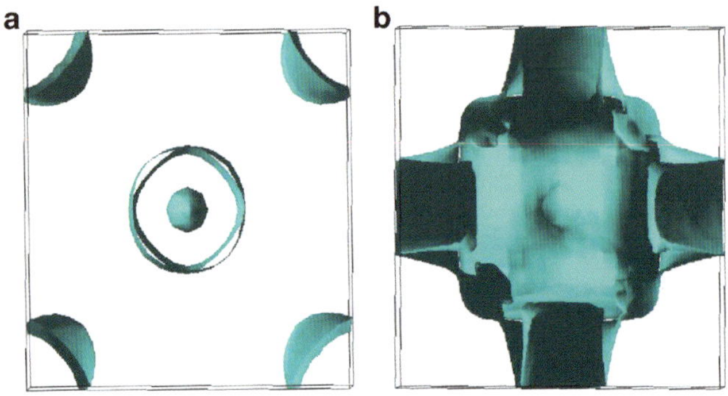

Fig. 4.8 Fermi surface of FeTe as function of doping δ: (**a**) $\delta \sim 0.0$, (**b**) $\delta \sim 0.5$ electrons per formula unit [323]

Ruderman–Kittel–Kasuya–Yosida (RKKY) type. From total energy values as calculated within the LDA and their mapping on the predictions of the localized model, the value of exchange parameters for both compounds have been extracted [329]:

$$\begin{aligned} &\text{FeTe:} \quad J_1 = 2.1\,\text{meV}/S^2, \quad J_2 = 15.8\,\text{meV}/S^2, \quad J_3 = 10.1\,\text{meV}/S^2; \\ &\text{FeSe:} \quad J_1 = 71\,\text{meV}/S^2, \quad J_2 = 48\,\text{meV}/S^2, \quad J_3 = 8.5\,\text{meV}/S^2. \end{aligned}$$

As has been shown in [65], in the localized $J_1 - J_2 - J_3$ model the SDW state is more favourable if $J_2 > 1/2\, J_1$, and for $J_3 > 1/2\, J_2$ the bicollinear structure (Fig. 4.8a) gains. The last condition is exactly satisfied for FeTe and not satisfied for FeSe.

As we will see later on, a more satisfactory explanation of the peculiar magnetic structure of FeTe follows from band structure calculations.

ARPES measurements of $Fe_{1+x}Te$ shows that Fermi surface contains hole- and electron- pockets and sheet centred in X point, though it is reconstructed from local intensity spectra [330, 331].

4.1.4 Electronic Structure of Doped Compounds

Under deviation from stoichiometry, the number of Fe atoms in an FeX compound exceeds that of chalcogen, a fact that can be expressed in two ways: either as an FeX_{1-x} compound with chalcogen vacancies, or as $Fe_{1+x}X$ with an excess of iron. At the same level of doping x, these situations correspond to two different compounds, since in the second case the excess Fe atoms enter interstitial positions in the lattice. The vacancy model FeX_{1-x} was studied in [332], and the model with excess Fe – in [324].

The authors of [332] calculated the band structure of FeX_{1-x} with $x = 0.125$ by an LSDA method. Since $0.125 = 1/8$, they used a periodic model of doped compound, considering a unit cell, which expands one site into eight, hence seven atoms of X and one vacancy. An interesting result is an appearance of the gap at the Fermi level for electrons with one (majority) spin projection, whereas for the other (minority) spin component there is no gap. Consequently, $FeSe_{0.85}$ comes out to be a semimetal.

It is obvious that the local states on those Fe atoms close to vacancies are different from those on far-from-vacancies Fe atoms. Let us denote the former as Fe1 and the latter as Fe2. Calculations of local spin density on Fe atoms yield an interesting result: the magnetic moment on the Fe1 atom, $2.14\,\mu_B$, is set antiparallel to magnetic moments $-1.10\,\mu_B$ on atoms Fe2. Therefore, a presence of a vacancy induces an antiparallel setting of Fe spins in its vicinity and at a distance. As is shown in [332], such a state is stable.

In another work [324], a model of excess Fe in the $Fe_{1+x}Te$ compound was used. The iron atoms labelled as Fe2 are situated in the plane occupied by the Te atoms and affect the state of the Fe1 atoms, those situated in the basal plane. The calculation for the $Fe_{1.125}Te$ system has been done within the LDA, assuming a periodic placement of excess Fe atoms, one per supercell – like in [332] where a vacancy in the chalcogen lattice per unit cell was added. It follows from the calculation that the magnetic moment is $2.4\,\mu_B$ on an Fe2 site and only $1.6\,\mu_B$ on Fe1. A large moment on excess Fe atoms must persist even when the antiferromagnetic ordering in the FeTe plane is suppressed by doping or applying external pressure.

It is noteworthy that the valence of excess Fe atoms is Fe^{+}, differently from Fe^{2+} for the atoms in the basal FeTe plane. This unusual valence has as a consequence that each excess Fe atom brings an extra electron carrier into the system, rapidly enlarging the electron pocket of the Fermi surface. This should lead to a rapid breakdown of nesting on the (π, π) vector and suppression of the antiferromagnetic order. However, as was shown in [320], the antiferromagnetic ordering survives up to quite high levels of doping.

Since both virtual crystal approximation and supercell approximation are not accurate enough for disordered systems, in [333] by using the coherent potential approximation (CPA) method the disordering effects in the FeSe system have been studied, at different substitutions of Se by S and Te, and of Fe by Co, Ni, Cu. The electronic structure calculations at different levels of doping have led the author to the following conclusions: (1) a small amount of excess Fe leads to a substantial lowering in energy of d bands forming the Fe3d states, whereas the vacancies in the Se sublattice affect the p-bands, formed by Se4p states; (2) a substitution of Se by S or Te enhances the nesting of the Fermi surface; (3) a substitution of Fe by Co, Ni or Cu results in a lowering of the energy of 3d bands, thus complicating the conditions for nesting. These results make a basis for an analysis of experimental data for the FeSe alloy with the abovementioned dopants.

In [334], the results of a detailed ARPES study on a good-quality single crystal $Fe_{1.03}Te_{0.7}Se_{0.3}$ ($T_c = 13\,K$) have been reported. Hole and electron pockets around Γ and M points, respectively, have been identified. The pockets are of nearly the same size, so that a good nesting occurs on the (π, π) wave vector. The revealed structure of the Fermi surface is analogous to that in the majority of

FeAs-compounds. At the same time, it is different from that for the selenium-non-doped $Fe_{1+x}Te$, in which the hole pockets near the X point have been found, and a bi-collinear magnetic structure takes place. In the same work, an isotrope superconducting gap on the hole surface around Γ has been determined from the ARPES data, of the size $\Delta = 4$ meV, so that $2\Delta/kT_c \sim 7$, thus indicating a strong coupling. Therefore, selenium-doped $FeTe_{1-x}Se_x$ compounds exhibit electronic properties analogous to those of doped FeAs-compounds.

We mention moreover a study, by inelastic neutron scattering, of spin fluctuations in the $Fe_{1+y}Te_{1-x}Se_x$ system. Earlier [335], an observation of spin resonance in this system within the superconducting phase has been reported, along with the already discussed resonance in doped AFe_2As_2-systems. However, the statistical weight of such fluctuations makes only a small part of the whole spin fluctuations spectrum. In this relation, in [336] the spin fluctuations have been studied in both a superconducting sample $Fe_{1.04}Te_{0.51}Se_{0.49}$ and a non-superconducting one, $Fe_{1.04}Te_{0.73}Se_{0.27}$, over a broad range of energies up to 300 meV. The spin spectrum has a pronounced quasi-two-dimensional character, with the dominance of spin fluctuations having incommensurate wave vectors in the vicinity of the $(\pi,0)$ wave vector of the stripe structure. Therefore, doped $Fe_{1+x}Te$ compounds exhibit static and dynamic magnetic properties similar to those of the FeAs-compounds.

4.1.5 Magnetic Structure of FeTe

In [323], an electronic structure calculation for FeTe has been reported, which apparently resolves the problem of magnetic ordering in this compound. In Fig. 4.8, a calculated Fermi surface (as its projection onto the ab plane) is shown for two levels of doping. The (a) panel corresponds to the stoichiometric composition and (b) to a doping of $\delta = 0.5$ electrons per formula units that corresponds to the $Fe_{1.068}Te$ composition. It is seen that the Fermi surface topology is very sensitive to the concentration of excess Fe near the stoichiometry. At $\delta = 0$, the situation standard for all FeAs-type compounds occurs: two hole pockets near Γ and two electron ones near the M point, so that the nesting takes place on the vector $\mathbf{Q} = (\boldsymbol{\pi}, \boldsymbol{\pi})$, which connects the hole and the electron sheets of the Fermi surface. However, already at $\delta = 0.5$ the Fermi surface changes considerably, the nesting on the $(\boldsymbol{\pi}, \boldsymbol{\pi})$ vector disappears, but another one appears on the vector $\mathbf{Q}' = (\boldsymbol{\pi}, \mathbf{0})$.

Due to such drastic modification of the Fermi surface, the spin fluctuation spectrum must also undergo quite considerable changes. The imaginary part of the Stoner spin susceptibility has been calculated:

$$\chi_0(\mathbf{q}) = \left(f(\varepsilon_{\mathbf{k}}) - f(\varepsilon_{\mathbf{k+q}})\right) / \left(\varepsilon_{\mathbf{k}} - \varepsilon_{\mathbf{k+q}} - \omega - i\delta\right). \tag{4.1}$$

The values of this function in the plane $q_z = 0$ are shown in Fig. 4.9). Under stoichiometry (Fig. 4.9a), the $\mathbf{Q} = (\pi, \pi)$ nesting is quite prominent, but at $\delta = 0.5$ it

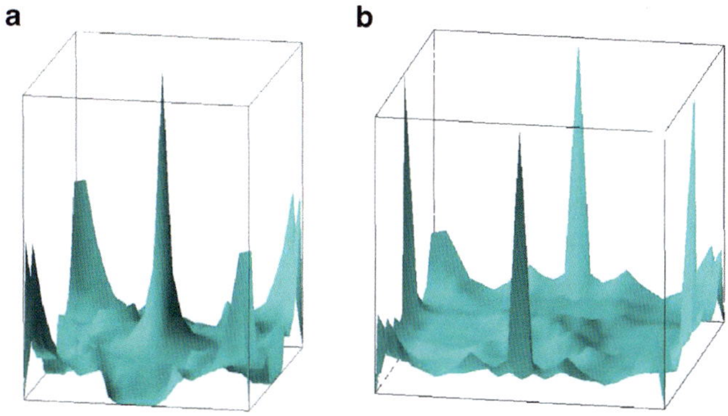

Fig. 4.9 Imaginary part of calculated Stoner susceptibility $\chi_0(q_z = 0)$ for FeTe: (**a**) $\delta \sim 0.0$, (**b**) $\delta \sim 0.5$ electrons per formula unit [323]

is suppressed and replaced by the $\mathbf{Q}' = (\pi, 0)$ one. The latter nesting leads to the magnetic structure shown in Fig. 4.4b in $Fe_{1.068}Se$.

On further increase of doping, the Fermi surface topology goes on changing, and the $(\pi, 0)$ nesting disappears. At $\delta \sim 1.1$, the doping corresponding to the $Fe_{1.141}Te$ composition, already an incommesurate spin structure emerges, which is detectable in experiment [322]. At this level of doping, the dispersion of electrons along the z direction becomes important that, namely, drives the spin out of he a, b plane.

In [323], the exchange interactions J_{1a}, J_{2a}, J_{1b} and J_{2b}, shown in Fig. 4.4, were calculated from first principles. Their magnitudes for $Fe_{1.068}Te$ turned out to be as follows (in meV):

$$J_{1a} = 7.6, \quad J_{1b} = +26.5, \quad J_{2a} = -46.5, \quad J_{2b} = 34.9, \tag{4.2}$$

the negative values corresponding to antiferromagnetic coupling. For LaOFeAs, the exchange interactions are different:

$$J_{1a} = -47.4, \quad J_{1b} = 6.9, \quad J_{2a} = -22.4\,. \tag{4.3}$$

Therefore, the largest interaction defined in LaOFeAs, that among the nearest neighbours, has the opposite sign in FeTe, that leads to a different type of magnetic structure. The magnitude of the Fe magnetic moment, $2.09\mu_B$, is in good agreement with the experimental value $1.97\mu_B$.

A recovery of the "lost" $(\pi, 0)$ nesting in FeTe explains the double stripe structure detected in experiment and supports the spin fluctuation nature of superconductivity in Fe chalcogenides, beyond to that already established for Fe arsenides.

Experiments of inelastic neutron scattering on single-crystalline $Fe_{1.07}Te_{0.75}Se_{0.25}$ and $FeTe_{0.7}Se_{0.3}$ samples revealed broad magnetic peaks in the vicinity of the $(\frac{1}{2}0\frac{1}{2})$ antiferromagnetic wave vector, thus indicating an incommensurate magnetic short-range order [337]. These peaks appear on cooling below 40 K. Their intensity grows on further cooling, and they persist in the superconducting phase. At larger Se concentration, the intensity of the peaks decreases, and superconductivity strengthens. This hints for a competition existing between the states characterized by short-range magnetic order and superconductivity, in the FeSe-system. The theory analysis concerning magnetic structure and lattice dispersion in FeSe and FeTe-compounds is addressed in [338, 339].

In conclusion, we refer to two works [340, 341] related to the study of superconductivity in the FeSe-system. In [340], a behaviour of the superconducting transition temperature under pressure in the $FeSe_x$ system ($x = 0.8$ and 1.0) was studied. In both cases, a step-like increase of T_c in the (T, P)-plane has been found. There is a local maximum of $T_c \sim 11$ K at $P \sim 1$ GPa and a rapid increase of T_c in the pressures range up to ~ 3 GPa, after which T_c reaches saturation at $T_c \sim 25$ K for $FeSe_{1.0}$ and $T_c \sim 21$ K for $FeSe_{0.8}$. In another work [341], a measurement of temperature behaviour of heat capacity in the $FeSe_x$ ($T_c \sim 8.8$ K) indicated that the superconducting gap does not possess zeros at the Fermi surface; however, the data obtained do not permit to make a distinction between the s and $s^{\pm}$ symmetries of the order parameter.

4.2 Compounds of the LiFeAs Type

4.2.1 Superconductivity

Following *Re*OFeAs and AFe_2As_2, a new compound belonging to the FeAs-group has been synthesized, LiFeAs, in which superconductivity with quite high T_c of 16–18 K [14,15,342] was found. Thorough structural and physical measurements on a single crystal [14] have shown that LiFeAs belongs to the class of compounds built of the FeAs layers, such as *Re*OFeAs and AFe_2As_2; the FeAs layers are separated by lithium planes (Fig. 4.10).

Differently from all other known undoped FeAs-compounds, LiFeAs does not exhibit any magnetic instability, and superconductivity exists in it without any doping. Charge carriers in it are electrons; upper critical field is $H_{c2} > 80$ T. LiFeAs can be doped by holes, via introducing Li vacancies. Thus, in Li_xFeAs at $x = 0.6$, T_c of 18 K was obtained [15]. In an isostructural NaFeAs compound, the superconductivity was well detected, albeit with a lower $T_c = 9$ K.

An important piece of information about the origins of anomalous properties of LiFeAs can be gained from the measurements of T_c under pressure [343]. It was found that T_c increases linearly as pressure grows, in the ratio 1.5 K/GPa. The same behaviour is typical for undoped $CaFe_2As_2$, $SrFe_2As_2$ and $BaFe_2As_2$. They become superconducting on achieving critical pressures of 0.23; 2.7 and 2.5 GPa, correspondingly [218, 219]. It is interesting that superconductivity in these compounds first rapidly grows beyond these critical values of pressure, but then T_c starts to gradually decrease with pressure.

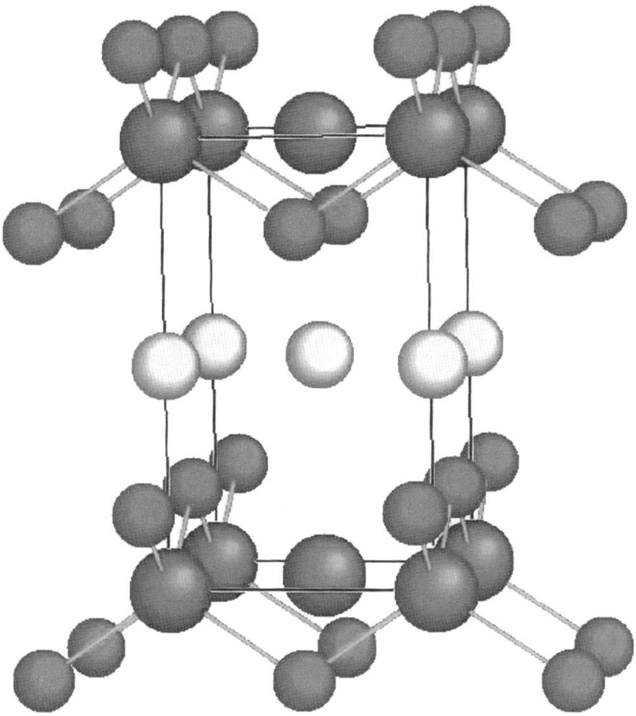

Fig. 4.10 Crystal structure of LiFeAs [344]

It should be noted that the lattice parameter and unit cell volume of LiFeAs are smaller than in other FeAs-systems – both *Re*OFeAs and AFe_2As_2 – due to a tiny ionic radius of Li. Consequently, the LiFeAs compound exists in quasi-compacted state, equivalent to being under applied external pressure. In analogy with the AFe_2As_2 compound, magnetic ordering in such compacted state is already suppressed, that is why a superconducting state could emerge. An application of a genuine pressure affects T_c in exactly the same way as it works in the AFe_2As_2 compounds, i.e. linearly decreases T_c as pressure grows. It would have been important to verify this picture in exposing LiFeAs to negative pressure, for example by depositing a film of this material onto a substrate with larger lattice parameter. Moreover, it would have been useful to perform electronic structure calculations of LiFeAs at different pressures.

4.2.2 Electronic Structure

Electronic structure of LiFeAs was calculated in [134, 344, 350]. The crystal structure belongs to the $P4/nmm$ space group, with the following atomic positions:

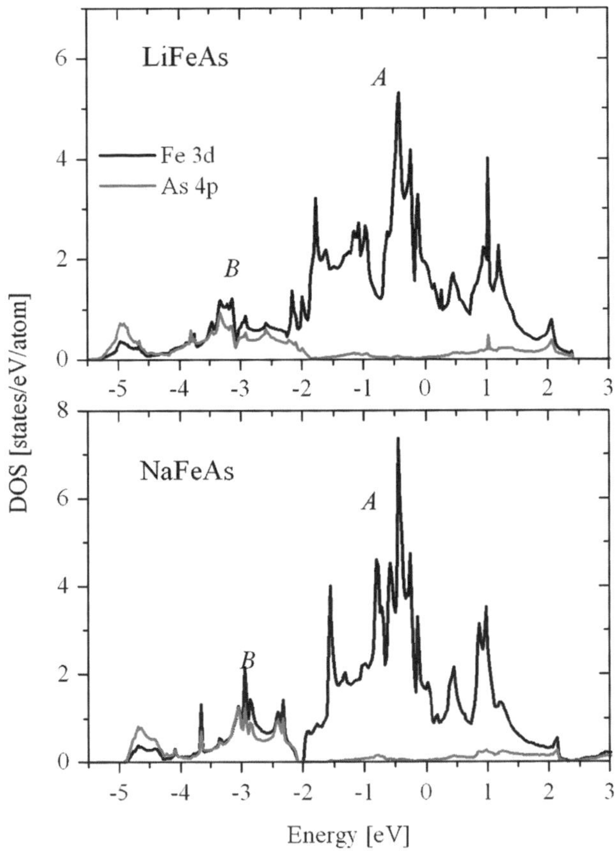

Fig. 4.11 Local density of states in LiFeAs and NaFeAs for Fe3*d* and As4*p* orbitals [346]

Fe(2*a*) (0,0,0); Li(2*b*) (0,0,1/2); As(2*c*) (0,1/2,z_{As}) and lattice parameters a = 3.76 Å; c = 6.28 Å [15]. These structure data have been used in a full LDA calculation of electronic structure: total and partial densities of states, energy band spectrum and the Fermi surface [344]. All these characteristics turned out to be quite close to those of previously studied *Re*OFeAs AFe_2As_2; hereby, an idea gets support that electronic properties of FeAs-type compounds are essentially formed by the FeAs layers.

Since the value of the z_{As} coordinate was not previously reported, the authors of [344] optimized it in a calculation by the full-potential LAPW method and obtained z_{As} = 0.21696. The abovementioned characteristics of electronic structures have been then reported as corresponding to this value. The Fermi surface of LiFeAs consists of two hole cylindrical sheets in the centre of the Brillouin zone and two electron sheets in its corners, like in all other previously calculated compounds. The difference is that in case of LiFeAs, the two sheets in Γ are much closer to each other.

For undoped compound NaFeAs, the density of states was calculated within the LDA in [345, 346], in parallel to the results for LiFeAS. The calculation results for Li- and Na-based compounds agree well (Fig. 4.11). As is other classes of FeAs-systems, the Fe3d states dominate in the region $-2\,\text{eV} < E < 2\,\text{eV}$, whereby the Fermi level falls onto the downhill slope of the major peak.

Detailed investigation of electronic structure of NaFeAs from the first principles was performed in [347].

In [345], an LSDA calculations have been done for several magnetically ordered states in both compounds. It turned out that the lowest energy corresponds to an SDW state, like in other FeAs-systems. The energy of this state is even lower than that of the non-magnetic one, which creates problems of consistency with existing experimental data in this relation. Namely, in [14, 15, 342, 348, 349] for stoichiometric LiFeAs and NaFeAs no traces of magnetic ordering was found, differently from the situation in all other FeAs-compounds. At the same time, superconductivity with not small values of T_c is present. For the moment, no obvious solution of this problem is found. It can be only speculated that in synthesized samples, some deviation from stoichiometry is present, due to which the magnetic ordering is already suppressed, and the superconducting state sets on. In any case, [15] gives notice that a sample synthesized had a nominal formula $Li_{1-x}FeAs$, i.e. with lithium deficiency. It is possible that synthesized NaFeAs samples were non-stoichiometric as well.

In view of the declared absence of magnetic order in LiFeAs along with the occurrence in it of superconductivity with $T_c = 18\,\text{K}$, an assumption was raised that this superconductivity might be induced by electron–phonon interaction. The authors of [345] calculated the phonon spectrum and matrix elements of electron–phonon interaction for LiFeAs and NaFeAs. On the basis of these data, the electron–phonon coupling parameter λ was calculated, to be 0.29 and 0.27 for LiFeAs and NaFeAs, correspondingly. Taking into account the density of states, $N(E_F) \approx 4\,(\text{eV})^{-1}$ for both compounds, they arrived at $T_c < 1\,\text{K}$, unambiguously indicating that the coupling mechanism in LiFeAs, like in other FeAs-compounds, is not a phononic one.

Electronic structure has been calculated for a hypothetical compound LiFeSb [351] in an assumption of it having the same crystal structure as LiFeAs. It is understandable that the calculated electronic structure turned out to be very close to that of LiFeAs; however, the trial compound was predicted to have SDW instability, differently from the real LiFeAs. Would it be possible to synthesize such compound and by means of doping to suppress the magnetic ordering in it, one could have expected an emergence of superconductivity.

Even as no magnetic instability has been detected in LiFeAs, a comprehensive study [352] of the other compound of this type, NaFeAs, indicated three different phase transitions near 52, 41 and 23 K. They can be attributed to structural, magnetic, and superconducting transitions, correspondingly, in similarity with all other FeAs-compounds, in which superconductivity has been found.

4.3 Compounds of the AFFeAs (A = Sr, Ca) Type

4.3.1 Primary Experimental Observations

In [353, 354], a synthesis of a new family of FeAs-systems, not containing oxygen whose role has been taken by fluorine, was reported for the first time. Two stoichiometric compounds, SrFFeAs and CaFFeAs, laid foundation of a new group of superconductors, belonging to the structural type of LaOFeAs. In these compounds, the standard FeAs layers are interlaced by SrF or CaF layers, forming, at room temperatures, a crystal structure with the $P4/nmm$ space group. Indeed, a doping of these systems led to an onset of superconductivity. Thus, the doping of CaFFeAs with cobalt gave T_c of about 22 K [355], and in $Sr_{1-x}La_xFFeAs$ the $T_c = 36$ K has been achieved [356].

Thorough studies of stoichiometric compound SrFFeAs [353,354] demonstrated that it is a metal with SDW magnetic structure, which appears at $T_s = 173 - 175$ K. Simultaneously with the magnetic ordering, a structural phase transition from tetragonal into orthorhombic phase occurs, as is the case in other compounds of the types *Re*OFeAs or AFe_2As_2. These transitions are indicated by well represented anomalies in temperature dependencies of electrical resistivity, magnetic susceptibility and specific heat. Moreover, the Mössbauer spectroscopy on the ^{57}Fe nuclei has shown [353] that at 175 K a magnetic splitting of the NMR line occurs which indicates an onset of magnetic ordering.

The measurements of the Hall constant turned out quite interesting. R_H (Fig. 4.12). $R_H = 0$ at room temperature and downwards till the magnetic ordering temperature, from where on (for lower temperatures) R_H rapidly increases and remains positive. This makes distinction of SrFFeAs from all other FeAs-compounds

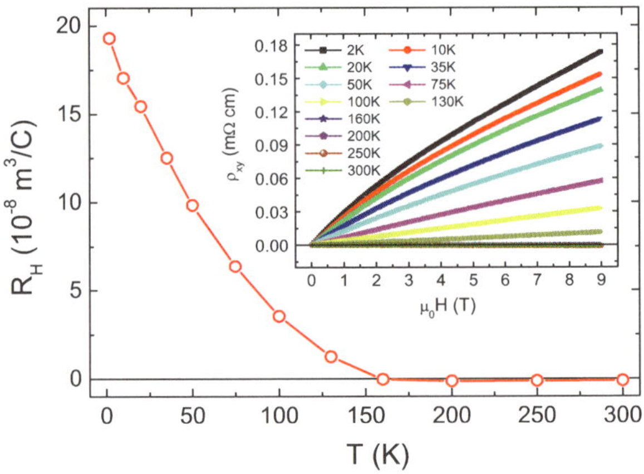

Fig. 4.12 Temperature dependence of the Hall constant for a sample of SrFFeAs [354]. Inset: Hall resistivity ρ_{xy} as function of magnetic field at different temperatures

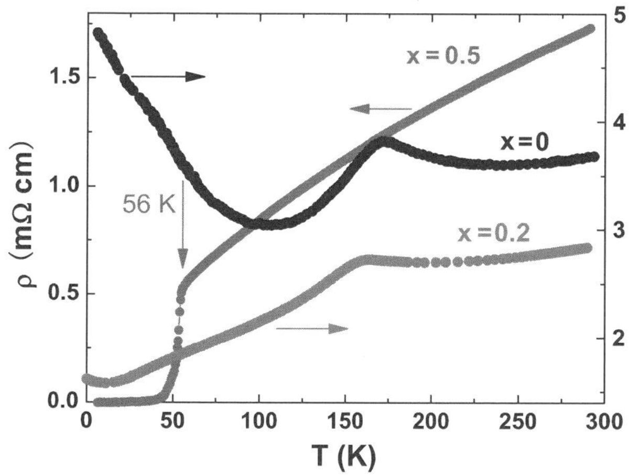

Fig. 4.13 Temperature dependence of electrical resistivity for three samples of the $Sr_{1-x}Sm_xFFeAs$ system [357]

in which R_H stays negative. Apparently, the following scenario of this behaviour of R_H can be put forward. The crucial point is an onset of the SDW ordering, which results in forming a gap at the Fermi surface, so that a hole pocket appears either below the gap, or, on the contrary, above it. Pushing this pocket away to beyond the Fermi level occurs gradually, as the gap moves with temperature. The electron pocket meanwhile remains pinned at the Fermi level, providing a net dominance of electronic carriers. In other aspects, the properties of SrFFeAs do resemble those of other FeAs-group compounds.

Wu et al. [357] reports an achievement of the recordly high $T_c = 56$ K in a sample of $Sr_{1-x}Sm_xFFeAs$ with the Sm composition $x = 0.5$. In Fig. 4.13, the temperature dependence of electrical resistivity is shown for three samples of this system, with $x = 0$; 0.2 and 0.5. In the stoichiometric compound SrFFeAs, the resistivity shows a typical anomaly at $T_s = 172$ K, due to magnetic ordering. Below T_s, the resistivity grows as temperature decreases, revealing a semiconductor type of conductivity. A doping with Sm to $x = 0.2$ pushes the T_s down to 163 K, while below T_s the semiconductor-type conductivity is suppressed. At Sm concentration of $x = 0.5$, a superconducting state is formed, that is confirmed by observation of a diamagnetic state emerging at $T = 53.5$ K in the field $H = 10$ Oe. The temperature of the onset of diamagnetic state corresponds to the middle of the transition into superconducting state, as judging by temperature dependence of resistivity.

It is interesting to compare the $\rho(T)$ behaviour of the previously studied compound $SmO_{1-x}F_xFeAs$ [75] at $x = 0.15$ and 0.20. In the normal phase, $\rho(T)$ of this compound behaves like that in $Sr_{1-x}Sm_xFFeAs$. However, in the latter the superconductivity does only appear at $x = 0.5$. A discovery of high-temperature superconductivity in the oxygenless AFFeAs system opens the perspectives not less important than its primary discovery in the, by now, well-studied *Re*OFeAs system.

We point out one more work [358] in which magnetic and structural phase transitions have been studied in the $Na_{1-\delta}FeAs$ (an analog of LiFeAs) compound, in which the superconductivity with $T_c = 23\,K$ occurs due to the deficiency of Na atoms. On cooling down from room temperatures, this compound undergoes a sequence of phase transitions: a structural one from tetragonal into orthorhombic phase at $T_S = 49\,K$, then the magnetic one with anomalously low value of the mean magnetic moment at Fe atom, $0.09 \pm 0.04\,\mu_B$, and, finally, the superconducting transition at $T_c = 23\,K$. These observations indicate that the superconductivity of NaFeAs, similarly to the case of other FeAs-compounds, occurs due to a closeness to antiferromagnetic ordering.

4.3.2 Electronic Structure

LDA calculations of the electronic structure of two compounds, SrFFeAs and CaFFeAs, have been done in [359, 360]. The overall conclusion from these studies is that these new compounds have electronic structure very close to that established for *Re*OFeAs and AFe_2As_2 systems; the reason of this similarity is, of course, the presence of FeAs-layers, which in both cases are responsible for the formation of electron spectra and the Fermi surface. Obviously, the closeness of crystal structure parameters is important as well.

For SrFFeAs, experimentally measured lattice parameters are as follows: $a = 3.9930\,Å$, $c = 8.9546\,Å$ [353], and for CaFFeAs: $a = 3.879\,Å$, $c = 8.601\,Å$ [354]. The Fe and F atoms in both compounds occupy the positions Fe(2b): (0.75, 0.25, 0.5); F(2a):(0.75, 0.25, 0). Sr and As are, in the SrFFeAs compound, at Sr(2c): (0.25, 0.25, 0.1598), As(2c):(0.25, 0.25, 0.6527) [353]. The A–F layers in both compounds (A = Sr, Ca) are equivalent to the *Re*O layers of LaOFeAs, hence a considerable similarity in their electronic structures can be expected.

Indeed, the calculations [359, 360] confirmed it. In both compounds (A = Sr, Ca), the Fermi level crosses the bands which stem from the t_{2g} orbitals of the Fe3d shell. The Fe3d states are situated in the energy range from −2 to 2.5 eV. The most important differences occur in the energy range between −5.5 and −2 eV, occupied by the As4p states. In LaOFeAs, the As4p and O2p states do strongly overlap. In SrFFeAs, the As4p states lie in the same energy interval −5.5 to −2 eV; however, the F2p states are completely separated from them and localized in the range −7.5 to −6 eV.

This difference in the profound part of spectrum does not affect the states in the vicinity of the Fermi level, that is why the band spectrum and the Fermi surface for AFFeAs compounds does not differ much from their counterparts in oxygen-containing *Re*OFeAs compounds (Fig. 4.14). A noticed difference is that the AFFeAs compounds are more two-dimension-like, therefore the sheets of their Fermi surfaces are almost cylindrical. The data given in [359] do completely

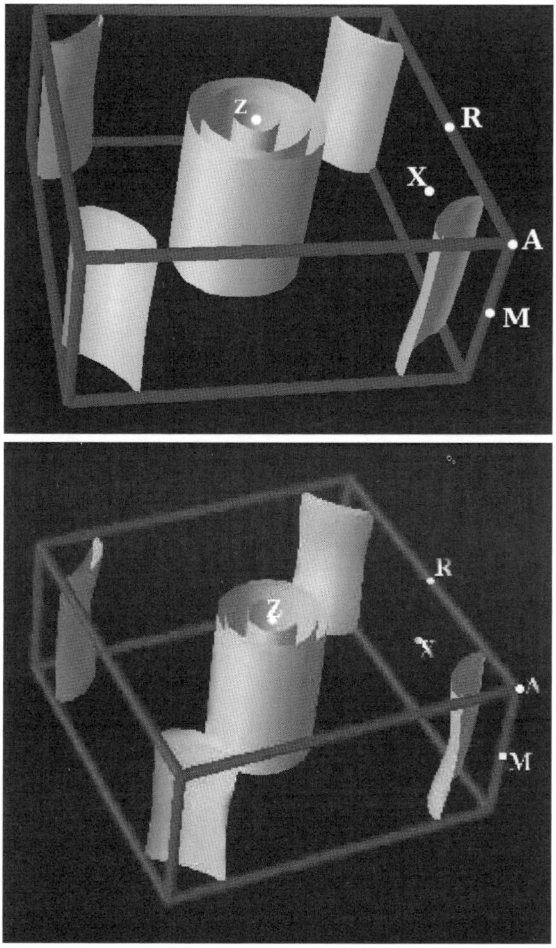

Fig. 4.14 Fermi surface of SrFFeAs (*left panel*) in comparison with that of LaOFeAs (*right panel*) [359]

agree with the calculations of [360]. The difference of Fermi surfaces of the two compounds is negligibly small [359, 360]. The calculated densities of states at the Fermi level after [359], 1.83 eV·Fe for CaFFeAs and 2.0 eV·Fe for SrFFeAs are somehow at variance with the results of [360], 1.895 eV·Fe for CaFFeAs and 1.540 eV·Fe for SrFFeAs.

Chapter 5
Theory Models

5.1 General Properties of Compounds from Different Classes of FeAs-Systems and Corresponding Theory Objectives

5.1.1 *Crystal and Magnetic Structures*

In the previous chapters, we considered various classes of the FeAs-systems: LaOFeAs (1111), $SrFe_2As_2$ (122), LiFeAs (111) and $Fe_{1+x}Se$ (11). Here, chemical formulae indicate typical representatives of each class, and the numbers give the relative compositions of constituent atoms. In this section, we mark the common properties of the FeAs-systems throughout different classes.

All compounds at room temperature have tetragonal crystal structure. As temperature drops, they undergo a structural phase transition into the orthorhombic phase. Thus, all compounds of the 1111 class belong to the $P4/nmm$ space group, which at low temperature is replaced by the orthorhombic one *Cmma*. Crystal structure of the 122 class possesses the $I4/mmm$ space group, which transforms into the orthorhombic one *Fmmm*. The 111 compounds have the $P4/nmm$ space group, similarly with the 11 compounds. At reduced temperatures, they become either monoclinic $P2_1/m$, or orthorhombic, depending on the level of doping.

In Table 5.1, the lattice constants of some compounds are given. In the upper line for each system, the values of a, b, c lattice parameters in tetragonal phase (at high temperatures) is given, and in the lower line – in the orthorhombic phase (at low temperatures), after [361].

The a, b parameters of tetragonal phase are of the order of ~4 Å for all compounds, and in the orthorhombic phase – by factor of $\sqrt{2}$ larger, because the primitive cell in that phase has to be rotated by 45° relative to the axes in the basal plane. Differently for the c parameter, which changes on turning from 1111 to 122 compounds, the latter comprising not one but two FeAs planes in its unit cell. In 111 and 11 compounds, the c parameter is smaller than in 1111, because the FeAs-layers in them are separated by monoatomic planes and not by multiatomic fragments as in LaOFeAs. A much larger value of the c parameter for the $(Sr_3Sc_2O_5)Fe_2As_2$ compound comes from a big size of the complex motif $Sr_3Sc_2O_5$, inserted between the FeAs-planes, in contrast to just Ba atom on its place in $BaFe_2As_2$.

Table 5.1 Lattice parameters of FeAs-compounds of various classes in tetragonal and orthorhombic phases [361]

Compound	$a(\mathring{A})$	$b(\mathring{A})$	$c(\mathring{A})$	Reference
LaOFeAs (175 K)	4.0301	$\equiv a$	8.7368	[59, 72, 73, 362, 363]
(4 K)	5.7099	5.6820	8.7265	
CeOFeAs (175 K)	3.9959	$\equiv a$	8.6522	[12, 364]
(30 K)	5.6626	5.6327	8.6382	
PrOFeAs (175 K)	3.977	$\equiv a$	8.6057	[365]
(5 K)	5.6374	5.6063	8.5966	
NdOFeAs (175 K)	3.9611	$\equiv a$	8.5724	[61, 366]
(0.3 K)	5.6159	5.5870	8.5570	
$CaFe_2As_2$ (175 K)	3.912	$\equiv a$	11.667	[220, 221, 275]
	5.542	5.465	11.645	
$SrFe_2As_2$ (300 K)	3.920	$\equiv a$	12.40	[276, 277, 367]
(150 K)	5.5695	5.512	12.298	
$BaFe_2As_2$ (175 K)	3.9570	$\equiv a$	12.9685	[273, 274, 278]
(5 K)	5.61587	5.57125	12.9428	
LiFeAs (215 K)	3.7914	$\equiv a$	6.3639	[14, 342]
$Fe_{1.068}Te$ (80 K)	3.81234	$\equiv a$	6.2517	[316, 320, 322, 368]
(5 K)	3.83435	3.78407	6.2571	
$(Sr_3Sc_2O_5)Fe_2As_2$	4.0678	$\equiv a$	26.8473	[369]
$(Sr_4V_2O_6)Fe_2As_2$	3.9302	$\equiv a$	15.6664	[178]
$(Sr_4Sc_2O_6)Fe_2As_2$	4.050	$\equiv a$	15.809	[170]

All FeAs-compounds are built out of FeAs-planes, which make an identical element in all these systems. The FeAs plane is, in fact, a triple sandwich whose middle layer is a quadratic lattice of Fe atoms, and equally quadratic layers of As above and below are displaced such that each Fe atom is surrounded by a tetrahedron of As atoms (Fig. 5.1).

A structural phase transition in undoped compounds, which occurs at the temperature T_s, is accompanied by magnetic ordering at the temperature T_N. In 1111 compounds, the structural transition is a first-kind phase transition, whereas the magnetic ordering occurs at a lower temperature, and is a second-kind phase transition. In the 122 compounds, the structural and magnetic transitions happen simultaneously and are phase transitions of the first kind.

Magnetic ordering, therefore, sets on in a distorted orthorhombic phase, so that within the plane of each FeAs-layer a collinear antiferromagnetic SDW structure is formed, as shown in Fig. 5.2. The spins of Fe atoms are directed along or against the a axis, in alternation; along the b axis, the spins are repeated without change, forming ferromagnetic chains in the plane. On moving along the c axis, the spins alternate, like along the a. Therefore, the corresponding structure has a wave vector $\mathbf{q}_M = \pi(101)$. If the spins along the c axis do not change, the wave vector would be $\mathbf{q}_M = \pi(100)$. In Fig. 5.3, the corresponding three-dimensional magnetic structure of 1111 and 122 compounds is shown.

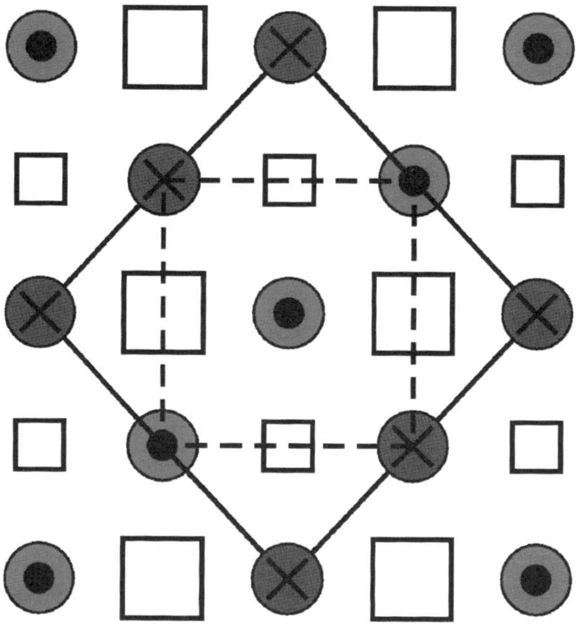

Fig. 5.1 Top view of a FeAs layer. The Fe atoms are marked by *dark circles*. The As atoms above the plane are indicated by larger *squares*, below the plane – by *smaller squares* [88]

Fig. 5.2 Magnetic ordering in the FeAs layer of 1111 and 122-type compounds. The *shaded region* indicates a unit cell of the orthorhombic phase [361]

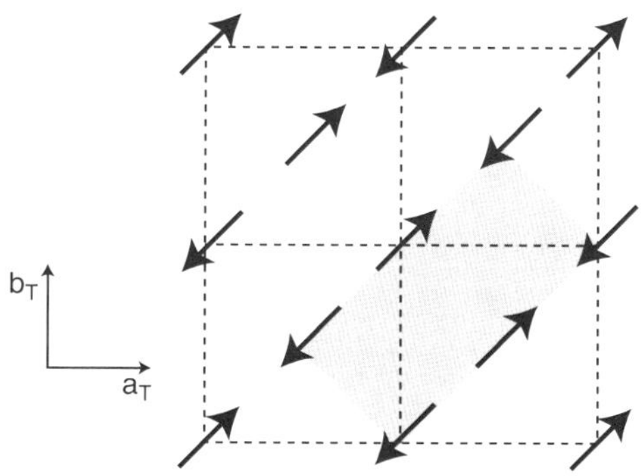

The data on magnetic structure in different classes of FeAs-systems are collected in Table 5.2. In 1111 compounds, the magnetic moment at Fe atom is small, revealing an itinerant nature of magnetism in these systems. In 122 compounds, the Fe magnetic moment is just below one Bohr magneton. This means that the 122 compounds exhibit a more localized type of magnetism. In the last column of Table 5.2, the wave vector of the magnetic structure $\mathbf{q}_M$ (in the units of π) is given. We see that in the 1111 compounds, the ordering along the c axis can be either ferromagnetic or antiferromagnetic, whereas in the 122 compounds it is antiferromagnetic only.

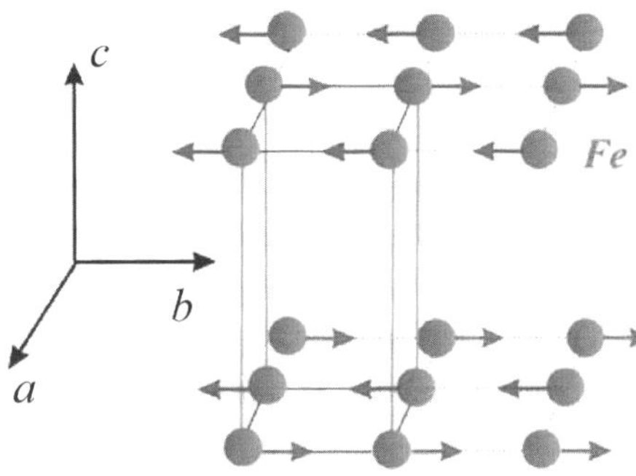

Fig. 5.3 Magnetic structure of 1111 and 122-type compounds. Only Fe atoms are shown. In the *ab* plane, the spin configurations and spin directions are identical for all compounds. Along the *c* axis, both ferromagnetic and antiferromagnetic spin arrangements are possible [61]

Table 5.2 Temperatures of structural T_s(K) and magnetic T_N(K) phase transitions in nondoped FeAs-compounds, along with the mean magnetic moment μ_{Fe} at Fe atom (in (μ_B) and the wave vector of the magnetic ordering, $\boldsymbol{q}_{Fe}$ in the Fe sublattice

Compound	T_s	T_N(Fe)	μ_{Fe}	q_{Fe}	
LaOFeAs	155	137	0.36	101	[59,72,73,362]
CeOFeAs	158	140	0.8	100	[12]
PrOFeAs	153	127	0.48	100	[365]
NdOFeAs	150	141	0.25	101	[61,366]
$CaFe_2As_2$	173	173	0.8	101	[220,221,275]
$SrFe_2As_2$	220	220	0.94	101	[276,277,367]
$BaFe_2As_2$	142	143	0.87	101	[274]
$Fe_{1.068}Te$	67	67	2.25	100	[322]

The physical foundations of realizing the SDW magnetic structure can be well clarified within the localized $J_1 - J_2$ model, as a result of competition (frustration) between antiferromagnetic exchange interactions between nearest neighbours to each Fe atom, on one side, and between next-nearest neighbours, on the other side. The most important problem for a theory is an explanation of small values of mean magnetic moments at Fe atoms, especially in the 1111 compounds, in other words – resolving the controversy in the relation of itinerant and localized aspects in the formation of magnetic moments in these compounds.

The next problem is a theoretical description of how does the antiferromagnetic ordering temperature T_N depend on doping; in particular, an evaluation of critical dopant concentration at which the magnetic order gets suppressed. These problems of magnetic ordering in the FeAs-systems are closely related to peculiarities of their band structure, notably, to the structure of their Fermi surfaces.

5.1.2 Peculiarities of the Electronic Structure

As revealed by all band structure calculations of FeAs-compounds, the states near the Fermi level are formed by the 3*d* orbitals of Fe atoms. Since in FeAs-systems of different classes (1111, 122, 111 and 11), the Fe atoms are situated in the FeAs-layers which are completely identical, the electronic structure of different

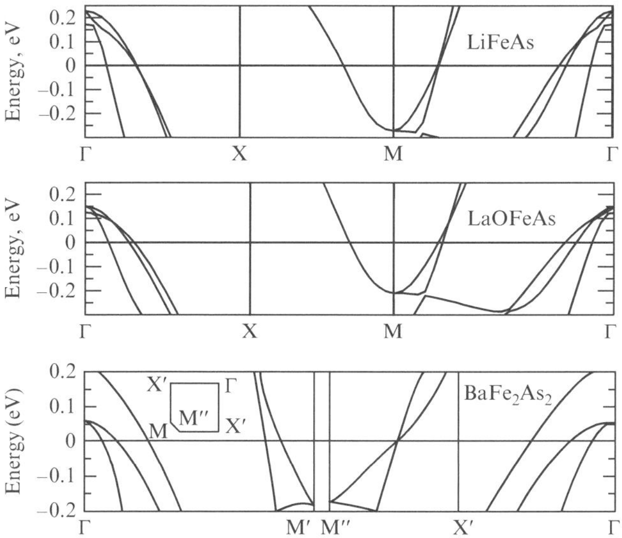

Fig. 5.4 Dispersion curves for LaOFeAs, $BaFe_2As_2$ and LiFeAs compounds, calculated in the LDA [344]

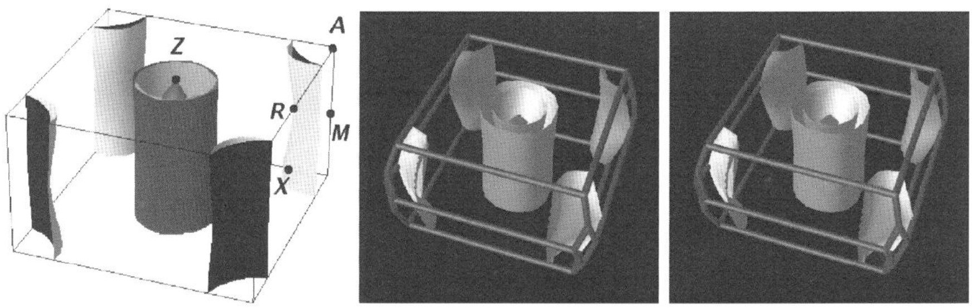

Fig. 5.5 Fermi surfaces for three compounds, belonging to different classes of the FeAs-systems: LaOFeAs, $BaFe_2As_2$ and LiFeAs [133, 344]

compounds turns out to be very close, and the differences in dispersed bands crossing the Fermi level not substantial; consequently, the Fermi surfaces in these compounds are similar. This is illustrated by Fig. 5.4, in which the LDA calculations of band dispersions in the vicinity of the Fermi level are given for three compounds from different classes – 1111, 122 and 111.

In all three cases, hole bands occur in the vicinity of Γ, and electron bands near the *M* point. They give rise to corresponding pockets of the Fermi surfaces, shown in Fig. 5.5.

The comparison of these results exhibits a universal character of the Fermi surface in different FeAs compounds: it consists of hole sheets in the middle of the Brillouin zone, and electron sheets in its corners. The sizes of these sheets do not differ much over various compounds. Substantial is a two-dimensional character of electron states (the sheets are weakly distorted cylinders). Another important observation is that the sizes of hole and electron sheets in each given compound are comparable, which indicates towards the nesting with the wave vector **Q**, connecting the Γ and M points of the Brillouin zone.

Therefore, for all compounds of the FeAs-type a unique schematic structure of the Fermi surface can be suggested, which is shown in Fig. 2.35. According to LDA calculations, one more hole sheet appears around Γ. This sheet is three-dimensional, i.e. not cylindrical, but its radius varies on moving along the Γ–Z line. From one compound to another, the size of pockets (both hole and electron ones) changes. The relative sizes of the sheets in Γ relative to those in M do change as well, but the topology of the Fermi surface is preserved, thus ensuring the similarity of the compounds' physical properties.

5.1.3 Asymmetry of the Electron/Hole Doping

On doping of initial stoichiometric compounds with other elements, additional charge carriers – either electrons or holes – may appear. Thus, in the *Re*OFeAs compounds a substitution of oxygen by fluorine or of iron by cobalt results in the electron doping. In 122 systems, e.g. $BaFe_2As_2$, a barium by potassium substitution corresponds to the hole doping. In case of electron doping, the Fermi surface sheets around the M point get expanded, on hole doping – those around Γ. A typical case is that in a pristine compound the electron and hole pockets are of comparable size, so that the nesting takes place which results in an amplification of magnetic susceptibility on the nesting wave vector and, as a result, in formation of a magnetically ordered SDW structure. On doping, the nesting breaks down, and the magnetic ordering gets suppressed. We note that the ultimate cause of this suppression is not so the disappearance of nesting as, more importantly, the rise of fluctuations in the spin system. The problem of destruction of magnetic order on doping is one of the most important tasks for the theory.

At first sight, it seems that both the electron and the hole dopings lead to the same effects: a suppression of long-range SDW order and an onset of superconducting state. However, this result is achieved at different levels of doping. For example, in $LaO_{1-x}F_xFeAs$ the emergence of the superconducting state begins as $x \simeq 0.1$, whereas in $Ba_{1-x}K_xFe_2As_2$ – at $x \simeq 0.4$. The differences were observed in other properties as well. Thus, the compounds with electron doping are rather itinerant magnets, whereas the hole-doped systems exhibit more properties of localized magnets, even if both the former and the latter represent the systems which are intermediate ones between strictly itinerant and strictly localized cases. The above difference is underlined by the fact that in $Ba_{1-x}K_xFe_2As_2$, the magnetic moment at the Fe atom makes $\simeq 0.8\,\mu_B$, and in $LaO_{1-x}F_xFeAs \simeq 0.3\mu_B$.

This difference is related, apparently, to the placement of the Fermi level in stoichiometric compounds. In both types of systems, E_F falls onto the descending edge of a peak in the Fe3*d* DOS, therefore on the hole doping the density of states at the Fermi level, $\rho(E_F)$, increases, and the hole pockets of the Fermi surface grow. Thats what, apparently, leads to differences in the behaviour of systems doped with electrons and with holes.

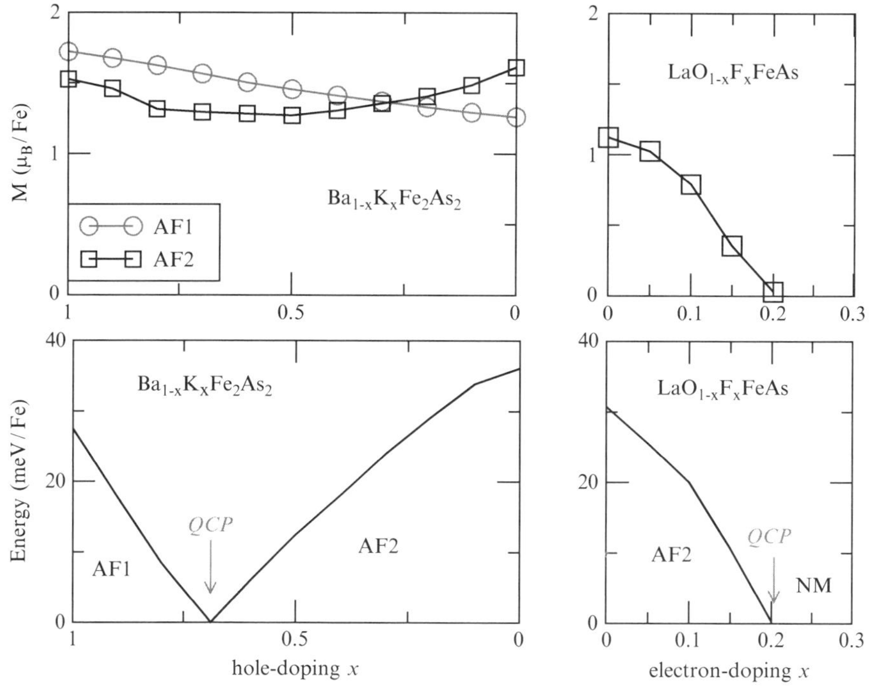

Fig. 5.6 Magnetic moments and the energy of magnetically ordered phases (with respect to the nonmagnetic one), depending on the hole or electron doping [370]

To trace the tendencies in the variation of physical properties with the level of doping, in [370] the LDA calculations have been done of electronic structure and magnetic state for two systems, $LaO_{1-x}F_xFeAs$ and $Ba_{1-x}K_xFe_2As_2$. The calculations comprised three states, the non-magnetic one (NM) and two antiferromagnetic (AF1, AF2), the AF2 state corresponding to the observed SDW, and AF1 – to chessboard antiferromagnetic ordering. The results of calculation are shown in Fig. 5.6.

In the compounds with the hole doping, in both magnetic structures the magnetic moment on an Fe atom varies weakly. In the systems with electron doping, the magnetic moment varies strongly and disappears at $x \simeq 0.2$ in the AF2 phase. The AF1 phase in these systems turns out as energetically unfavourable. The phase diagrams for both systems are shown in the bottom part of the figure. In the case of electron doping, the energies of two phases, AF2 and NM, become equal at $x \simeq 0.2$, as the local magnetic moment disappears. In the hole-doped system, the AF1 and AF2 energies become equal at $x \simeq 0.7$, while from the side of lower x the AF2 phase is more energetically favourable. The $x \simeq 0.2$ and $x \simeq 0.7$ points, marked in the figures as QCP, are quantum critical points. For the electron-doped system, they correspond to a transition from SDW into non-magnetic phase, and for the hole-doped system – to a phase transition between two magnetically ordered phases.

In the first case, we have a situation corresponding to experimental data for $LaO_{1-x}F_xFeAs$, and in the second one – a prediction of the magnetic phase transition in $Ba_{1-x}K_xFe_2As_2$. The last result is not confirmed experimentally that apparently reveals a neglected role of quantum fluctuations in the vicinity of the critical point.

A possibility of magnetic phase transition in this system can be seen from a simplified $J_1 - J_2$ Heisenberg model for this compound, exhibiting properties rather consistent with a localized description. From the mean field approximation of this model, it follows that the total energy of the AF1 phase is $E_1 = -2J_1 + 2J_2$, and of the AF2 phase – $E_2 = -2J_2$ per Fe atom. Consequently, at $J_2 = 1/2\ J_1$ the two magnetically ordered phases become degenerate in energies, in which case one can expect strong quantum fluctuations. In the electron-doped LaOFeAs system, as the calculations indicate, $J_2 \gg 1/2\ J_1$, but in hole-doped systems J_1 and J_2 are of the same order of magnitude. From a comparison of the energies obtained from LDA calculations for $Ba_{1-x}K_xFe_2As_2$ with those following from the $J_1 - J_2$ model, one can extract the variation of the J_1 and J_2 exchange interactions with x. It is remarkable that, following from such mapping, the equality $J_2 = 1/2\ J_1$ occurs at $x \simeq 0.7$, the very quantum critical point indicated in the figure. Thus, the phase diagram obtained from LDA calculations for $Ba_{1-x}K_xFe_2As_2$ is consistent with a general view on this system as a localized magnet. It is obvious that all the results obtained in [370] is rather crude approximations can only pretend to deliver a qualitative level of comparison between electron-doped and hole-doped systems, but nevertheless the thus revealed tendencies concerning the electron-hole symmetry in the FeAs-compounds deserve attention.

5.1.4 Problems of Symmetry of the Superconducting Order Parameter

Whereas in what regards crystalline, magnetic and electronic structure of FeAs-systems, a consensus is achieved in their understanding, the symmetry of the superconducting order parameter and the mechanisms of electron pairing present problems far from being resolved. Different experimental techniques lead to contradicting results. This regards notably the number of superconducting gaps for a compound given and, in particular, the symmetry of the superconducting order parameter. The latter is, to a large extent, determined by whether the gap at the Fermi surface contain zeros (in isolated points or along the lines of zeros).

In general, one can sort out two groups of experimental methods; one measuring the magnitude of the superconducting order parameter, and the other its phase, which is closely related to the presence or absence of zeros in the superconducting order parameter. The first group comprises thermodynamical methods dealing with measurements of the electron heat capacity, London penetration depth of magnetic field, or NMR in which the rate of spin–lattice relaxation, $1/T_1$, is measured. Close to thermodynamical methods are tunnel measurements, notably on point contacts with Andreev's reflection. The second type of methods allows to measure the phase of the order parameter, primarily, from the behaviour of the current in Josephson currents in the magnetic field. Such experiments allowed to unambiguously probe the symmetry of the order parameter in cuprates and identify it as the $d_{x^2-y^2}$ symmetry. In what concerns the FeAs-systems, such experiments have not yet been numerous; however, a considerable amount of theoretical suggestions already has been done, which undoubtedly will be realized, and are expected to become informative.

The majority of experiments on different classes of FeAs-compounds have been done by thermodynamical and tunnel methods. Unfortunately, the data collected are often contradictory, and their interpretation ambiguous. In a number of experiments, superconductor order parameter without nodes on the Fermi surface have been reported [105, 111, 114, 116, 248, 371, 372]. In other experiments, it seemed that the nodes of the order parameter have been detected [26, 97, 98, 101, 104, 108, 109, 373–376]. It should be noted that the claims of the existence of zeros in superconducting order parameter from NMR data (an observation of a power-low dependence of the type $1/T_1 \sim T^3$) are absolutely not conclusive, because a similar behaviour can follow for a superconductor with the $s^{\pm}$-symmetry of the order parameter in the presence of non-magnetic impurities in the sample, as will be discussed in detail in Sect. 5.4.3.

What is so far definitely established on the subject of superconducting pairing in the FeAs-compounds? Merely the fact that the pairing is singlet. The measurement of the Knight shift on the Co-doped $BaFe_2As_2$ compound indicated without doubt [187] a complete suppression of magnetic susceptibility below T_c for all orientations of the magnetic field relative to the single crystal axes, which fact is incompatible with triplet superconductivity. For other compounds, the measurements have been done on polycrystals only that gives a picture averaged over directions, which, nevertheless, again agrees with singlet superconductivity only. Therefore, only three possibilities for the symmetry of the superconducting order parameter remain: a conventional s-pairing (with several gaps, as predominantly under discussion) that of the $s^{\pm}$-type, and of the d-type. A definite choice between them, at present, seems difficult.

Thermodynamic methods do primarily lead to a conclusion about the nodal order parameter (i.e. with zeros at the Fermi surface). However, for example [255, 377] dealing with heat capacity argue towards an absence of zeros of the order parameter. The temperature dependence of heat capacity in the 122 compounds cannot be adjusted to theory curves corresponding to the case of one gap only, whereas an assumption about two (or more) gaps demands a larger number of parameters and becomes therefore less reliable.

Tunnel experiments, e.g. PCAR, reveal immediately the structure of the density of states in the superconducting phase; however, the spectra obtained are strongly sensitive to the properties of the contact surface, which is difficult to control.

As has been already mentioned, Josephson contacts of different types would be able to provide a more specific information about the symmetry of the superconducting order parameter. Thus, in the setting when the tunnel current be measured along the c axis, in case of d-symmetry the resulting current would cancel by symmetry. However, in experiment [261] a considerable current has been measured that speaks against the d symmetry.

Let us notice that change of the magnetic flow in a loop made of two superconductor: niobium and $NdO_{0.88}F_{0.12}FeAs$ is first time measured in [378]. These results prove that the pairing in $NdO_{0.88}F_{0.12}FeAs$ is spin-singlet with even-parity and that there is sign change in the superconducting order parameter.

A general conclusion from the researches done so far is that one cannot draw any unambiguous conclusion concerning the symmetry of the superconducting order parameter in the FeAs-compounds. In the following, theoretical, chapters it will be shown that different electron models of the FeAs-compounds lead to the $s^{\pm}$-superconducting state as the most energetically optimal one in the conditions when the hole and electron sheets of the Fermi surface exist with the nesting vector close to the wave vector of the SDW structure. We will repeatedly come back to the question by which experiments this suggestion can be confirmed.

5.1.5 Isotopic Effect

Already on earlier stages of studies of FeAs-systems, it was established that electron–phonon interaction is small and cannot yield that large T_c values as have been found in these new superconductors. Calculations of [17] indicated the values of the electron–phonon coupling constant $\lambda_{ph} \approx 0.21$. The more so surprising was a large isotopic effect, measured in [379] for the $SmO_{1-x}F_xFeAs$ compound ($x = 0.15$) and in $Ba_{1-x}K_xFe_2As_2$ ($x = 0.4$). The index of isotopic effect α ($T_c \sim M^{-\alpha}$) on substituting ^{56}Fe by ^{54}Fe turned out to be $\alpha \approx 0.34 - 0.37$. Even more astonishing was a detection of inverse isotope effect with $\alpha = -0.18 \pm 0.03$ in $Ba_{1-x}K_xFe_2As_2$ ($x = 0.4$), see [380].

Even if generally accepted that this is the interaction of electrons with spin fluctuations that is responsible for superconducting pairing, it should be taken into account that magnetic and structural instabilities in initial stoichiometric compounds appear in close vicinity to each other. Therefore, spin and structural fluctuations in doped superconductors are closely related, and through this coupling various lattice effects can be revealed in superconducting properties.

In relation with a discovery of direct and inverse isotopic effect, in [381] the superconducting transition temperature has been calculated within the model with $s^{\pm}$ symmetry of the order parameter, by solving the Eliashberg equations for a superconductor with hole and electron pockets at the Fermi surface. A draft solution of equations in the spirit of the BCS approximation lead to the following expressions for T_c and the isotopic coefficient α:

$$T_c = 1.14\omega_{\text{ph}} \exp(\frac{1 + \lambda^{+}_{\text{AF}} + \lambda^{+}_{\text{ph}}}{\Lambda - \lambda^{-}_{\text{ph}}}), \tag{5.1}$$

$$\alpha = 1/2[1 - \frac{1 + \lambda^{+}_{\text{AF}} + \lambda^{+}_{\text{ph}}}{1 + \lambda^{+}_{\text{AF}}}(\frac{\Lambda}{\Lambda - \lambda^{-}_{\text{ph}}})^2], \tag{5.2}$$

$$\text{where} \quad \Lambda = \frac{\lambda^{-}_{\text{AF}}}{1 - \lambda^{-}_{\text{AF}}/1 + \lambda^{+}_{\text{AF}} \ln \frac{\omega_{\text{AF}}}{\omega_{\text{ph}}}}.$$

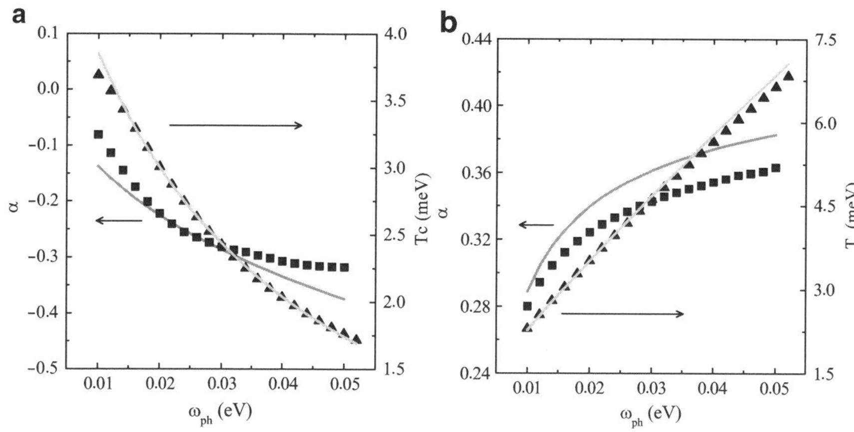

Fig. 5.7 T_c and α, as functions of ω_{ph}. (**a**) $\omega_{AF} = 10\,meV$, $\lambda_{AF}^{inter} = 0.99$, $\lambda_{ph}^{intra} = 0.73$; (**b**) $\omega_{AF} = 20\,meV$, $\lambda_{AF}^{inter} = 1.67$, $\lambda_{ph}^{intra} = 0.22$ [381]. *Triangles* and *squares* correspond to numerical solutions of the Eliashberg equations, *solid lines* trace the calculations along the formulae (5.1) and (5.2)

Here, ω_{AF} and ω_{ph} are limiting frequencies of spin fluctuations and phonons, respectively, whereas $\lambda_{AF}^{\pm}$ and $\lambda_{ph}^{\pm}$ are linear combinations of the corresponding properties, related to the pairing within the same pocket and between different pockets:

$$\lambda_{AF}^{\pm} = \lambda_{AF}^{inter} \pm \lambda_{AF}^{intra}, \quad \lambda_{ph}^{\pm} = \lambda_{ph}^{inter} \pm \lambda_{ph}^{intra}.$$

Numerical solutions of the Eliashberg equations, not using the BCS simplification, are shown in Fig. 5.7, where T_c and α are depicted as functions of ω_{ph}.

In Fig. 5.7a, dominating is the property λ_{ph}^{intra}, therefore $\lambda_{ph}^{inter} = 0$ is set, leading to a positive α, that is the conventional isotopic effect. In Fig. 5.7b, the situation is reversed: the pair interaction λ_{ph}^{inter} is dominating, while $\lambda_{ph}^{intra} = 0$, in which case the inverse isotopic effect occurs. Turning to experiments [379,380], where on the same compound $Ba_{1-x}K_xFe_2As_2$ ($x = 0.4$) in one case $\alpha > 0$ and in the other $\alpha < 0$ was measured, the above concept can be contested, unless some hidden difference exists between the samples on which these measurements have been done. This controversy demands for further investigations.

5.2 Role of Electron Correlations

5.2.1 Dynamical Mean Field Theory (DMFT)

The results outlined above of electronic structure calculations for FeAs-compounds within the LDA do not take into account the effects of electron correlations, which are usually in place in systems containing d- or f-elements. There are many experimental observations indicating a presence of such correlations in FeAs-compounds. Thus, electrical resistivity in the normal phase turns out to be too large [8, 60]. In optics, the Drude peak is absent [60], and a redistribution of spectral weight of the conductivity occurs as the temperature changes.

Compounds with strong electron correlations are efficiently described by the dynamical mean-field theory (DMFT), based on the idea that in systems with d and f electrons, the inter-electron Coulomb interaction is localized at the atomic scale; therefore, in such systems the spatial correlations can be neglected, but, as far as possible, the dynamical fluctuations should be taken into account. This approach turns out to be exact in the limiting case of infinite dimension of space ($d \to \infty$) [382], but works well in realistic cases $d = 3$ and even $d = 2$. One can make acquaintance with DMFT following the reviews [383, 384] and the monograph [385].

The DMFT approach is equally well applicable to the SKS models as to real compounds. It takes into account the crystal structure of compounds and the atomic orbitals of the elements they are made of. It is instructive to discuss the main ideas of the method on a simplest, yet extremely comprehensive, non-degenerate Hubbard model, given by the Hamiltonian

$$H = \sum_{ij\sigma} t_{ij}\, c^{+}_{i\sigma} c_{j\sigma} + U \sum_{i} n_{i\uparrow} n_{i\downarrow}. \tag{5.3}$$

Here, the first term describes the transfer (hopping) of electrons from the site i of the lattice to the site j, and the second term – the local Coulomb interaction of electrons at the same site. Therefore, a localized nature of interactions is built into the model from the very beginning. $c_{i\,\sigma}$ ($c^{+}_{i\,\sigma}$) are the Fermi operators of destruction (creation) of an electron at the site i with spin $\sigma = \uparrow, \downarrow$, and $n_{i\,\sigma} = c^{+}_{i\,\sigma}\, c_{i\,\sigma}$ is the operator of the number of electrons in the state ($i\,\sigma$).

The Hubbard model contains two parameters: $W = 2zt$ (t being the transfer matrix element between the nearest neighbours, z – the number of nearest neighbours), and U. To these two parameters, one can yet add the electron concentration n, or chemical potential μ.

A condition $U \geq W$ corresponds to the case of strong correlations, for handling which the DMFT has been namely developed, as it does not use any perturbation theory in interaction strength. The mathematical apparatus of the DMFT is the method of temperature Green's functions (GF) [386]. The electron Green's function is defined by the relation

$$G^{\sigma}_{ij}(\tau - \tau') = -\langle \hat{T}\, c_{i\sigma}(\tau)\, c^{+}_{j\sigma}(\tau') \rangle, \tag{5.4}$$

where

$$c_{i\sigma}(\tau) = e^{(H-\mu N)\tau}\, c_{i\sigma}\, e^{-(H-\mu N)\tau}$$

is the $c_{i\,\sigma}$ operator in the Heisenberg representation with thermodynamical "time" τ comprised within the interval $0 \leq \tau \leq \beta = 1/T$, $\hat{T}$ is the operator or chronological arrangement over the axis of imaginary time; $\langle \cdots \rangle$ stands for statistical averaging,

N is the operator of full number of electrons. The Fourier component of the GF over the τ parameter, $G_{ij}(i\,\omega_n)$ (5.4), depends on discrete imaginary frequencies $i\,\omega_n = i\,(2n+1)\,T$, $n = 0, \pm 1, \pm 2, \ldots$. An analytical continuation from the imaginary axis into the real one, $i\,\omega_n \to \omega + i\,\delta$, leads to the GF which describes the properties of system [386].

The one-particle Green's function obeys the Dyson equation

$$G = G_0 + G_0\,\Sigma\,G, \tag{5.5}$$

where G_0 is the GF of non-interacting electrons, and Σ – the so-called self-energy describing the effects of electron–electron interaction.

In the Fourier representation over both lattice sites $i - i'$ and time variable $\tau - \tau'$, the GF $G(\mathbf{k}, i\omega_n)$, and the self-energy $\Sigma(\mathbf{k}, i\omega_n)$, become functions of momentum $\mathbf{k}$ and discrete frequencies $i\omega_n$. The spin index σ will be left out for the following, as we will deal exclusively with the paramagnetic phase. In the $d \to \infty$ limit, the self-energy does not depend on $\mathbf{k}$; the relation

$$\Sigma(\mathbf{k}, i\,\omega_n) = \Sigma(i\omega_n) \tag{5.6}$$

makes the pivotal approximation of DMFT [385]. In this approximation, the solution of the Dyson equation (5.5) is as follows:

$$G(\mathbf{k}, i\,\omega_n) = \frac{1}{i\,\omega_n + \mu - \varepsilon_{\mathbf{k}} - \Sigma(i\,\omega_n)}, \tag{5.7}$$

where $\varepsilon_{\mathbf{k}}$ are Fourier components of the hopping matrix elements t_{ij}.

On summing up the previous relation over momenta, we arrive at the expression for $G(i\,\omega_n)$:

$$G(i\,\omega_n) = \sum_{\mathbf{k}} G(\mathbf{k}, i\,\omega_n) = \sum_{\mathbf{k}} \frac{1}{i\,\omega_n + \mu - \varepsilon_{\mathbf{k}} - \Sigma(i\,\omega_n)}, \tag{5.8}$$

which is nothing else as the diagonal matrix element $G_{jj}(i\,\omega_n)$ of the GF (5.4). For a spatially homogeneous system, the site index j can be omitted.

On the other side, we can consider $\Sigma(i\,\omega_n)$ as the self-energy part of the GF $G_{loc}(i\,\omega_n)$ of some auxiliary problem on the same lattice, which problem is defined via the action $S[\mathcal{G}_0]$, being in its turn a functional of certain function $\mathcal{G}_0(\tau - \tau')$:

$$S[\mathcal{G}_0] = -\int_0^{\beta} \mathrm{d}\tau \int_0^{\beta} \mathrm{d}\tau' \sum_{\sigma} c_{\sigma}^{+}(\tau)\,\mathcal{G}_0^{-1}(\tau - \tau')\,c_{\sigma}(\tau') + U \int_0^{\beta} \mathrm{d}\tau\, n_{\uparrow}(\tau)\, n_{\downarrow}(\tau). \tag{5.9}$$

With the help of action (5.9), we can calculate an electron GF of the auxiliary single-impurity problem to be denoted $G_{\mathrm{loc}}(i\,\omega_n)$.

The condition of self-consistency between the lattice problem and the impurity problem would be:

$$G(i\,\omega_n) = G_{\mathrm{loc}}(i\,\omega_n), \tag{5.10}$$

from which $\mathcal{G}_0(\tau-\tau')$ can be found. It plays the role of an effective dynamical field which acts on an electron situated on a certain site, from all other sites of the lattice. The above self-consistency condition takes the form

$$G_{\mathrm{loc}}(i\,\omega_n) = \sum_{\mathbf{k}} \frac{1}{i\,\omega_n + \mu - \varepsilon_{\mathbf{k}} - \Sigma(i\,\omega_n)}; \tag{5.11}$$

it connects $\mathcal{G}_0(i\,\omega_n)$ and $\Sigma(i\,\omega_n)$. At the same time, all the components of the single-impurity problem are related via the Dyson equation

$$\mathcal{G}_0^{-1}(i\,\omega_n) = G_{\mathrm{loc}}^{-1}(i\,\omega_n) + \Sigma(i\,\omega_n), \tag{5.12}$$

that is a different formulation of the previous (5.5). Therefore, we have two equations (5.11) and (5.12) on two independent properties, $\mathcal{G}_0^{-1}(i\,\omega_n)$ and $\Sigma(i\,\omega_n)$, from which they can be determined, once a way to obtain the GF of the single-impurity problem, $G_{\mathrm{loc}}(i\,\omega_n)$, is given. These are the main equations of the DMFT, reducing the problem of calculating the GF on the lattice to the problem of calculating the GF of an effective impurity problem. For resolving the latter, efficient numerical algorithms have been developed: the quantum Monte Carlo method (QMC), numerical renormalization group (NRG), iterative perturbation theory (IPT) and others, see [384, 387].

The DMFT analysis of non-degenerate Hubbard model had shown a substantial reconstruction of the electron spectrum as an effect of strong Coulomb interaction. For $U \ll W$, a conventional Fermi-liquid picture takes place, with well-defined spectrum of quasiparticle states, without a damping on the Fermi surface. As U grows, the Fermi-liquid persists, but the weight of quasiparticle states determined by the formula

$$Z = \left[1 - U\frac{d}{d\omega}\mathrm{Re}(\omega)\right]^{-1}_{\omega=0} \simeq \frac{m}{m^*}, \tag{5.13}$$

decreases, and the electron effective mass m^* gets larger. The system enters the regime of strongly correlated metal. In case of half-filled band ($n = 1$), as U reaches a certain critical value $U_c \sim W$, a gap opens at the Fermi level, and the Mott transition into dielectric state occurs. In the vicinity of this transition from the metallic side, a three-peak structure emerges in the density of states: a narrow quasiparticle peak of the width of about ZW appears at the Fermi level, and below and above it – broad Hubbard bands, centred at energies of about $\pm U/2$, which correspond to non-coherent states. As U approaches U_c, the weight of the quasiparticle peak is gradually displaced towards the Hubbard bands, and the Mott splitting of the initial electron band occurs. On deviation from the half-filled band occupation, the three-peak structure becomes asymmetric, because the electron-hole symmetry is broken. As such deviation gets larger, the quasiparticle peak and one of Hubbard bands merge together; the system becomes more metallic.

We note moreover that the quasiparticle peak at the Fermi level appears due to the scattering of electrons on fluctuations of the spin density. This is a manifestation of the Kondo effect, or, more precisely, of the Suhl–Abrikosov resonance.

5.2.2 LDA+DMFT Calculation for ReOFeAs Compounds

An inclusion of electron correlations by DMFT in calculations for the $LaO_{1-x}F_x$ FeAs compound has been done in [18, 388] soon after a discovery of HTSC in the latter. LDA calculations of electronic structure, thoroughly discussed above, indicate a dominance, in the vicinity of the Fermi level, of Fe3d atomic states which spread over the energy range from −2 to 2 eV. The Fermi level $\omega = 0$ falls onto a steep descent of the density of states of d orbitals, $\rho_d(\omega)$, so that all five orbitals, including t_{2g}-states (d_{xz}, d_{yz},$d_{x^2-y^2}$) and e_g-states ($d_{3z^2-r^2}$, d_{xy}), yield comparable contributions into $\rho_d(\omega)$, and hence they all must be included into the DMFT treatment.

The LaOFeAs compound has to be described by a multi-orbital Hubbard model, taking into account, along with the Coulomb interaction, the Hund's exchange interaction between electrons on the same atom. Therefore, the term in the Hamiltonian responsible for inter-electron interactions should be adopted in the following form:

$$H_{\text{int}} = U \sum_{i,a} n_{ia\uparrow} n_{ia\downarrow} + U' \sum_{i,a\neq b} n_{ia} n_{ib} - J_H \sum_{i,ab} \mathbf{S}_{ia} \mathbf{S}_{ib}, \tag{5.14}$$

where a and b numerate orbitals. The inter-orbital Coulomb interaction is typically expressed in terms of U, and the exchange integral via the relation $U' = U - 2\,J_H$ [384, 387]. The electron GF then becomes a matrix with the elements

$$G^{\sigma}_{ij,ab}(\tau - \tau') = -\langle \hat{T}\, c_{ia\sigma}(\tau)\, c^{+}_{jb\sigma}(\tau') \rangle, \tag{5.15}$$

and all the equations of DMFT consistently acquire matrix form. The method unifying the LDA approach in a calculation for a given compound with the DMFT, and taking into account, on this basis, of strong electron correlations, got a notation "LDA+DMFT" [384, 387]. In this method, similarly to the above described case of the non-degenerate Hubbard model, the problem of an electron spectrum on the lattice is reduced to a solution of a single-impurity system, but now with degenerate atomic orbitals.

As soon as LDA calculations for LaOFeAs indicated a quasi-two-dimensional structure of the electronic states, it suffices to consider a single atomic layer with its square lattice of Fe atoms. Haule, Shim and Kotliar [18] performed LDA+DMFT calculations with five d orbitals and interaction parameters

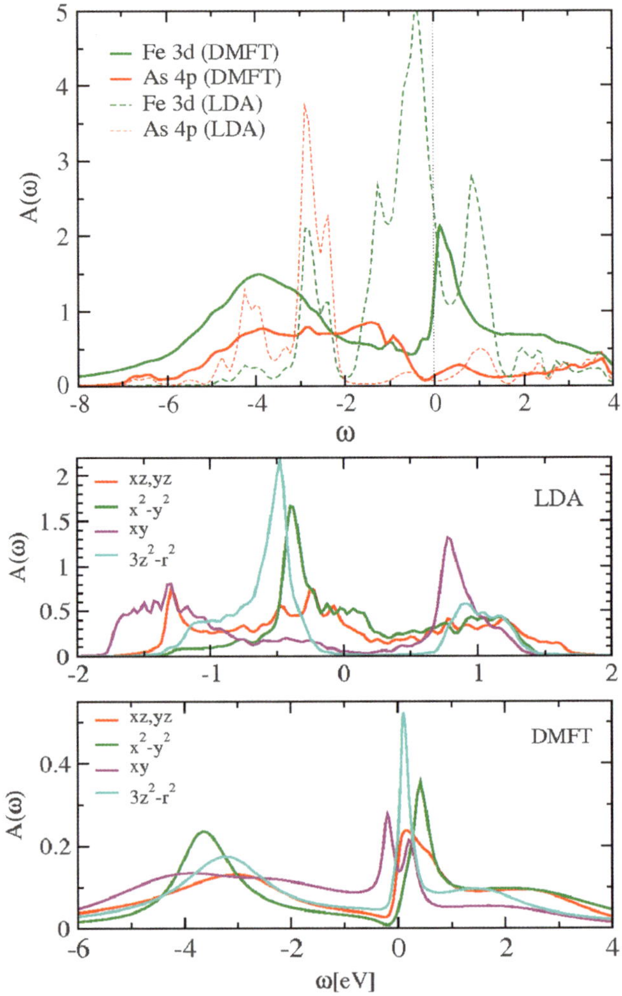

Fig. 5.8 Density of states of the LaOFeAs compounds: Fe3*d* and As4*p* according to DMFT (**a**); Fe3*d*-partial densities of states for different *d*-orbitals according to LDA (**b**) and DMFT (**c**) [18]

$$U = 4\,\text{eV}, \quad J_H = 0.7\,\text{eV}. \tag{5.16}$$

According to the nominal electron configuration of Fe^{2+} ions in this compound, the number of localized electrons per atom is $n = 6$. The calculation results are depicted in Fig. 5.8, in which the density of Fe3*d* states from an LDA+DMFT [384, 387] calculation is shown. In the upper panel, along with the $\rho(\omega)$ for Fe3*d* states also the $\rho(\omega)$ for the As4*p* states are given, which hybridize with Fe3*d*. It is seen that correlation modifies the spectra of both Fe3*d* and As4*p* states.

In the LDA, the whole spectrum of Fe3d states is comprised within a rather narrow interval, from −2 to 2 eV. With correlations taken into account, it spreads from −8 to 4 eV. Far below the Fermi level, a broad Hubbard band of incoherent states, with its maximum at about −4 eV, is situated. The Fermi level is positioned now at the left steep side of the quasiparticle peak. The contributions into $\rho(\omega)$ of d-orbitals is shown in Fig. 5.8c. It is seen that the quasiparticle peak splits for different orbitals, and the splitting due to crystal field gets amplified under the inclusion of correlations. The analysis shows that the damping of the quasiparticle states ($\sim \operatorname{Im}\Sigma(\omega)$) is large, and the effective mass calculated after the relation (5.13) is by 3–5 times larger than the free electron mass. Nevertheless, the system remains metallic, even if a poor metal with low concentration of carriers – electrons and holes, localized in the **k**-space in the vicinity of M and Γ points, as it was predicted in LDA calculations.

A calculation of optical conductivity $\sigma(\omega)$ revealed an absence of the Drude peak and an appearance of a broad peak, corresponding to incoherent transitions from the lower Hubbard band into the peak of quasiparticle states just above the Fermi level. These both features of $\sigma(\omega)$ witness that under the conditions (5.16), the system is a poor metal. The values of parameters chosen (5.16) correspond to strong correlations, because U surpasses the d-band width, $W = 4$ eV. However, the critical value U_{c} of the order of W applies to the non-degenerate Hubbard model only. In case of N_d-fold degeneracy of an electron band, the critical value of the Coulomb potential is given by the formula [389]:

$$U_{\mathrm{c}} \simeq \sqrt{N_d}\,U_{\mathrm{c}}^0 - N_d\,J_H, \tag{5.17}$$

where U_{c}^0 is the critical value corresponding to the non-degenerate Hubbard model. For d states $N_d = 5$, therefore U_c does somehow exceeds W at the parameter values of (5.16). Strictly speaking, the formula should be applied for estimation of the critical U value in LaOFeAs with care, because the d-orbitals are split by the crystal field, and this splitting enhanced by correlation. Therefore, a priori it remains unclear how far the compound under consideration is from the metal–insulator (MI) transition. The authors of [18] performed a separate DMFT calculation with $U = 4.5$ eV and found out that a gap opens at the Fermi level. It means that LaOFeAs, for which the parameter of interaction is (5.16), are near to the Mott transition; consequently, the other *Re*OFeAs compounds should belong to those with strong electron correlation. The stoichiometric compound itself, as well as the doped compositions, represents a strongly correlated metal.

We note that in the work being discussed [18], the single impurity model was solved by the QMC method with continuous time [384, 387], for the temperature $T = 116$ K. Almost simultaneously, another work appeared [388], in which by LDA+DMFT the density of states has been calculated for the same compound LaOFeAs and for the same parameter values (5.16) as in [18], but the solution of the impurity problem was achieved by the standard variety of QMC, the Hirsch-Fye method [384,385,387]. The calculations were done for the value of $\beta = 10\,(\mathrm{eV})^{-1}$ that corresponds to the temperature by an order of magnitude higher than that used in [18].

The overall structure of the spectrum (the presence of the Hubbard bands and the quasiparticle peak in the vicinity of the Fermi level) is the same as in [18]; however, the fine structure of the quasiparticle states is markedly different. The renormalized densities of the t_{2g} states, d_{xz}, d_{yz} and $d_{x^2-y^2}$, indicate a pseudogap below the Fermi level, whereas both e_g-states, $d_{3z^2-r^2}$ and d_{xy}, exhibit a quite metallic density of states at the Fermi level. It turned out that the t_{2g} states are filled to almost half occupation, whereas the occupation of the e_g states is substantially higher. These results resemble a situation with a selective MI transition in the multi-orbital Hubbard model, when part of orbitals has a typically metallic density of states whereas the other part – a dielectric one, i.e. with a gap at the Fermi level. As a result, such a material is metal, but one with peculiar properties.

To find out how far the system is from the Mott MI transition, the authors of [388] calculated $\rho(\omega)$ for the value of $U = 5\,\text{eV}$, and did not obtain any gap at the Fermi level. According to their results, LaOFeAs is not close to the MI transition, i.e. this compound is in the regime of moderate, but not strong, correlations. The discrepancy between the results of [18] and [388] might be related to different choice of a solver for the single-impurity problem: the standard QMC vs. the continuous-time QMC. This controversy stimulated further studies of the role of correlations in *Re*OFeAs-compounds within the LDA+DMFT approach.

A further analysis to this end can be found in [390,391]. In these works, the electron spectrum and optical conductivity of two compounds, $Re\text{O}_{1-x}\text{F}_x\text{FeAs}$ ($Re =$ La, Sm), have been calculated using the same parameter values (5.16) as in the preceding works [18,388]. The auxiliary problem was being solved by the IPT method, which is less accurate than the QMC by reasonably efficient when applied to multi-orbital systems, and moreover has no limitation on the side of low temperatures, in contrast to QMC.

Based on their calculation results, the authors of the works [390,391] suggest that the *Re*OFeAs compounds are rather substances characterized by strong electron correlations. Thus, with the use of the same parameter values (5.16), the LDA+DMFT calculations done within the five-bands model [18, 388, 390] led to different conclusions. Whereas [18, 390] claim that the *Re*OFeAs compounds belong to strongly correlated ones, the authors of [388] suggest that the role of electron correlations in them is weak, or, at least, moderate. A further step in elucidating this controversial situation was done in [392].

5.2.3 LDA+DMFT Calculation on an Extended Basis

Anisimov et al. [392] arrived at a conclusion that the problem of the importance of correlations in the *Re*OFeAs compounds has to be solved not on the limited basis, with only Fe3d-states taken into account, but in an extended one, including also As4p and O2p states, because of the importance of the Fe3d hybridization with the latter. A guiding idea to this came from the analysis of appropriate interaction parameters, U and J_H. The best way to determine them would be from a first-principles calculations. An analysis of such calculations previously done in cuprates and transition metal oxides indicates that the parameters of screened Coulomb interaction, calculated within the random phase approximation, do strongly depend on the basis chosen.

Such calculations of parameters U and J_{H} for LaOFeAs, with two different (in what concerns basis sets) setups for the solution of the LDA problems, reveal quite different results: in the case when only Fe3d orbitals participate, $U \simeq 0.5\,\mathrm{eV}$ and $J_{\mathrm{H}} \simeq 0.5\,\mathrm{eV}$, whereas with the extended basis, including As4p and O2p orbitals, $U \simeq 3 \div 4\,\mathrm{eV}$ and $J_{\mathrm{H}} \simeq 0.8 \div 1.0\,\mathrm{eV}$. This means that in the calculations of the electron spectrum in the LDA+DMFT scheme, the extended basis should be used as well.

With such a basis, already in the LDA the partial density of states is markedly different from that which follows from a calculation with reduced basis, i.e. in the five-band model (Fig. 5.9).

Along with the spectrum comprised in the range $-2 \leq \omega \leq 2\,\mathrm{eV}$, which is very close to that obtained with the reduced basis, in the density of Fe3d states a region is seen $-3 \leq \omega \leq -2\,\mathrm{eV}$, left unshaded in the bottom panel of Fig. 5.9, formed by hybridization with As4p states. Below them, a weak contribution is present (filled black) due to hybridization with O2p states. These hybridization effects do strongly affect the above values of U and J_{H} parameters, extracted from the corresponding band structures.

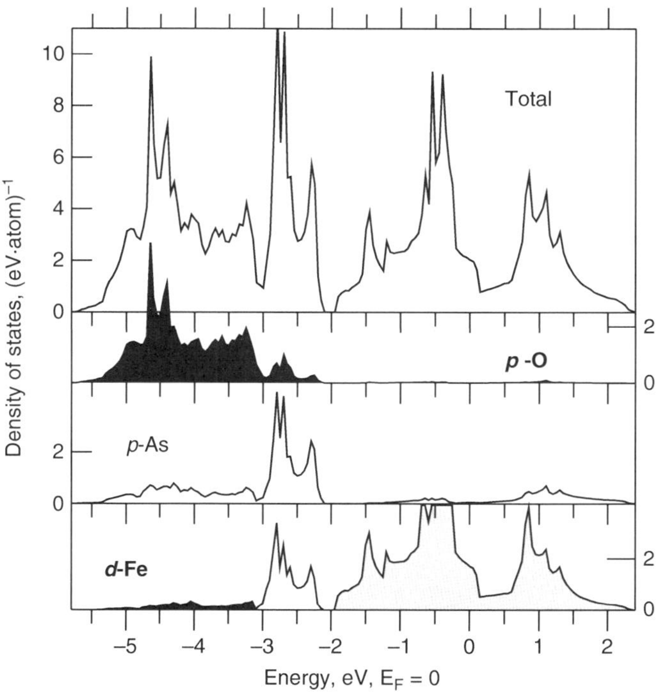

Fig. 5.9 Full and partial densities of states in LaOFeAs, calculated within the LDA in the extended basis [392]

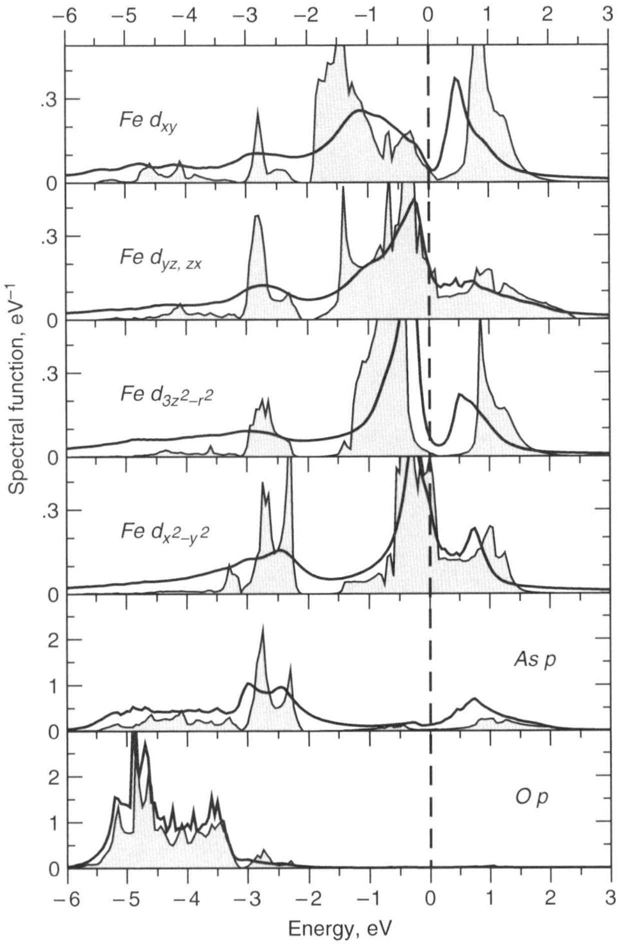

Fig. 5.10 Partial densities of states with electron correlations taken into account in a LDA+DMFT calculation, in an extended basis with the parameter values $F^0 = 3.5$ eV, $J_H = 0.81$ eV (*solid lines*). *Shaded areas* correspond to an LDA calculation [392]

Partial densities of states, as calculated within the LDA+DMFT with the extended basis, are shown in Fig. 5.10. The F^0 specified in the caption is the Slater parameter, through which the Coulomb parameter is expressed, as $U = F^0 - \frac{1}{2} J_H$. Thus, a realistic U value turns out to be 3.1 eV. We see that the spectra obtained look like those with a weak correlation. Indeed, the upper and lower Hubbard bands are missing. In the energy range $-2 \lesssim \omega \lesssim 2$ eV, the spectra are qualitatively similar to LDA results, following from the five-band model. The reason for correlation effects to be weak, given a relatively high value of the Coulomb U parameter, is a strong hybridization between Fe3d and As4p states (refer to the peak in the spectral function in the region $-3 \lesssim \omega \leq -2$ eV. This hybridization results in an additional screening of Coulomb interactions between the Fe3d orbitals.

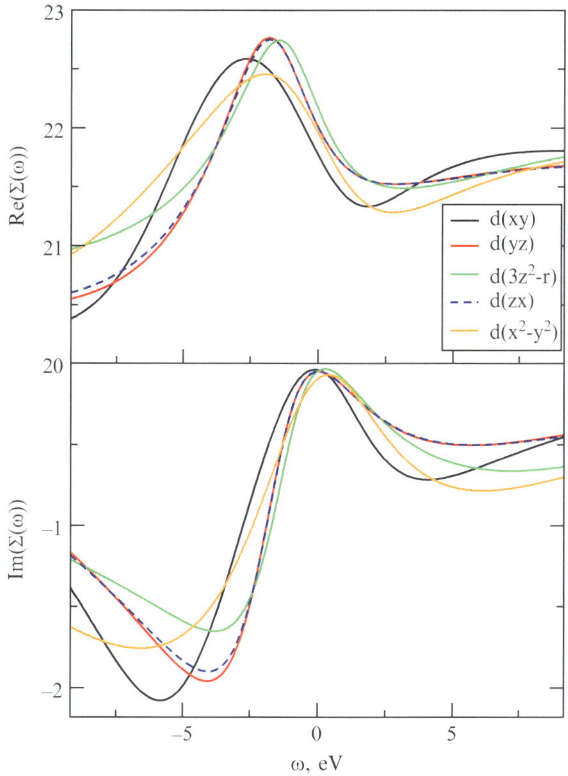

Fig. 5.11 Real and imaginary parts of the self-energy $\Sigma(\omega)$, calculated in LDA+DMFT under the same conditions as in [392]

Characteristics of quasiparticle states in the vicinity of the Fermi level can be found from the real and imaginary parts of the self-energy $\Sigma(\omega)$, Fig. 5.11. Thus, the weight of quasiparticle states is given by (5.13), from which the Z values of 0.56, 0.54, 0.45, 0.56 follow for the orbitals d_{xy}, (d_{xz}, d_{yz}), $d_{3z^2-r^2}$ and $d_{x^2-y^2}$, correspondingly, and, consequently, $m^*/m \sim 1/Z < 2$. These numbers of Z are substantially larger than $0.2 \div 0.3$, obtained in [18], and indicate a not so strong correlation in the LaOFeAs compound. The imaginary part of $\Sigma(\omega)$ in the vicinity of the Fermi level is small, even if different from zero. Taken together, the deviations from the Fermi-liquid behaviour are not big, at variance with the conclusions of [18, 390, 391]. Therefore, the bulk of characteristics of LaOFeAs, calculated by LDA+DMFT with the parameter values of U and J_H evaluated with an extended basis, indicates that this compound should be attributed to the class of moderately correlated systems [392].

In LaOFeAs, each Fe ion is tetrahedrically coordinated by four As ions, with a slight tetragonal distortion. In the tetrahedral point group T_d, the five d orbitals split by the crystal field into the low-energy doublet $(3z^2 - r^2, xy)$, corresponding to the e_g irreducible representation, and the high-energy triplet $(x^2 - y^2, xz, yz)$, which belongs to the t_{2g} representation. Calculations of the orbital energies of the Wannier functions revealed that the crystal field splitting, $t_{2g} - e_g$, is very small:

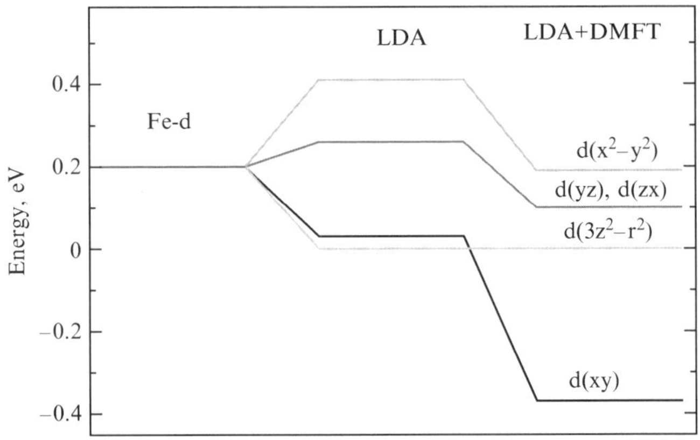

Fig. 5.12 Splitting of the Fe3d-orbitals obtained in the LDA+DMFT calculation [392]

$\Delta_{\mathrm{cf}} \approx 0.25$ eV. A small tetragonal distortion of a tetrahedron leads to an additional splitting of t_{2g} and e_g levels with the following orbital energies, the energy of the lowest $d_{3z^2-r^2}$ orbital taken as a reference: $\varepsilon_{d3z^2-r^2} = 0$; $\varepsilon_{dxy} = 0.03$; $\varepsilon_{dxz,yz} = 0.26$; $\varepsilon_{dx^2-y^2} = 0.41$ eV. The correlation has, as a result, not only a narrowing of bands, but also a shift of Fe3d orbital energies. An addition of $\mathrm{Re}\,\Sigma(0)$ to the LDA orbital energies give, again with respect to $\varepsilon_{d3z^2-r^2,dxy} = 0.37$; $\varepsilon_{dxz,yz} = 0.10$; $\varepsilon_{dx^2-y^2} = 0.20$ eV (see Fig. 5.12). We note that the actual band shifts are smaller because of the p–d hybridization.

A similar calculation has been done for another type of compounds, $BaFe_2As_2$, at the same values of interaction parameters, $U = 3.1$ eV, $J_{\mathrm{H}} = 0.81$ eV [393]. The spectral density of Fe3d and As4p states is shown in Fig. 5.13. A comparison with Fig. 5.10 shows the similar type of behaviour of electronic structure affected by electron correlations in the compounds with a single FeAs layer and two layers in the unit cell. In neither case are manifestations of strong electron correlations (Hubbard bands of incoherent states) visible. At the same time, a narrowing of 3d bands and, correspondingly, an increase of the effective mass occurs (see Table 5.3). As in the LaOFeAs compound, in the double-layer $BaFe_2As_2$ an increase of mass due to correlations makes $m^*/m \approx 2$ that allows to attribute the FeAs-systems to moderately correlated ones.

5.2.4 Comparison with Experiment

In Fig. 5.14, full spectral density of Fe3d and As4p states, calculated for the $BaFe_2As_2$ compound, is shown in comparison with photoemission spectra [394]. The calculated spectrum reproduces the experimental data fairly well, including the relative placement of d and p bands, and a large contribution of the d states at the top of the valence band (Fig. 5.15).

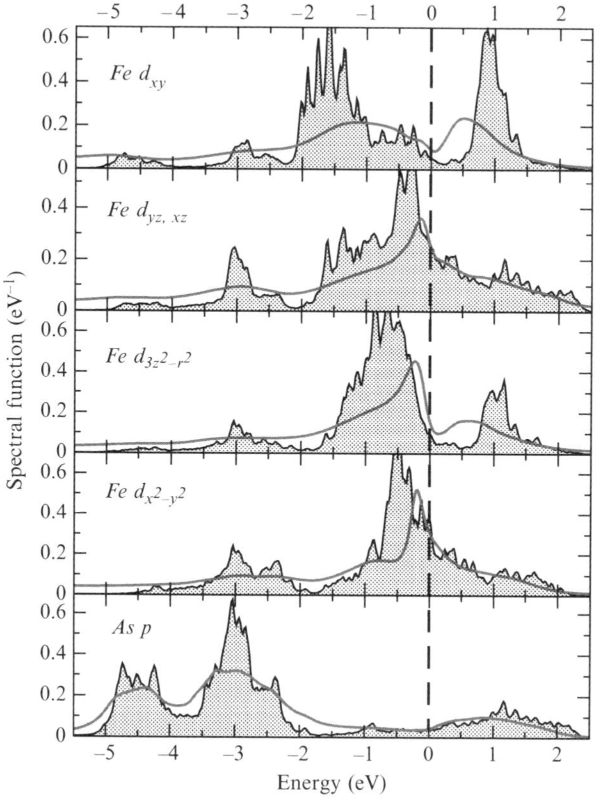

Fig. 5.13 Partial densities of states of $BaFe_2As_2$ from LDA+DMFT calculation on extended basis with parameter values $U = 3.1$ eV, $J_H = 0.81$ eV (*solid lines*). *Shaded areas* correspond to the LDA calculation [393]

Table 5.3 Energy gain of three magnetically ordered phases relative to the nonmagnetic phase, after [471]

	Checkeboard	Stripe	Double-stripe
$BaFe_2As_2$	16	94	0.6
FeTe		207	230

The calculation of the spectral density $A(\mathbf{k}, \omega)$ for the same compound permitted to trace the dispersion curves along the maxima of $A(\mathbf{k}, \omega)$ depending on the wave vector $\mathbf{k}$. The curves calculated in the vicinity of Γ and X points are shown in upper panels of Fig. 5.16. The results obtained are in good agreement with the ARPES data [395] shown in the bottom panels of the same figure.

A comparison of LDA+DMFT one-particle spectral functions, calculated for a strong enough Coulomb interaction ($U = 3.1$ eV) and taking into account the Fe3d–As4p hybridization [392], with different experimental spectra of LaOFeAs is given in Figs. 5.16 and 5.17. In Fig. 5.16, the calculated Fe3d spectral function

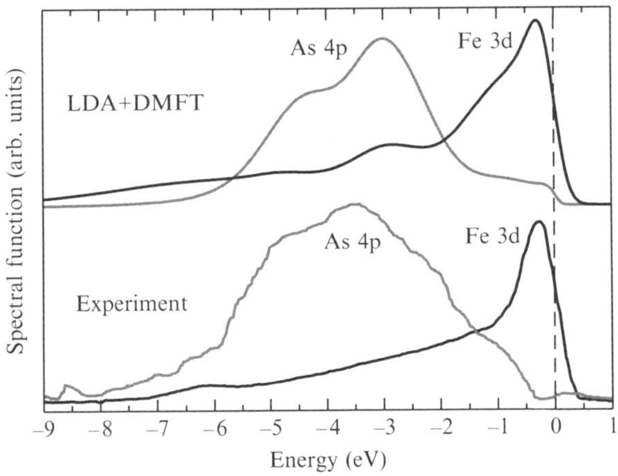

Fig. 5.14 Spectrum of Fe3d and As4p states in the $BaFe_2As_2$ compound as calculated in the LDA+DMFT approach [393], in comparison with PES data [394]

is compared with the FeL_3 X-ray emission spectrum (XES) of LaOFeAs [396], measured under resonance conditions, i.e. when the energy of incident photons is sufficient for an excitation of the L_3 level only. The energy is hence not sufficient to excite the L_2 level, and out of two possible transitions, $3d4s \longrightarrow 2p_{3/2}$ and $3d4s \longrightarrow 2p_{1/2}$, only the first one is realized that helps to avoid an overlap of L_3 and L_2 spectra. By force of dipole selection rules, the FeL_3 XES reflects the distribution of the occupied Fe3d states, since the (also permitted) contribution from the Fe4s states is small, and can be neglected. A comparison of theoretical Fe3d spectral function with the experimental spectra in Fig. 5.16 reveals their agreement in what regards the placement of the main Fe3d peak in the vicinity of the Fermi level, and also of a low-energy peak (at ~2.5 eV) due to the hybridization of Fe3d and As4p states. This low-energy peak is less clearly seen in the experimental spectrum, and at least a part of difference may be due to the fact that in the calculated spectrum, the energy dependence of the matrix element of the transition probability has not been taken into account.

Measurements of resonant X-ray emission FeL_3 spectra in other FeAs-compounds: $CaFe_2As_2$ [397], NaFeAs and LiFeAs [346] indicated their big similarity with the LaOFeAs spectrum (Fig. 5.18). Between all these FeAs-systems, there is similarity in the positioning, with respect to the Fermi level, of the intensity maximum of the Fe L_3 XES, and of the low-energy band genetically related to the Fe3d–As4p hybridization. At the same time, the Fe L_3 XES do not reveal any traces of the lower Hubbard band, predicted by LDA+DMFT calculations, in which under assumption of strong correlations ($U = 4.0$ eV), the Fe3d–As4p hybridization was not taken into account [18, 390].

In Fig. 5.17, the LDA+DMFT spectral function of the total density of states is compared with experimental photoemission valence-band spectrum of LaOFeAs [398]. In a photoemission spectrum, the kinetic energy of all electrons emitted from

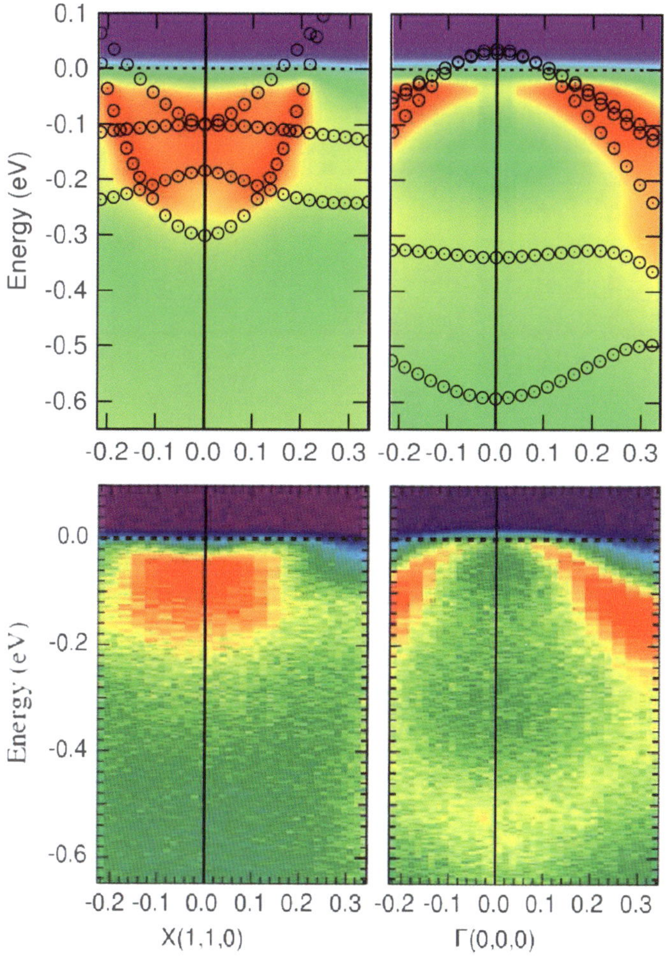

Fig. 5.15 Comparison of the dispersion curves of the $BaFe_2As_2$ compound as calculated in the LDA+DMFT approach [393] with the ARPES data of [395]

the valence band is detected, and, because of the absence of any symmetry-related selection, the resulting spectrum corresponds to a distribution of the total density of states (up to the differences in the atomic ionization cross-sections for different elements). In this relation, note that in the experimental photoemission spectrum a contribution due to O2p states near the bottom of the valence band at $-3 \div -6$ eV, which did not appear in the Fe L_3 XES because of a very weak hybridization between the Fe3d and O2p states, is clearly seen (Fig. 5.10). We note again a good agreement between experimental and theoretical spectra. Therefore, the photoemission spectra of FeAs-compounds can be reasonably explained on the basis of an LDA+DMFT calculation, with the hybridization of Fe3d and As4p spectra taken into account.

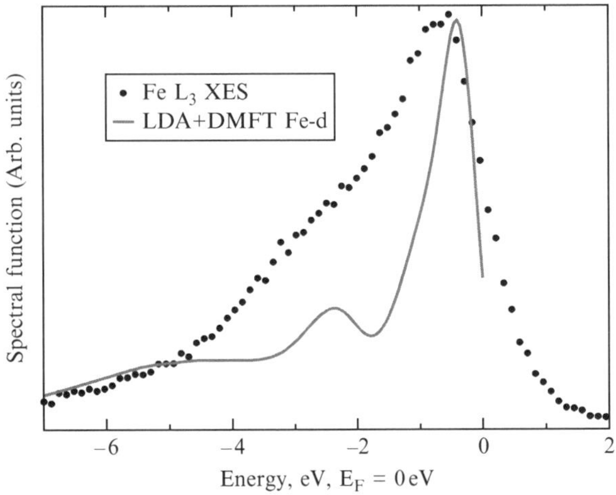

Fig. 5.16 Comparison of the calculated Fe3d LDA+DMFT spectral function (*solid line*) with an experimental FeL_3 X-ray emission spectrum of LaOFeAs [392]

Another experimental argument in favour of such conclusions comes from the studies of X-ray photoemission spectra (XPS) of the Fe2p core levels in FeAs-compounds. It has been established in earlier studies that in the metallic Fe and in those compounds which are characterized by itinerant character of Fe3d electrons, the Fe2p XPS have simple shape with two peaks, separated by the spin-orbit splitting of the Fe $2p_{1/2} - 2p_{3/2}$ levels. On the other side, in case of localized Fe3d electrons, e.g. in FeO, the Fe2p XPS contain pronounced low-intensity satellites [399]. The experiments revealed an absence of such satellites in the Fe2p XPS of FeAs-compounds, namely for LaOFeAs [398], $CaFe_2As_2$ [397], LaOFeP [92], see Fig. 5.19 and $FeSe_x$ [400], that is one more point for an itinerant character of Fe3d electrons in these systems.

This conclusion matches the results of measured intensities ratio within non-resonance Fe $L_{2,3}$ XES in LaOFeAs, $CaFe_2As_2$, NaFeAs and LiFeAs (Fig. 5.20). Under non-resonant excitation conditions, both L_2 and L_3 levels get ionized, and in the spectrum two lines appear, at the energies ~705 and ~718 eV, which correspond to the Fe L_3 ($3d4s \longrightarrow 2p_{3/2}$ transition) and Fe L_2 ($3d4s \longrightarrow 2p_{1/2}$ transition) emission spectra, correspondingly. For free atoms, the intensity relation between the L_2 L_3 emission lines, $I(L_2)/I(L_3)$, depends on the statistical population of the $2p_{1/2}$ and $2p_{3/2}$ levels only, and, consequently, has to be 1/2. In metals and other systems with itinerant 3d electrons, along with the radiative transitions ($3d4s \longrightarrow 2p_{1/2,3/2}$), also a non-radiative one of the type $L_2L_3M_{4,5}$ may occur, i.e. when under an ionization of the L_2 level, the vacancy is filled by an electron from the L_3 level, ejecting an Auger electron out of the $M_{4,5}$ shell. Such a process does noticeably reduce the $I(L_2)/I(L_3)$ intensity ratio [401]. As follows from Fig. 5.20, the $I(L_2)/I(L_3)$ intensity ratio remains the same for all abovementioned FeAs compounds, and stays closer to the relation in elemental Fe than in FeO.

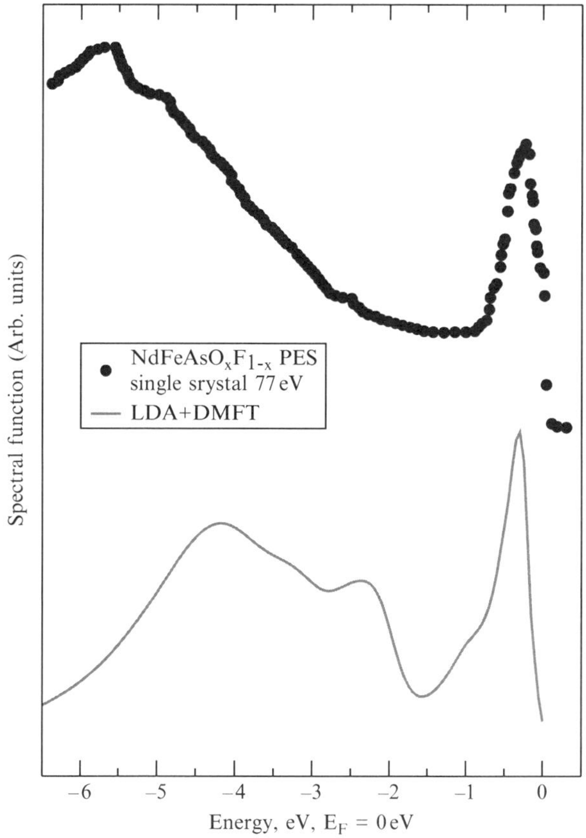

Fig. 5.17 Calculated full spectral function (*solid line*) and experimental photoemission spectrum of LaOFeAs [392]

Summarizing, a comparison of X-ray emission and photoelectron spectra of LaOFeAs against two lines of LDA+DMFT calculations have shown that the experiment is closer to those calculations which take into account the hybridization of Fe3*d* and As4*p* states that substantially weakens the electron correlations. This indicates that the FeAs-compounds do not belong to strongly correlated systems but are, rather, those with weak or moderate electron correlations [392].

The experimental evidence given above supports the conclusions from the LDA+ DMFT calculations concerning a moderate role of electron correlations in the FeAs-compounds. In these calculations, no re-distribution of electron states is observed which would lead to a formation of Hubbard incoherent bands, but an enhancement of the quasiparticles' effective masses occurs, on an average, in the relation $m^*/m \approx 2$. The same conclusion is done in a detailed study [402].

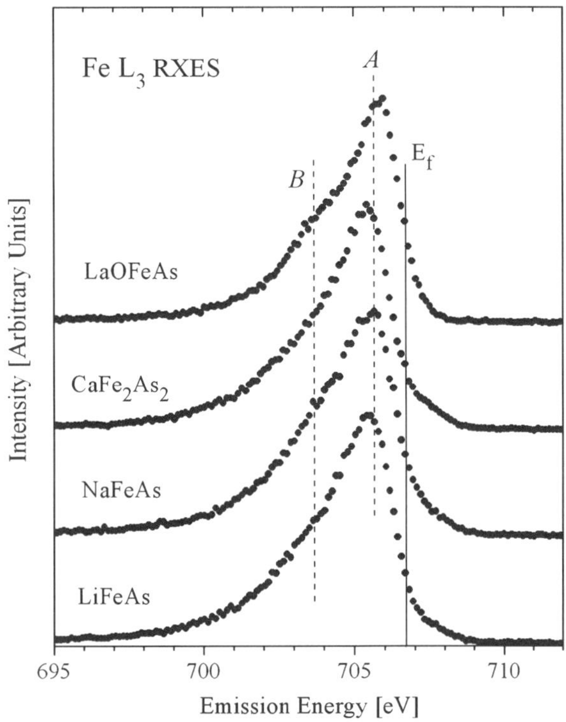

Fig. 5.18 FeL_3 XES LaOFeAs [396], $CaFe_2As_2$ [397], LiFeAs and NaFeAs [346]

5.3 A Minimal Two-Orbital Model

5.3.1 *Formulation of the Model*

So far, the question is not yet definitely resolved about whether the FeAs-systems are strongly or weakly correlated ones, therefore it makes sense to consider both limits: $U \gg W$ and $U \ll W$. In the first case, the localized aspect of electron and spin states is dominating, in the second case – the itinerant one. The two limits demand for a use of different mathematical tools. In the case of strong Coulomb interaction, one can pass from the initial Hamiltonian, expressed via conventional Fermi operators, to an effective Hamiltonian, in which the exchange interaction appears between spins at different sites. Thus, the simplest non-degenerate Hubbard model at $U \gg W$ transforms into the $t - J$ model, for which special methods of analysis are developed since long. An interest for the $t - J$ model was stimulated by the fact that it is considered as the basic model for the description of electronic properties of the high-temperature superconducting cuprates. A peculiarity of the $t - J$ model is an appearance of incoherent states and a distortion of the Fermi-liquid picture. In the opposite case $U \ll W$, the Coulomb interaction can be treated within the standard perturbation theory, and the properties of systems are

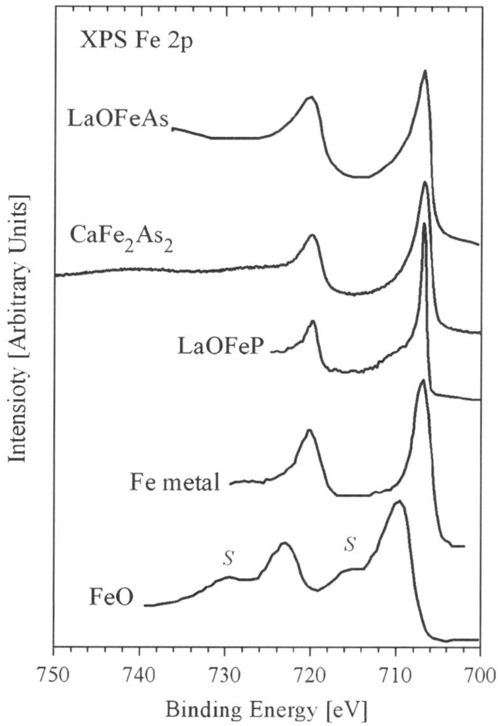

Fig. 5.19 *XPS* Fe 2*p* LaOFeAs [398], $CaFe_2As_2$ [397], LaOFeP [92], FeO [399] and Fe [403]

close to Fermi-liquid properties. This means that in the vicinity of the Fermi level, a quasiparticle description with a weak damping of quasiparticles is possible.

In relation to FeAs-systems, the both approaches may a priori be allowed, so long as the question about the relation of parameters U and W in them is not definitely resolved. Apparently, they are of the same order, but it makes interest to consider two limiting cases of strong and weak Coulomb interaction, to clarify the tendencies of the localized and itinerant aspects of their description. In the following, we consider both approaches, starting from that of a weak Coulomb interaction.

In both approaches, to simplify a theory analysis based on analytical methods, it makes sense to use a minimal model of FeAs-compounds. This minimal electron model for the FeAs-systems follows from the crystal structure of the FeAs layers. The Fe atoms form a square lattice; below and above these planes, the planes of As atoms are situated, also forming square lattices, so that the As atoms are either above or below the centre of each placket formed by four Fe atoms. As a result, each Fe atom is surrounded by a tetrahedron of As atoms (Fig. 5.21).

Concerning Fig. 5.21, an explanation is due. The t_{2g}-orbitals d_{xy}, d_{yz} and d_{xz} make four-petal structures, situated in the xy, yz and xz planes, correspondingly. Therefore, the d_{xz} and d_{yz} orbitals extend in the planes normal to the basal xy-plane. In Fig. 5.21, a projection of the d_{xz} and d_{yz} orbitals onto the basal plane is drawn. The four-petal motif in a certain lattice site is, in fact, composed of two

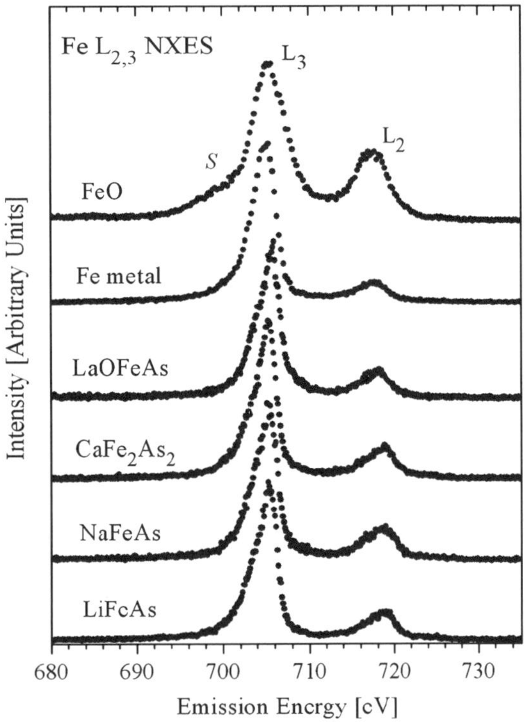

Fig. 5.20 Fe L_3 NXES of some FeAs-systems [392]

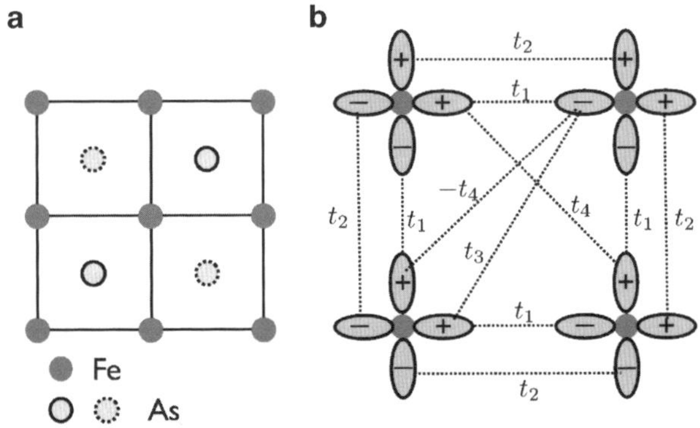

Fig. 5.21 Crystal and electron structure of the FeAs layer. (**a**) Positions of Fe atoms (in the plane) and As atoms above and below the Fe-plane (*solid* and *dotted circles*, correspondingly); (**b**) orientation of the d_{xz} and d_{yz} orbitals of the Fe atoms and matrix elements of the electron hopping between the nearest and the next-nearest Fe neighbours [404]

degenerate orbitals, d_{xz} and d_{yz}. The "+" and "−" indicate the sign of the wave function in a certain part of a given petal of the orbital. The dotted lines connecting different petals of the nearest or next-nearest neighbours indicate a corresponding matrix element of hopping. Its magnitude depends on the signs of the wave function, so that e.g. t_1 and t_2 values are different.

As follows from the band structure calculations discussed above, the major contribution to the electron density of states near the Fermi level comes from the 3d states of Fe atoms. Among the five d orbitals, two degenerate ones, d_{xz} and d_{yz}, should be singled out, which are shown in Fig. 5.21. An overlap of orbitals determines the magnitude of hopping matrix elements between them, for the nearest and next-nearest neighbours. If only these two orbitals are taken into account, a two-orbital model of the FeAs-compounds (more precisely, of an FeAs-plane) results, described by a Hamiltonian

$$H_0 = \sum_{\mathbf{k}\sigma} \psi^+_{\mathbf{k}\sigma} T(\mathbf{k}) \psi_{\mathbf{k}\sigma}, \tag{5.18}$$

where $\psi^+_{\mathbf{k}\sigma} = \left(c^+_{x\,\mathbf{k}\,\sigma}, c^+_{y\,\mathbf{k}\,\sigma}\right)$ is a two-component spinor, composed of creation operators for a $c^+_{x\,\mathbf{k}\,\sigma}$ electron in the d_{xz} orbital and $c^+_{y\,\mathbf{k}\,\sigma}$ – in the d_{yz} orbital, and $T(\mathbf{k})$ is a two-row matrix, made out of matrix elements of the transitions shown in Fig. 5.21:

$$T(\mathbf{k}) = \begin{pmatrix} \varepsilon_x(\mathbf{k}) - \mu & \varepsilon_{xy}(\mathbf{k}) \\ \varepsilon_{xy}(\mathbf{k}) & \varepsilon_y(\mathbf{k}) - \mu \end{pmatrix}. \tag{5.19}$$

Here,

$$\left.\begin{aligned} \varepsilon_x(\mathbf{k}) &= -2t_1 \cos k_x - 2t_2 \cos k_y - 4t_3 \cos k_x \cos k_y \\ \varepsilon_y(\mathbf{k}) &= -2t_2 \cos k_x - 2t_1 \cos k_y - 4t_3 \cos k_x \cos k_y \\ \varepsilon_{xy}(\mathbf{k}) &= -4t_4 \sin k_x \sin k_y \end{aligned}\right\}. \tag{5.20}$$

The Hamiltonian (5.18), being a quadratic form of Fermi operators, can be diagonalized by a canonical transformation from the $c_{s\mathbf{k}\sigma}$ ($s = x, y$) operators to the operators $\gamma_{s\mathbf{k}\sigma}$:

$$c_{s\mathbf{k}\sigma} = \sum_{\nu=\pm} a^s_\nu(\mathbf{k})\, \gamma_{\nu\mathbf{k}s}, \tag{5.21}$$

with

$$\begin{aligned} a^x_-(\mathbf{k}) &= \left[\frac{1}{2}\left(1 + \frac{\varepsilon_-(\mathbf{k})}{\sqrt{\varepsilon^2_-(\mathbf{k}) + \varepsilon^2_+(\mathbf{k})}}\right)\right]^{1/2} = a^y_+, \\ a^x_+(\mathbf{k}) &= \left[\frac{1}{2}\left(1 - \frac{\varepsilon_-(\mathbf{k})}{\sqrt{\varepsilon^2_-(\mathbf{k}) + \varepsilon^2_+(\mathbf{k})}}\right)\right]^{1/2} = -a^y_+. \end{aligned} \tag{5.22}$$

Taking into account the chemical potential, the Hamiltonian of the model acquires the form:

$$H_0 = \sum_{\mathbf{k}\sigma} \sum_{\nu=\pm} E_\nu(\mathbf{k})\, \gamma^+_{\nu\,\mathbf{k}\sigma} \gamma_{\nu\,\mathbf{k}\sigma}, \tag{5.23}$$

where the band energies are

$$E_{\pm}(\mathbf{k}) = \varepsilon_{+}(\mathbf{k}) \pm \sqrt{\varepsilon_{-}^{2}(\mathbf{k}) + \varepsilon_{xy}^{2}(\mathbf{k})} - \mu, \tag{5.24}$$

where $\varepsilon_{\pm}(\mathbf{k}) = \frac{1}{2}\left[\varepsilon_x(\mathbf{k}) \pm \varepsilon_y(\mathbf{k})\right]$.

We introduce a one-electron Matsubara Green's function:

$$\mathcal{G}^{\sigma}(\mathbf{k}, \tau) = -\langle \hat{T}_{\tau}\, \psi_{\mathbf{k}\sigma}(\tau)\, \psi_{\mathbf{k}\sigma}^{+}(\tau)\rangle, \tag{5.25}$$

where all notations are standard ones: $\psi_{\mathbf{k}\sigma}(\tau)$ is a $\psi_{\mathbf{k}\sigma}$-operator in the Heisenberg representation with Hamiltonian H and imaginary time $0 \leqslant \tau \leqslant \beta = 1/T$:

$$\psi_{\mathbf{k}\sigma}(\tau) = e^{H\tau}\, \psi_{\mathbf{k}\sigma}\, e^{-H\tau}; \tag{5.26}$$

$\hat{T}_{\tau}$ is the operator of ordering over the variable τ, and $\langle\cdots\rangle$ stands for statistical averaging.

In view of the fact that the Hamiltonian (5.18) does not contain interaction, the equation of motion for the GF (5.25) is closed, and after the Fourier expansion over the Matsubara frequencies

$$g^{\sigma}(\mathbf{k}, \tau) = \sum_{n} e^{i\omega_n\tau}\, g^{\sigma}(\mathbf{k}, i\omega_n) \tag{5.27}$$

we arrive at an explicit expression for the electron GF in the form of a 2×2 matrix:

$$\mathcal{G}^{\sigma}(\mathbf{k}, i\,\omega_n) = \frac{1}{\big(i\,\omega_n - E_{+}(\mathbf{k})\big)\big(i\,\omega_n - E_{-}(\mathbf{k})\big)} \begin{pmatrix} i\omega_n - \varepsilon_x(\mathbf{k}) - \mu & -\varepsilon_{xy}(\mathbf{k}) \\ -\varepsilon_{xy}(\mathbf{k}) & i\omega_n - \varepsilon_y(\mathbf{k}) - \mu \end{pmatrix}. \tag{5.28}$$

5.3.2 *Band Structure of the Spectrum*

The poles of the Green's function, $E_{+}(\mathbf{k})$ and $E_{-}(\mathbf{k})$, determine the spectrum of quasiparticles within the model under consideration. There are two branches, the relative position of which and the shape of dispersion curves depends on the magnitude of the matrix elements $t_1, \ldots, t_4$. They have to be adjusted from the condition for the formulae (5.24) to give a spectrum consistent with numerical calculations for the FeAs-compounds. In [404], the following values have been chosen:

$$t_1 = -1; \quad t_2 = 1.3; \quad t_3 = t_4 = -0.85 \tag{5.29}$$

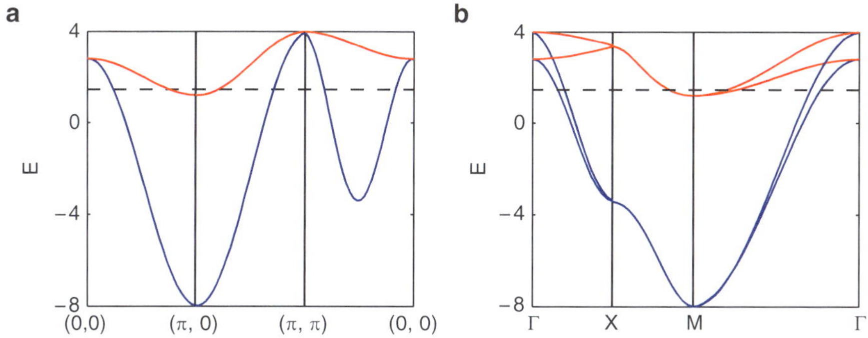

Fig. 5.22 Band structure in the two-orbital model with the parameter values (5.29) in the extended (**a**) and reduced (**b**) Brillouin zone [404]

(in dimensionless units, with $|t_1| = 1$). Then from a condition of half band occupation (one electron per Fe atom) the resulting value for the chemical potential is $\mu = 1.5$.

The spectrum (5.24), as calculated with the parameters thus chosen, is depicted in Fig. 5.22 in the extended and reduced Brillouin zones. The extended BZ corresponds to a choice of primitive cell with one Fe atom, whereas the reduced one – with two Fe atoms. The dashed line in Fig. 5.22a and b corresponds to the Fermi level, $\mu = 1.45$. We see that the Fermi level is crossed by two branches of hole quasiparticles in the vicinity of Γ and two branches of electron ones close to the M point. The Fermi surface is shown in Fig. 5.23 within the extended and reduced zone schemes. As we see, the model considered yields two hole pockets in the vicinity of Γ and two electron ones around the M point, therefore, it can serve as a minimal model of FeAs-compounds.

We introduced so far the kinetic part only of the system's energy. To describe various ordered phases in this model, it is necessary to include an interaction into the Hamiltonian – first of all, the Coulomb repulsion of electrons on the same site, or exchange interaction on different sites. The types of interaction are, primarily, determined by a conclusion to which of the limits is the FeAs-system under consideration closer, to the itinerant or the localized one. In the first case, it is assumed that the one-site Coulomb interactions are smaller than the widths of the d band. The opposite limit corresponds to an assumption about the closeness of the Mott–Hubbard transition and to an existence of well-defined orbitals on the sites occupied by Fe atoms. The model in which the Coulomb repulsion dominates reduces to taking into account the effective exchange interactions between the nearest and next-nearest neighbours only, and hence can be called the $t - J_1 - J_2$ model. The opposite model with weak Coulomb interaction will be called the itinerant model. Without so far going deeply into the question about what does rather occur in reality, we shall consider both limits separately, first of all, from the point of view of a possibility to form a superconducting state. The minimal two-orbital model will be used for an analysis of superconductivity within the weak coupling approximation in Sect. 5.4.

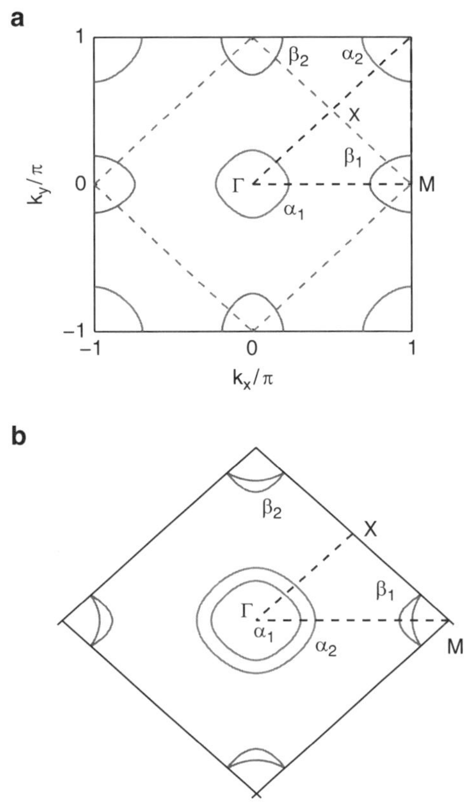

Fig. 5.23 Fermi surface in the two-orbital model with the hopping parameters (5.29), plotted in the extended (**a**) and reduced (**b**) Brillouin zone [404]

5.3.3 Mean Field Approximation

After that we have seen an ability of the two-orbital model to reproduce the main features of the band spectrum and Fermi surface in the FeAs-compounds, we turn to a possibility to describe magnetic ordering. For this, it is necessary to add to the Hamiltonian (5.18), which describes the hopping of electrons over the lattice, the H_{int} term of interaction between the electrons:

$$H_{\text{int}} = U \sum_{i\alpha} n_{i\alpha\uparrow} n_{i\alpha\downarrow} + \left(U' - \frac{J}{2}\right) \sum_{i} n_{ix} n_{iy} + J \sum_{i} c^{+}_{ix\uparrow} c^{+}_{ix\downarrow} c_{iy\downarrow} c_{iy\uparrow} - 2J \sum_{i} \mathbf{S}_{ix} \mathbf{S}_{iy}. \qquad (5.30)$$

In the two-orbital model, two orbitals d_{xz} and d_{yz} are included, which will be numbered by an index $\alpha = x, y$ (not to mix up with conventional x, y, z coordinates). The expression (5.30) is a standard one for the Coulomb interaction and takes into account the rotational symmetry of the latter [405]. The first two terms here take into account the Coulomb interaction of electrons on the same and on two different orbitals, respectively; the third term describes a transition of a pair from one orbital to another one, and the last term – the Hund's exchange.

In [406] which will be outlined in a minute, the $H_{\rm int}$ term is treated in the mean field approximation. For this, the order parameter is introduced, which specifies a spin ordering with the wave vector $\mathbf{q} = (\pi, 0)$:

$$\langle c^{+}_{i\alpha\sigma}\, c_{i'\alpha'\sigma'}\rangle = \left(n_\alpha + \frac{\sigma}{2}\cos\mathbf{q}\,\mathbf{r}_i \cdot m_\alpha\right)\delta_{ii'}\,\delta_{\alpha\alpha'}\,\delta_{\sigma\sigma'}. \quad (5.31)$$

The mean field approximation consists in a replacement of the initial Hamiltonian (5.30) by a quadratic one, $H^{\rm MF}$, whereby a pair of Fermi operators is replaced by the mean value of (5.31). As a result, we get:

$$\begin{aligned} H^{\rm MF} &= -UN\sum_\alpha\left(n_\alpha^2 - \frac{1}{4}m_\alpha^2\right) - 4\left(U' - \frac{J}{2}\right)Nn_xn_y + \frac{1}{2}JNm_xm_y \\ &+ \sum_{\mathbf{k}\sigma}\left(U\,n_x + 2\left(U' - \tfrac{J}{2}\right)n_y\right)c^{+}_{\mathbf{k}x\sigma}\,c_{\mathbf{k}x\sigma} + \sum_{\mathbf{k}\sigma}\left(Un_y + 2\left(U' - \tfrac{J}{2}\right)n_x\right)c^{+}_{\mathbf{k}y\sigma}\,c_{\mathbf{k}y\sigma} \\ &- \frac{1}{4}\sum_{\mathbf{k}\sigma}(U\,m_x + J\,m_y)\left(c^{+}_{\mathbf{k}x\sigma}\,c_{\mathbf{k+q}x\sigma} + c^{+}_{\mathbf{k+q}x\sigma}\,c_{\mathbf{k}x\sigma}\right) \\ &- \frac{1}{4}\sum_{\mathbf{k}\sigma}(U\,m_y + J\,m_x)\left(c^{+}_{\mathbf{k}y\sigma}\,c_{\mathbf{k+q}y\sigma} + c^{+}_{\mathbf{k+q}y\sigma}\,c_{\mathbf{k}y\sigma}\right). \end{aligned} \quad (5.32)$$

The four parameters n_x, n_y, m_x and m_y are determined from the minimization of energy at half-filling of the band states. This leads to the values of the order parameters $n_x = n_y = 1/2$. As concerns the magnetic order parameters m_x and m_y, they are functions of U and J. It follows from the equations of energy minimization that for a fixed J value, two critical values of U exist, U_{c_1} and U_{c_2}. For $U < U_{c_1}$, $m_x = m_y = 0$. In the range $U_{c_1} < U < U_{c_2}$, non-zero solutions of the equations for m exist; simultaneously, a gap opens in some parts of the Brillouin zone, but the Fermi level still crosses the bands, i.e. the system remains a metal. At $U > U_{c_2}$, the equations of minimizations do still have non-zero solutions, but a gap develops throughout on the Fermi level, and the system turns into an insulator.

Therefore, an intermediate regime $U_{c_1} < U < U_{c_2}$ exists, in which a magnetic ordering occurs, and the system is metallic. The phase transition at $U = U_{c_2}$ is a continuous one. The values of U_{c_1} and U_{c_2} do strongly depend on the matrix elements of hopping. If adjusted from fitting of the dispersion curves, as obtained from the model Hamiltonian, to calculated LDA curves, the order of magnitude of U_{c_1} and U_{c_2} makes several eV. However, if they are considered as the Koster–Slater parameters, their estimated values come out by an order of magnitude smaller, even if a qualitative picture of the electron spectrum and the relations $U_{c_1}(U)$, $U_{c_2}(U)$ remain close in both approaches. We outline some results calculated in [406] when using the hopping parameters estimated as the Koster–Slater ones [407].

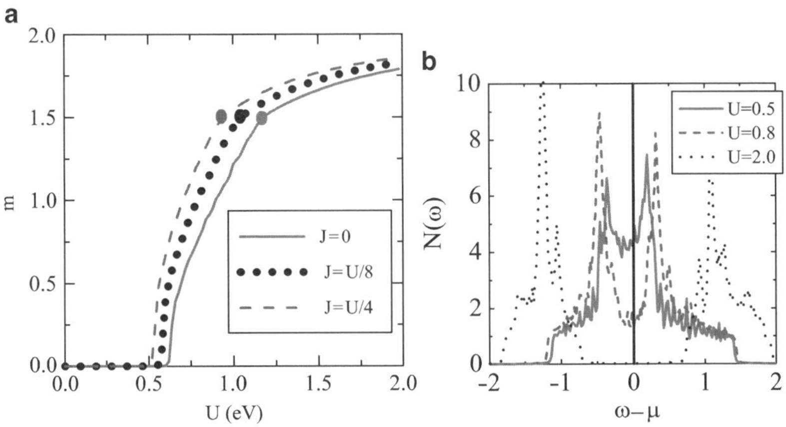

Fig. 5.24 Calculation results for the two-orbital model in the mean-field approximation [406]: (**a**) magnetic order parameter as function of U; (**b**) density of states corresponding to different U values

In Fig. 5.24, the calculated values of the magnetic order parameter $m = m_x + m_y$ and the densities of states are shown for three different regimes. As is seen from Fig. 5.24, the magnitude of the mean spin moment does abruptly change from U_{c_1} ($\simeq 0.5\,\text{eV}$) to U_{c_2} ($\simeq 1\,\text{eV}$). The density of states (Fig. 5.24b), calculated for $J = U/4$ at three values of the Coulomb parameter, reveals three regimes: $U < U_{c_1}$ (solid line), $U_{c_1} < U < U_{c_2}$ (dashed line) and $U > U_{c_2}$ (dotted line), of which the former two correspond to a metallic state, and the later one – to an insulator. In the intermediate state, a pseudogap appears which gradually approaches zero as U goes to U_{c_2}.

Manifestations of the magnetic ordering in the electron spectrum are seen from Fig. 5.25. The (b) panel there of corresponds to the intermediate regime, in which a gap opens between two bands at some fragments of the spectrum, but the Fermi level cuts the dispersion curves. In the (c) panel, an insulating state is seen; in this case, the dispersion curves are split due to an onset of ferromagnetic ordering.

Therefore, the mean field approximation permits to formulate a two-orbital model of the SDW ordering with the wave vector $\mathbf{q} = (\pi, 0)$ in the metallic phase. The magnitude of the order parameter (mean value of spin at each Fe atom) is determined by the strength of the Coulomb potential U, or, more specifically, by the deviation of the latter from its critical value, U_{c_2}. Obviously, the conclusions drawn from the mean field approximation must be verified by other, more accurate methods. The calculations [407], done by exact diagonalization method for small clusters [408], using the cluster variation method [409], do confirm, for the two-orbital model, the above discussed picture of spin ordering in the magnetic phase of FeAs-compounds. This is namely the picture which is observed experimentally.

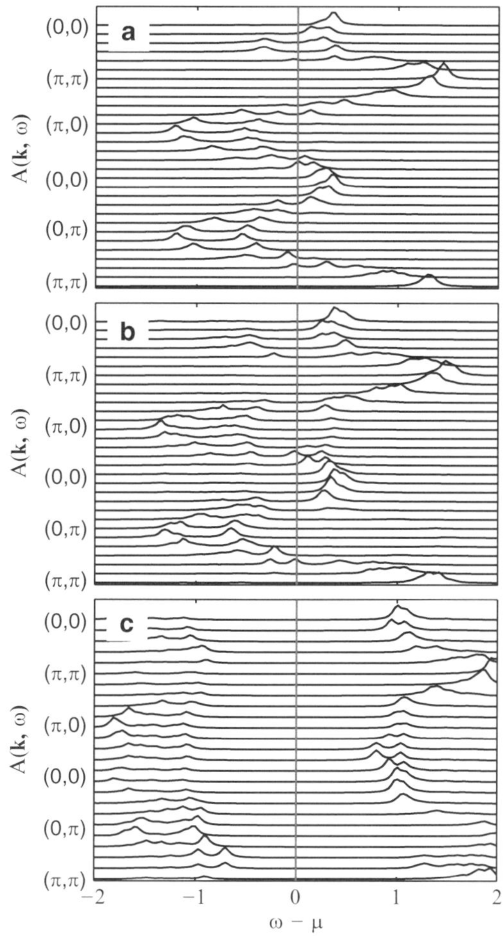

Fig. 5.25 Spectral function $A(\mathbf{k}, \omega)$ of the two-orbital model calculated along high-symmetry directions of the Brillouin zone for $J = U/4$, at three different values of the Coulomb potential: (**a**) $U = 0.5\,\text{eV}$, (**b**) $U = 0.8\,\text{eV}$, (**c**) $U = 2.0\,\text{eV}$ [406]

5.3.4 Numerical Calculation for Small Clusters

The approaches outlined for a description of the main physical properties of FeAs-systems, following from the two-orbital model, exhibit an efficiency of the latter. However, the question of whether it is indeed the minimal model for these compounds, in the sense of being able to provide a detailed description of properties of these materials, needs special studies. This was an ambition of [410], where numerical calculations with a model Hamiltonian have been performed on small clusters, using the Lanczos technique. Such calculations permit to extract information about the model at any values of the electron interaction parameters, without any use of small parameters, an immanent limitation in whatever form of the perturbation theory, including the FLEX calculations.

For such calculations, the two-orbital model is particularly appealing, since the calculation difficulties do rapidly increase with the number of interacting orbitals. The calculations done in [410] on a cluster of the $\sqrt{8} \times \sqrt{8}$ size, including two orbitals only, d_{xz} and d_{yz}, on each centre, indicated that the model does well reproduce the LDA results of undoped compounds of the LaOFeAs type at the values of hopping parameters as obtained by two ways: calculated within the Koster–Slater approach and from fitting the LDA calculations to the tight-binding model with two orbitals. The both approaches yield qualitatively consistent results in what regards the band spectrum and the Fermi surface. Thus, the size of the hole sheet near Γ and of the electron sheets around the X point of the extended Brillouin zone are close to those obtained in LDA calculations of compounds, where all five Fe3d orbitals were used.

The two-orbital model yields a correct magnetic SDW structure, as observed in experiment. Of the strongest interest are the predictions of the two-orbital model concerning the structure of the superconducting order parameter, in particular, the zeros of the order parameter at the Fermi surface. Experiments provide controversial results to this point, therefore it is important to know which are the predictions given by an exactly (numerically) solvable two-orbital model.

In the work cited, a realization of a singlet superconducting order parameter of different symmetries has been analyzed, within the two-orbital model. The complete classification of possible order parameters on the basis of representations of the D_{4h} point group was given in [411].

The most interesting are two order parameters which describe pairing at neighbouring Fe atoms. They are formed by the following operators of the singlet pairing:

$$\widehat{\Delta}_1^+ = \frac{1}{2N}\sum_{ia\mu}\left(c^+_{i\bar{a}\uparrow}\, c^+_{i+\mu a\downarrow} - c^+_{ia\downarrow}\, c^+_{i+\mu\bar{a}\uparrow}\right), \tag{5.33}$$

$$\widehat{\Delta}_2^+ = \frac{1}{2N}\sum_{ia\mu}\left(c^+_{ia\uparrow}\, c^+_{i+\mu a\downarrow} - c^+_{ia\downarrow}\, c^+_{i+\mu a\uparrow}\right). \tag{5.34}$$

Here, the a index numbers the d_{xz} and d_{yz} orbitals (it can be considered, consistently with the above definitions, as $a = x,\ y$); $\mu = \mathbf{x},\ \mathbf{y}$ are vectors connecting two atoms along the directions x and y of the quadratic lattice; N is the number of sites on the lattice (in the cluster chosen).

The $\widehat{\Delta}_1^+$ order parameter accounts for a pairing occurring on different orbitals, and $\widehat{\Delta}_2^+$ – on identical orbitals, situated at neighbouring lattice sites. According to [411], $\widehat{\Delta}_1^+$ transforms according to the B_{2g} irreducible representation, and $\widehat{\Delta}_2^+$ – according to the A_{1g} irreducible representation of the D_{4h} point group. The authors of [410] studied the realization of these two order parameters in the cluster, using the following approach: to the total number of electrons, corresponding to a half occupation and imitating an undoped compound, two more electrons were added which were coupled into pairs according to either (5.33) or (5.34) constructions. Then, the mean values of $\left\langle \widehat{\Delta}_1^+ \right\rangle$ and $\left\langle \widehat{\Delta}_2^+ \right\rangle$ were calculated in the mean field approximation at different values of the Hamiltonian parameters U and J. In this way, the regions of realization of these two order parameters on the (U, J) plane have been determined.

The calculations indicate that the B_{2g}-symmetry of the order parameter is energetically favourable in the region of intermediate values of the Coulomb interaction U, whereas the A_{1g}-symmetry is realized at $U \gtrsim 2.8\,\mathrm{eV}$, when a gap is about to open at the Fermi surface, and the system is at on the verge of becoming an insulator. The two superconducting states considered dominate over other types of singlet superconducting pairing. It is interesting that one of realizations of a singlet order parameter, with A_{1g}-symmetry and corresponding to a pairing of electrons situated at the next-nearest neighbouring Fe atoms, does exactly correspond to the $s^{\pm}$ coupling, in which the dependence of the order parameter on the wave vectors is described by the relation

$$\Delta_{s\pm}(\mathbf{k}) = \Delta_0 \cos k_x \cos k_y, \tag{5.35}$$

first introduced in [264]. However, this pairing is realized out of the physical range of (U, J) parameters. As concerns, the broadly used form of the $s^{\pm}$-coupling in which the signs of the order parameter are opposite on the hole and electron sheets, but $\Delta(\mathbf{k})$ does not depend on the wave vector, the authors of [410] argue that such an idea of a $s^{\pm}$ superconducting order parameter does not quite comply with the symmetry requirements.

Therefore, the calculations done on a small cluster [410] indicated that the two-orbital model does lead to the B_{2g}-symmetry of the superconducting order parameter for the values of U and J being within a realistic region. A detailed analysis has moreover revealed that such order parameter must have zeros on the electron pockets of the Fermi surface. At present, the ARPES data do not allow to draw an unambiguous conclusion about the zeros of the superconducting order parameter. More precise experimental studies must show whether these predictions of the two-orbital model will hold. Such experiments are very important, because the two-orbital model is appealing by its simplicity and a possibility it provides to perform numerical calculations of high accuracy. Should the further experiments enter in contradiction with the two-orbital model, it will be necessary to learn how should the latter be amended, to finally arrive at a conclusion: what is the minimal model for the FeAs-compounds?

In the following, multi-orbital models will be considered, up to the 5-orbital model which includes all degenerate $3d$ states of the Fe ion. Obviously, such a model is formally more adequate to reality; however, its efficiency is limited by the fact that the related numerical calculations become very involved, and it becomes difficult to single out those components of the theory which determine an essential behaviour of real FeAs-systems.

5.4 Multi-Orbital Model

5.4.1 Formulation of the Model

We have considered the two-orbital model of FeAs-systems. The mean field approximation in the spirit of the BCS theory leads to a possibility to realize, in a system, the superconducting states with different symmetries of the order parameter. One of such possibilities to realize is a so-called $s^{\pm}$ symmetry of the order parameter. Its distinguishing feature is that on both hole and electron sheets of the Fermi surface, an isotrope superconducting gap Δ appears, which, however, has opposite signs on these two sheets. If the order parameter is considered as a function of momentum over the Brillouin zone, it changes sign when coming from one sheet onto the other (see Fig. 5.26). Since the sizes of both hole and electron sheets in FeAs compounds are small, the line of zeros of the order parameter of the $s^{\pm}$ symmetry passes beyond the limits of the Fermi surface, and these zeros do not lead to specific dependencies of certain properties of a superconductor, as is, e.g. the case for the $s_{x^2-y^2}$ symmetry of the order parameter.

For the first time, a suggestion that, namely, this order parameter manifests itself in the FeAs-compounds has been put forward in two works [90, 117] independently. It should be pointed out that an idea of the $s^{\pm}$ symmetry of the order parameter has been introduced long before the discovery of superconductivity in FeAs systems, in relation to theoretical description of superconductivity in a number of other systems suggesting a multi-orbital treatment [412–415]. It was demonstrated by the authors of these works that in such a superconductor, the temperature dependence of spin–lattice relaxation rate $1/T_1$ might follow a quasi power-function (and not exponential, as in the BCS theory) law, therefore the experimental results may be understood in a different way than when assuming the d symmetry of the order parameter. In the works by Mazin et al. [90] and moreover Kuroki et al. [117, 416], the authors did not proceed from the minimal two-orbital model, but from a more realistic representation of the electronic structure of the FeAs compounds, namely that the states of the Fe3d electrons are formed by all five orbitals: d_{xz}, d_{yz}, $d_{x^2-y^2}$, $d_{3z^2-r^2}$ and d_{xy}. The band spectrum calculated is shown in Fig. 5.27. The Fermi surface within this five-band model, consisting of two hole pockets α_1 and α_2 around Γ and two electron pockets β_1 and β_2 near the M point, is shown in Fig. 5.28.

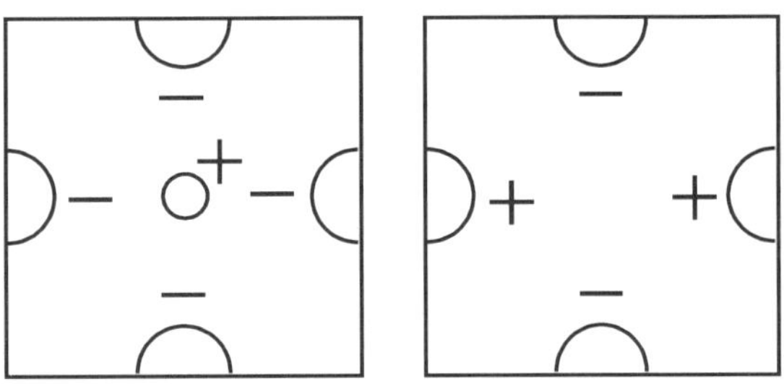

Fig. 5.26 Symmetry of the superconducting order parameter of the (**a**) $s^{\pm}$-type and (**b**) $d_{x^2-y^2}$-type [117]

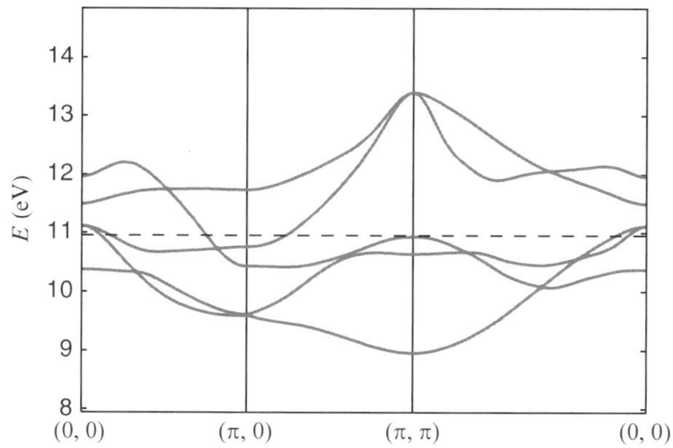

Fig. 5.27 Dispersion curves for the five-band model [117]

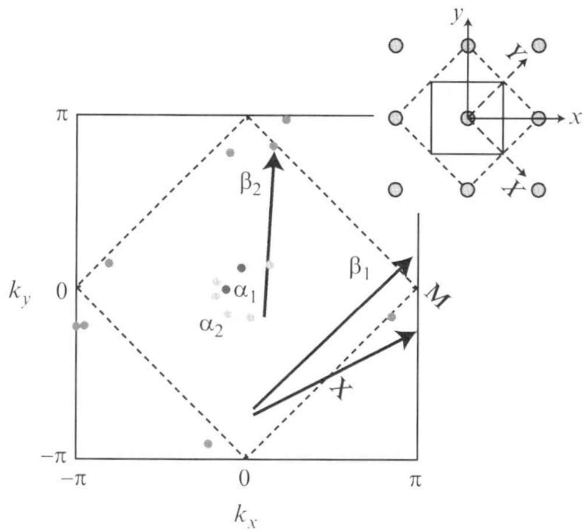

Fig. 5.28 Fermi surface of the $LaO_{1-x}F_xFeAs$ compound, $x = 0.1$, calculated within the model with five d orbitals (without taking into account the interaction between the electrons), after [117]. In the inset, the primitive unit cell with one Fe atom is shown (*solid line*), along with the extended one, with two Fe atoms (*dashed line*). In the main figure, the *solid line* marks the extended Brillouin zone, and the *dashed line* – the reduced one. *Arrows* indicate the nesting vectors [117]

5.4.2 *Equations for a Superconductor in the Fluctuation Exchange (FLEX) Approximation*

We use the Eliashberg approach, in which the retarded interaction of electrons is taken into account via charge and spin fluctuations. This interaction does, of course, come about due to the underlying Coulomb interaction. If the RPA or the FLEX approximation is used, the effective interaction of electrons can be expressed via dynamic susceptibilities of spin and charge.

For the one-band model of superconductivity, the FLEX approximation has been elaborated in [417] and generalized over the multiband superconductor in [415]. We write down the main equations for the electron GF of the multiband model:

$$H = \sum_{ij\sigma} t_{ij}^{ab} c_{iab}^{+} c_{jb\sigma} + U \sum_{ia} n_{ia\uparrow} n_{ia\downarrow}. \tag{5.36}$$

The electron GF, see (5.25), is a matrix over orbital indices a and b. The Dyson equation is a matrix one as well:

$$\{\hat{G}(\mathbf{k}, i\,\omega_n)\}^{-1} = \{\hat{G}^0(\mathbf{k}, i\,\omega_n)\}^{-1} - \hat{\Sigma}(\mathbf{k}, i\,\omega_n). \tag{5.37}$$

In the FLEX approximation, the self-energy is given by the following expression

$$\hat{\Sigma}_{ab}(k, i\,\omega_n) = \frac{T}{N} \sum_q G_{\alpha\beta}(k-q)\, V_{\alpha\beta}^1(q), \tag{5.38}$$

where the fluctuation–exchange interaction is expressed by the formula

$$V_{\alpha\beta}^{(1)}(q) = \frac{3}{2} U^2 \chi_{\alpha\beta}^s(q) + \frac{1}{2} U^2 \chi_{\alpha\beta}^c(q) - U^2 \chi_{\alpha\beta}^{\text{irr}}(q). \tag{5.39}$$

Here,

$$\hat{\chi}^s(q) = \hat{\chi}^{\text{irr}}(q)\{1 - U\,\hat{\chi}^{\text{irr}}(q)\}^{-1}, \tag{5.40}$$

$$\hat{\chi}^c(q) = \hat{\chi}^{\text{irr}}(q)\{1 + U\,\hat{\chi}^{\text{irr}}(q)\}^{-1} \tag{5.41}$$

represent the spin and charge susceptibilities, correspondingly. Both are expressed via their irreducible parts

$$\hat{\chi}_{\alpha\beta}^{irr}(q) = -T \sum_k G_{\alpha\beta}(k+q)\, G_{\alpha\beta}(k). \tag{5.42}$$

In all the above expressions (5.37)–(5.42), k and q are four-component vectors, e.g. $k = (\mathbf{k}, i\,\omega_n)$.

In the superconducting state, the GF contains an anomalous part $\Phi_{\alpha\,\beta}(k)$ and is a matrix of the doubled dimension relative to the GF for a normal state. A linearized equation on $\Phi_{\alpha\,\beta}(k)$ determines the temperature T_c of the superconducting transition. A standard derivation of the Eliashberg equation for the GF of a superconductor leads to the following linearized equation for $\Phi_{\alpha\,\beta}(k)$, for the singlet pairing:

$$\lambda\,\Phi_{\alpha\beta}(k) = T \sum_q \sum_{\alpha'\,\beta'} V_{\alpha'\beta'}^{(2)}(k-q)\, G_{\alpha'\alpha'}(q)\, G_{\beta\beta'}(q)\, \Phi_{\alpha'\beta'}(q), \tag{5.43}$$

where

$$V^{(2)}_{\alpha'\beta'}(q) = \frac{3}{2}U^2\chi^s_{\alpha\beta}(q) - \frac{1}{2}U^2\chi^c_{\alpha\beta}(q) + U \tag{5.44}$$

is the potential of the pairing interaction in the FLEX approximation. The pairing of electrons occurs via spin and charge fluctuations. Comparing (5.39) and (5.44), we note that the susceptibility of charge fluctuations enters the $V^{(1)}$ and $V^{(2)}$ with opposite signs. In case of triplet pairing, these contributions enter $V^{(1)}$ and $V^{(2)}$ with the same sign.

λ in the linear integral equation (5.43) determines the eigenvalues of the $\Phi_{\alpha\beta}(\mathbf{k}, i\,\omega_k)$ matrix. The superconducting transition temperature is defined by the condition: max $\lambda = 1$. Therefore, the problem of determining T_c reduces to a numerical solution of (5.43) and finding out its maximal eigenvalue.

Kuroki et al. [117, 418] performed a numerical solution of (5.43) within the RPA for a pairing potential, i.e. when the calculation of susceptibility is done with the Green's functions in which the self-energy corrections are neglected, and moreover neglected is the contribution due to charge fluctuations in (5.44). For the five-orbital model, the GF $G_{\alpha\beta}$ is a 5×5 matrix. The momentum space (k_x, k_y) was given by the 32 × 32 lattice; the number of Matsubara frequencies ω_n was 1024. The calculations have been done with the parameters which include the Coulomb repulsion between different orbitals, the Hund's exchange J_H, and the J' parameter:

$$U = 1.2\,\text{eV},\ U' = 0.9\,\text{eV},\ J = J' = 0.15\,\text{eV},\ n = 6.1,\ T = 0.02\,\text{eV}. \tag{5.45}$$

A small value of U was intentionally chosen, in view of preventing a dominance of the tendency towards antiferromagnetic ordering, even if such a value of U is definitely much less than it would be appropriate for the *Re*OFeAs compounds. The calculation results are depicted in Fig. 5.29; the spin susceptibility $\chi^s(\mathbf{k}, 0)$ has peaks near the points $(\pi, \pi/2)$, $(\pi/2, \pi)$ and (π, π). This is a manifestation of

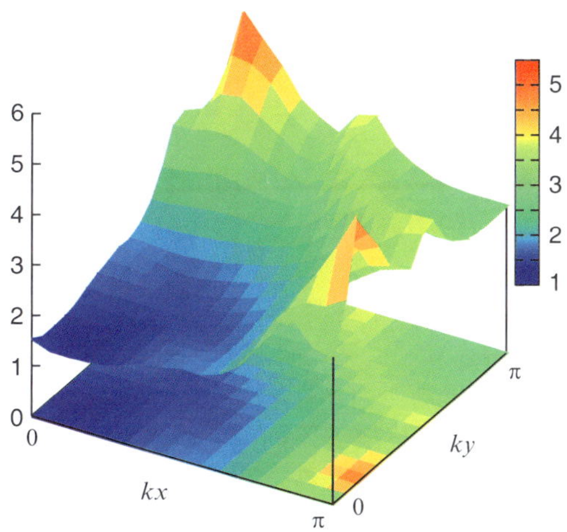

Fig. 5.29 Spin susceptibility within the RPA, for $i\omega_n = 0$, calculated for the parameter values (5.45) of the five-orbital model [117]

the Fermi surface (Fig. 5.28) characterized by the nesting vectors $(0, \pi)$ and $(\pi, 0)$, which connect the hole and electron pockets α and β, and moreover the vectors $(\pi, \pi/2)$ and (π, π), connecting the β_1 and β_2 pockets. This reveals the Fermi surface (Fig. 5.28) in which the nesting vectors $(0, \pi)$ and $(\pi, 0)$, connecting the hole and electron pockets α and β, as well as the $(\pi, \pi/2)$ and (π, π) vectors, connecting the β_1 and β_2 pockets, are present.

The solution of (5.43) for the superconducting order parameter has a symmetry shown in Fig. 5.26a. For the parameter values of (5.45), notably $T = 0.02\,\text{eV}$, this corresponds to the eigenvalue $\lambda = 0.94$. This means that T_c must be lower than 0.02 eV in the temperature units, but the solution should exist. It is remarkable that the hole and electron sheets of the Fermi surface are connected by the nesting vectors $(0, \pi)$ or $(\pi, 0)$, for which the spin susceptibility exhibits peaks. Another possible solution of (5.43) appears at parameter values different from those of (5.177), for example, at higher electron concentration ($n = 6.3$), when the hole pocket disappears and, consequently, the nesting $(0, \pi)$ or $(\pi, 0)$ does not work, but the nesting on the vectors $(\pi, \pi/2)$ and (π, π), connecting the electron sheets β_1 and β_2, still persists. There is namely for this case the solution shown in Fig. 5.26b was obtained. It represents an order parameter of the $d_{x^2-y^2}$-symmetry. The spin susceptibility then has the peaks at the above vectors.

Therefore, the solution outlined of the Eliashberg equation demonstrates the possibility of two different symmetries, $s^{\pm}$ and $d_{x^2-y^2}$, of the superconducting order parameter. Which of those will be realized depends on the interaction parameters. It is assumed that in the doped *Re*OFeAs compounds, the state with the $s^{\pm}$ symmetry will appear, differently from cuprates, where the $d_{x^2-y^2}$ symmetry is realized. Apparently, this is a consequence of the fact that the Coulomb interaction parameter U between the Cu atom in cuprates is higher than that between the Fe atoms in the FeAs-compounds. The $d_{x^2-y^2}$ symmetry is better suited for reducing somehow a repulsion between electrons in a Cooper's pair, even if their attraction via spin fluctuations is shaped by the same Coulomb interaction (see also [419]).

In the following, it will be shown which specific properties follow from the $s^{\pm}$ symmetry of the order parameter, judging by which an existence of this symmetry could have been confirmed for the FeAs-compounds from experimental data.

5.4.3 Properties of Superconductors with the $s^{\pm}$ Symmetry of the Order Parameter

First of all we shall show, on the basis of the BCS model, how exactly does a superconducting state of the $s^{\pm}$ symmetry appears in the case when the Fermi surface has a hole and an electron pockets. We proceed from the Hamiltonian

$$
\begin{aligned}
H = &\sum_{\mathbf{k}a\sigma} \varepsilon_{\mathbf{k}}^{a}\, c_{\mathbf{k}a\sigma}^{+} c_{\mathbf{k}a\sigma} \\
&+ \sum_{\mathbf{k}\mathbf{k}'\mathbf{q}} \sum_{aa'} V_{a,a'} c_{\mathbf{k}+\frac{\mathbf{q}}{2}a\uparrow}^{+} c_{-\mathbf{k}+\frac{\mathbf{q}}{2}a\downarrow}^{+} c_{-\mathbf{k}'+\frac{\mathbf{q}}{2}a'\downarrow} c_{\mathbf{k}'+\frac{\mathbf{q}}{2}a'\uparrow},
\end{aligned} \tag{5.46}
$$

in which $a = (h, e)$ numbers hole and electron pockets, and $V_{aa'}$ is an effective interaction of electrons via spin fluctuations. Its exact form is irrelevant, therefore we drop its dependence on the wave vector. In the mean field approximation, the Hamiltonian (5.46) reduces to the following:

$$H_{\mathrm{MF}} = \sum_{\mathbf{k}a\sigma} \varepsilon_{\mathbf{k}}^{a}\, c_{\mathbf{k}a\sigma}^{+} c_{\mathbf{k}a\sigma} + \sum_{\mathbf{k}a} \left[\Delta_a c_{\mathbf{k}a\uparrow}^{+} c_{-\mathbf{k}a\downarrow}^{+} + h.c. \right]. \tag{5.47}$$

Here, the Δ_h and Δ_e parameters are given by the expressions:

$$\Delta_h = V_{hh} \sum_{\mathbf{k}} \left\langle c_{-\mathbf{k}h\downarrow} c_{\mathbf{k}h\uparrow} \right\rangle + V_{he} \sum_{\mathbf{k}} \left\langle c_{-\mathbf{k}e\downarrow} c_{\mathbf{k}e\uparrow} \right\rangle, \tag{5.48}$$

$$\Delta_e = V_{ee} \sum_{\mathbf{k}} \left\langle c_{-\mathbf{k}e\downarrow} c_{\mathbf{k}e\uparrow} \right\rangle + V_{eh} \sum_{\mathbf{k}} \left\langle c_{-\mathbf{k}h\downarrow} c_{\mathbf{k}h\uparrow} \right\rangle. \tag{5.49}$$

By choosing the Hamiltonian of the form (5.47), we take into account a pairing of electrons situated within the same pocket. Let us consider a simple situation when the interaction within a pair belonging to the same pocket can be neglected, i.e. we assume $V_{hh} = V_{ee} = 0$. Then (5.48) and (5.49) reduce to the following two:

$$\Delta_h = -V_{he} \sum_{\mathbf{k}} \frac{\Delta_e}{\sqrt{\varepsilon_{\mathbf{k}}^2 + |\Delta_e|^2}} \tanh \frac{1}{2T} \sqrt{\varepsilon_{\mathbf{k}}^2 + |\Delta_e|^2}, \tag{5.50}$$

$$\Delta_e = -V_{he} \sum_{\mathbf{k}} \frac{\Delta_h}{\sqrt{\varepsilon_{\mathbf{k}}^2 + |\Delta_h|^2}} \tanh \frac{1}{2T} \sqrt{\varepsilon_{\mathbf{k}}^2 + |\Delta_h|^2}. \tag{5.51}$$

It is known that an effective electron interaction in the Cooper pairs via spin fluctuations has a repulsive character, i.e. $V_{eh} = V_{he} > 0$. In this case, the (5.50) and (5.51) have solutions if the signs of Δ_h and Δ_e are opposite. Let us write down the linearized equations for the Δ_h and Δ_e order parameters:

$$\left.\begin{aligned} |\Delta_h| &= V_{he}\, N_e(0)\, |\Delta_e| \ln\left(\frac{2\gamma\omega_0}{\pi T_c}\right) \\ |\Delta_e| &= V_{he}\, N_h(0)\, |\Delta_h| \ln\left(\frac{2\gamma\omega_0}{\pi T_c}\right) \end{aligned}\right\}. \tag{5.52}$$

From them, an expression follows for the superconducting transition temperature:

$$T_c = \frac{2\gamma\,\omega_0}{\pi} \exp\left(-1/\,V_h \sqrt{N_h(0) N_e(0)}\right). \tag{5.53}$$

Here, ω_0 is the cutoff energy of the pairing interaction, $\gamma = 1.78$.

The formula (5.53) is a generalization of the BCS formula over the case of two-component order parameter [420]. In [420], it has been obtained under an assumption that $V_{he} < 0$, i.e. is of attractive character, in which case the both order parameters have the same sign. A simple calculation outlined above [421] shows that in the case of repulsion ($V_{eh} > 0$), the signs of the order parameter on different sheets of the Fermi surface are opposite, i.e. we deal with a superconducting state possessing the $s^{\pm}$ symmetry of the order parameter.

If considering Δ_h and Δ_e as manifestations of a same function of wave vector over the whole Brillouin zone, a line of zeros of the order parameter, if the latter has the $s^{\pm}$ symmetry, must exist some place between the hole and electron sheets of the Fermi surface. This state has the features of the conventional BCS state. In particular, the solution of the non-linear equation for the gap Δ leads to the same relation $2\Delta = 3.52\,T_c$ as for the isotropic s state in the weak coupling approximation. A presence of the line of zeros between the sheets of the Fermi surface does, however, result in a specific behaviour of some experimentally observable properties of matter, which are easy to confuse them as a manifestation of zeros of the gap function at the Fermi surface, with other symmetries of the order parameter, e.g. of the d or p type.

Between the s and $s^{\pm}$ states, an important difference exists in the case when a superconductor contains non-magnetic impurities. They do not affect T_c in the case of the order parameter having the s symmetry. In the same time, their effect on the $s^{\pm}$-state is the same as that of magnetic impurities in a conventional superconductor, i.e. they suppress T_c. For the $s^{\pm}$ state, the impurity potential $U_{\text{imp}}(\mathbf{q})$ has the intraband component $U_{\text{imp}}(0)$ and the interband one, $U_{\text{imp}}(\mathbf{Q})$. The latter juggles fermions with Δ and $-\Delta$ gaps, i.e. it acts as a magnetic impurity. In [422], an effect of both impurity potentials, $U_{\text{imp}}(0)$ and $U_{\text{imp}}(\mathbf{Q})$, on T_c has been studied. In the case $U_{\text{imp}}(\mathbf{Q})/U_{\text{imp}}(0) \geqslant 1$, the T_c gets suppressed by impurities, but for $U_{\text{imp}}(\mathbf{Q})/U_{\text{imp}}(0) < 1$, T_c practically does not change with the impurity concentration. In FeAs-compounds, according to estimates of [422], $U_{\text{imp}}(\mathbf{Q})/U_{\text{imp}}(0) \simeq 0.5$ that explains a stability of the superconducting state of the $s^{\pm}$ symmetry towards the presence of impurities. The impurity-related scattering does also change the behaviour of magnetic susceptibility in the superconducting phase.

The dynamic susceptibility in the RPA is given by the standard formula:

$$\chi_s(\mathbf{q},\,\Omega) = \frac{\chi_s^0(\mathbf{q},\Omega)}{1-\Gamma_{sdw}^{(r)}\,\chi_s^0(\mathbf{q},\Omega)}. \tag{5.54}$$

The $\chi_s^0(\mathbf{q},\,\Omega)$ property for a $s^{\pm}$ superconductor in the pure limit (without impurities) for $\mathbf{q} \approx \mathbf{Q}$ behaves asymptotically as

$$\left.\frac{\operatorname{Im}\chi_s^0(\mathbf{q},\Omega)}{\Omega}\right|_{\Omega\to 0} \sim (\mathbf{q}-\mathbf{Q})^2, \tag{5.55}$$

i.e. it is small. This smallness results in exponential smallness of the inverse relaxation time $1/T_1$, measurable in NMR experiments, by force of the known relation [423, 424]:

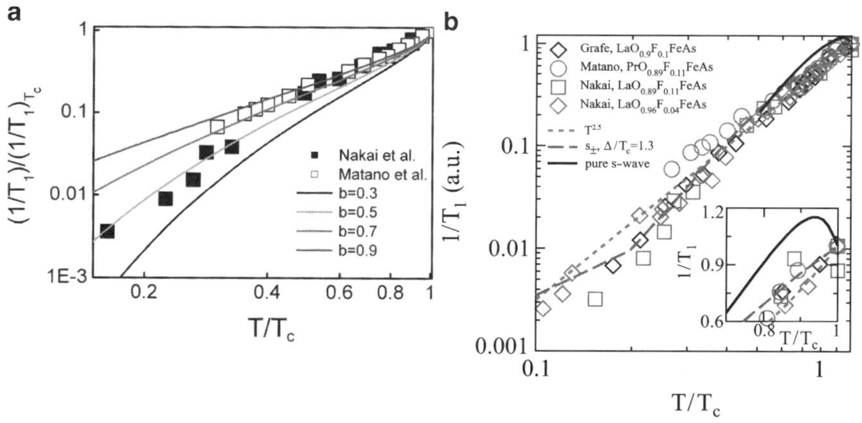

Fig. 5.30 Temperature dependence of $1/T_1$ for a superconductor of the $s^{\pm}$-type with non-magnetic impurities, as calclated in [423] (**a**) and [424] (**b**), in comparison with experimental NMR data

$$1/T_1 \sim T \sum_{\mathbf{q}} \left. \frac{\operatorname{Im} \chi_s(\mathbf{q}, \Omega)}{\Omega} \right|_{\Omega=0} \sim T \sum_{\mathbf{q}} \chi_s^2(\mathbf{q}, 0) \left[\frac{\operatorname{Im} \chi_s^0(\mathbf{q}, \Omega)}{\Omega} \right]_{\Omega=0}. \qquad (5.56)$$

Since $\chi_s^0(\mathbf{q}, 0)$ has maximum values in the vicinity of $\mathbf{q} = \mathbf{Q}$, $1/T_1$, because of the asymptotic (5.55), is exponentially small. However, in the presence of nonmagnetic impurities the expression (5.56) does not anymore exhibit exponential smallness. A numerical calculation shows, rather, a power-low rise of $1/T_1$ with temperature. The power-low exponent $1/T_1 \sim T^{\alpha}$ decreases with the rise of the $b = 2U_{\text{imp}}(\pi)/\Delta$ parameter, i.e. the strength of fermion scattering from one to the other sheet of the Fermi surface. Numerical results for $1/T_1$ along with the experimental data are shown in Fig. 5.30. At low T values, $1/T_1$ is exponentially small, but in the intermediate temperature region $T \lesssim T_c$ is closer to the power-law behaviour with the exponents α between 2 and 3. For superconductors with the d symmetry of the order parameter, $1/T_1 \sim T^3$, due to an existence of a line of zeros at the Fermi surface. The power-law variation of $1/T_1$, observed in the NMR experiments on FeAs-systems, used to be interpreted as a manifestation of the existence of gap zeros on the Fermi surface. However, as was shown in the work cited, similar behaviour can be induced by a presence of impurities in the superconductor of the $s^{\pm}$ type, where two gaps of opposite signs exist on two sheets of the Fermi surface.

We see that the spin–lattice relaxation rate $1/T_1$ in a superconductor with the $s^{\pm}$ symmetry of the order parameter, in the presence of non-magnetic impurities, exhibits a power-law dependence on T, differently from the exponential one, predicted by the BCS theory for a conventional superconductor with the s symmetry of the order parameter. There is a further anomaly in the temperature behaviour of $1/T_1$ in a superconductor of the $s^{\pm}$-type – namely, an absence of the Hebel–Slichter peak in the vicinity of T_c. This is detected in NMR experiments on FeAs-compounds along with the power-law behaviour of $1/T_1$ below T_c.

The Hebel–Slichter peak appears in $1/T_1$ due to the presence, in the expression of the spin susceptibility in a superconductor, of a so-called coherence factor

$$1 + \frac{\Delta_{\mathbf{k}} \Delta_{\mathbf{k}'}}{E_{\mathbf{k}} E_{\mathbf{k}'}},$$

where $\Delta_{\mathbf{k}}$ is the superconducting gap, and $E_{\mathbf{k}} = \sqrt{(\varepsilon_{\mathbf{k}} - \mu)^2 + |\Delta_{\mathbf{k}}|^2}$ is the energy of a quasiparticle state in a superconductor. In case of a conventional superconductor of the s-type, the product $\Delta_{\mathbf{k}} \Delta_{\mathbf{k}'} > 0$, as a result of which on an onset of the superconducting state, an enhancement of $1/T_1$ occurs, and the Hebel–Slichter peak emerges. For a superconductor of the $s^{\pm}$-type, the signs of gaps on the hole and electron sheets of the Fermi surface are opposite, due to which a suppression of the coherence factor follows, and the Hebel–Slichter peak does not appear. An omission of its observation in NMR experiments, on measuring the temperature dependence of spin–lattice relaxation, provides an evidence in favour of the $s^{\pm}$-symmetry of the order parameter.

The both calculations [423, 424] have been performed simultaneously and independently. The minimal model described in the previous section allows, with the help of natural assumptions concerning the parameters of the Hamiltonian, chosen so as to match the LDA calculations of the electronic structure of FeAs compounds, to find the solutions of the equations for superconductor with two gaps of opposite signs on the hole and electron sheets of the Fermi surface, and to provide a novel interpretation of experimental results on the measurements of the nuclear relaxation in the given row of compounds.

We mark in conclusion a number of works [419, 425–427], in which, within the minimal two-band model (see the next section), on the basis of electron pairing via spin fluctuations, also a possibility of the $s^{\pm}$-symmetry of the order parameter has been demonstrated. In these works, the parameter values are given under which the $s^{\pm}$-pairing is more energetically favourable than the d-pairing.

5.4.4 Three-Orbital Model

We outlined the results of the solution of the linearized equation for a superconductor within the RPA, when spin and charge susceptibilities are calculated with the help of bare Green's functions, which do not take inter-electron interactions into account. It seems of great interest to calculate these GF self-consistently, which idea namely makes the essence of FLEX. In [428], such calculations have been performed for a case of three-orbital model. The results obtained allow us to understand the important role played by inter-electron interactions in the formation of quasiparticle spectrum in the normal phase of a metal, in the spin susceptibility, and, finally, in the shaping of the pairing interaction.

The Hamiltonian of the model has to be chosen in the maximally general form, used in calculations of electronic structure of strongly correlated systems. For a multi-orbital model, it has to be written down in the form:

$$H = \sum_{\mathbf{k}} \Psi_{\mathbf{k}}^{+} T_{\mathbf{k}} \Psi_{\mathbf{k}} + H_{\text{int}}, \tag{5.57}$$

where $\Psi_{\mathbf{k}}^{+}$ is a many-component spinor, composed out of the Fermi operators of creation of electrons at the orbitals chosen, which are numbered by the indices a, a':

$$\begin{aligned} H_{\text{int}} = & \frac{1}{2}U \sum_{ia\sigma\neq\sigma'} c_{ia\sigma}^{+} c_{ia\sigma'}^{+} c_{ia\sigma'} c_{ia\sigma} + \frac{1}{2}U' \sum_{ia\neq a'\sigma\sigma'} c_{ia\sigma}^{+} c_{ia'\sigma'}^{+} c_{ia'\sigma'} c_{ia\sigma} \\ & + \frac{1}{2}J \sum_{ia\neq a'\sigma\sigma'} c_{ia\sigma}^{+} c_{ia'\sigma'}^{+} c_{ia\sigma'} c_{ia'\sigma} + \frac{1}{2}J' \sum_{ia\neq a'\sigma\neq\sigma'} c_{ia\sigma}^{+} c_{ia\sigma'}^{+} c_{ia'\sigma'} c_{ia'\sigma}. \end{aligned} \tag{5.58}$$

Here, the first and the second terms represent the Coulomb interaction of electrons at a given site, belonging to the same orbital and to different orbitals, correspondingly; the third term describes the exchange interaction of electrons at the site, and the last term – pair hoppings of electrons from one orbital onto the other.

The FLEX equations (5.38)–(5.42) do in principle remain unchanged, but the matrix structure of interactions becomes more complicated. If in the model which includes L orbitals the electron GF G_{mn} and the self-energy Σ_{mn} are $L\times L$ matrices, the interaction and susceptibility are the matrices of the $L^2 \times L^2$ size, so that the FLEX equations are now written down as follows:

$$\Sigma_{mn}^{(k)} = \frac{T}{N} \sum_{q} \sum_{\mu\nu} V_{n\mu,m\nu}(q)\, G_{\mu\nu}(k-q), \tag{5.59}$$

$$\begin{aligned} V_{n\mu,m\nu}(q) = & \frac{1}{2}\Big[3U^s \chi^s(q) U^s + U^c \chi^c(q) U^c \\ & -\frac{1}{2}(U^s + U^c)\,\chi^0(q)\,(U^s + U^c) + 3U^s - U^c\Big]_{\mu m,n\nu}, \end{aligned} \tag{5.60}$$

$$\chi^s(q) = \Big[1 - \chi^0(q) U^s\Big]^{-1} \chi^0(q), \tag{5.61}$$

$$\chi^c(q) = \Big[1 + \chi^0(q) U^c\Big]^{-1} \chi^0(q), \tag{5.62}$$

$$\chi^0_{\mu m,n\nu}(q) = -\frac{T}{N} \sum_{k} G_{n\mu}(k+q)\, G_{m\nu}(k). \tag{5.63}$$

Here, all indices n, m, ν, μ are of the same standing and run 1 through L. The U^s and U^c matrices have dimensions $L^2 \times L^2$, and their elements depend on the parameters entering the H_{int}: U, U', J, J'.

We can moreover write down the linearized equation for a superconductor, that is a generalization of (5.43) and (5.44):

$$\lambda \, \Phi_{mn}(k) = -\frac{T}{N} \sum_{q} \sum_{\alpha\beta} \sum_{\mu\nu} V^{s}_{\alpha m, n\beta}(q) \cdot G_{\alpha\mu}(k-q) \, G_{\beta\nu}(q-k) \, \Phi_{\mu\nu}(q), \tag{5.64}$$

$$V^{s}(q) = \frac{3}{2} U^{s} \, \chi^{s}(q) U^{s} - \frac{1}{2} U^{c} \, \chi^{c}(q) \, U^{c} + \frac{1}{2} \left(U^{s} + U^{c} \right). \tag{5.65}$$

Here, $V^{s}(q)$ represents pairing interaction in the singlet channel.

In [428], the FLEX equations were solved for the three-orbital model, so that G and Σ are 3×3 matrices, while U^{s} and U^{c} (as well as V, V^{s}, χ^{s}, χ^{c}) are 9×9 matrices. The three-orbital model, comprising the t_{2g}-orbitals (d_{xz}, d_{yz}, d_{xy}) was formulated in [429], where the $T_{\mathbf{k}}$ matrix was expressed via the hopping parameters $t_1, \ldots, t_7$ between nearest and next-nearest Fe atoms, and the transitions between all possible orbitals taken into account. The expressions for the matrix elements U^{s} and U^{c} in terms of the parameters of the Hamiltonian H_{int} are given in [428].

The hopping parameters were chosen from the condition that the three-orbital model (without taking the inter-electron interaction into account) gives the electronic structure as calculated within the LDA [81]. Thus, the following parameter values have been taken:

$$t_1 = -1.0; \; t_2 = 0.7; \; t_3 = -0.8; \; t_4 = -0.3; \; t_5 = 0.2; \; t_6 = 0.6; \; t_7 = -0.35, \tag{5.66}$$

whereas, in the energy units, $t_1 = 0.4\,\text{eV}$. For the number of d-electrons per Fe atom $n = 4$, which corresponds to underdoped three-orbital model, the chemical potential equals $\mu = 1.5$.

The equations for the self-energy Σ and the GF G were solved for the following choice of the interaction parameters. First of all, the symmetry conditions $J' = J$ and $U = U' + 2J$ were taken into account, and the value $U = 3.0\,\text{eV}$ taken. The exchange parameter J remained free and was varied within the limits

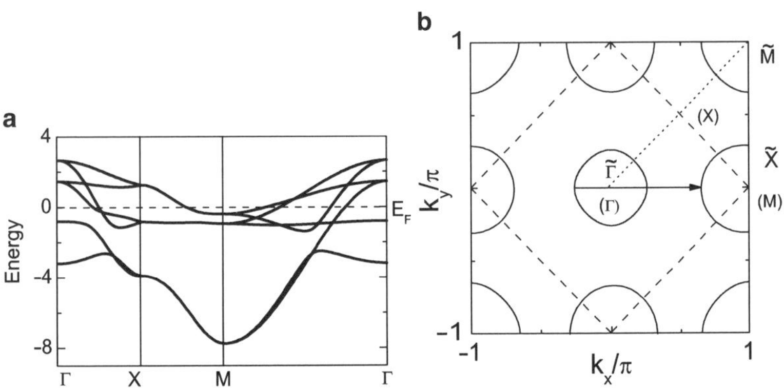

Fig. 5.31 Band structure of non-doped compound in the three-orbital model [428]

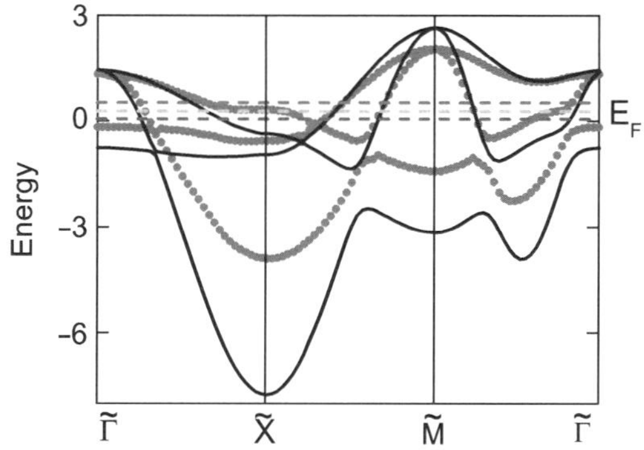

Fig. 5.32 Band structure in the three-orbital model, with electron–electron interaction taken into account. *Solid lines* are bare spectrum of Fig. 5.31, *dashed lines* – the spectrum renormalized by interaction, for the parameter values: $U = 3.0\,\text{eV}$, $J = 0.2\,U$ [428]

$0 \leqslant J \leqslant 0.5$. In Fig. 5.31a, the calculated band structure of underdoped compound with the hopping parameters (5.45) and zero interaction parameters is shown. The Fermi surface corresponding to these bands is depicted in Fig. 5.31b. As it could be expected, there is a hole sheet near Γ and the electron one near the M point. In Fig. 5.31b, the symmetry points are marked in the reduced Brillouin zone (Γ, M, X) and in the extended one ($\widetilde{\Gamma}$, $\widetilde{M}$, $\widetilde{X}$). An arrow connecting two sheets of the Fermi surface indicates the nesting vector $(0,\ \pi)$.

The solutions of self-consistent FLEX equations are shown in Fig. 5.32. The largest renormalization of the spectrum occurs in the vicinity of the $\widetilde{X}$ point $(0,\ \pi)$. This point corresponds to the nesting vector. An expression for the spin susceptibility $\chi^s(\mathbf{q}, 0)$ shows sharp peaks at the wave vectors $\mathbf{q} = (0,\ \pm\pi)$ and $\mathbf{q} = (\pm\pi,\ 0)$, which exactly correspond to the nesting vector. This result shows that a large renormalization of the electron spectrum at $\widetilde{X}$ occurs due to a scattering of electrons on the spin fluctuations with the wave vectors $(\pm\pi,\ 0)$ and $(0,\ \pm\pi)$. The renormalization factor (the relation of the d-band width with the electron interaction taken into account to the width of the trial band) does strongly depend on the magnitude of exchange interaction. It increases from 1.4 to 2.3 as J rises from $0.18\,U$ to $0.5\,U$. At $J \simeq 0.18\,U$, a rapid rise of the renormalization factor starts.

An analysis of the partial weight of the density of states $A(\mathbf{k},\,\omega) = -\frac{1}{\pi}\,\text{Im}\,G(\mathbf{k},\,\omega)$ shows that the d_{yz} and d_{xy}-orbitals dominate in the electron pockets, whereas the d_{xz}-orbitals yield the main contribution in the hole pocket. This shows that the spin fluctuations are mostly formed due to interorbital *particle–hole*-excitations. As is seen from the Hamiltonian H_{int}, the exchange Hund's term takes into account the interorbital intra atomic transitions, therefore the spin fluctuations grow with an increase of J. The peaks of spin susceptibility at the $(\pm\pi,\ 0)$ and $(0,\ \pm\pi)$ do also increase with the rise of J. Such structure of spin fluctuations is confirmed by neutron diffraction experiments.

We note now the three horizontal dashed lines in Fig. 5.32, which indicate the placement of the Fermi level for three cases: an underdoped compound (middle line), with 10% electron doping (upper line) and 20% hole doping (lower line). The corresponding values of the Fermi energy were obtained by solving the FLEX equations self-consistently. It is seen from the figure that the electron doping enlarges the pocket centred near the $\widetilde{X}$ point and shrinks the pocket centred around Γ. On the hole doping, the situation is opposite.

Finally, the solution of (5.64) with the renormalized electron Green's functions yields the following description of superconducting gaps in the **k**-space. The gap function Δ_{hh} is localized at the hole pocket and has a small contribution of the opposite sign in the points $(\pm\pi, 0)$ and $(0, \pm\pi)$. Inversely, the gap function Δ_{ee} is localized on the electron pockets, centred at $(\pm\pi, 0)$ and $(0, \pm\pi)$, and has a weak component, also of the opposite sign, in Γ. This occurs in such a way that the values of main contributions in Δ_{hh} and Δ_{ee} are of opposite signs. Therefore, the FLEX solution for the three-orbital model leads to the $s^{\pm}$-symmetry of the superconducting order parameter.

5.5 Detailed Analysis of the 5-Orbital Model

5.5.1 *The Hamiltonian of the Model*

The five-orbital model incorporating all d-orbitals of the Fe atoms in the LaOFeAs compound, which contribute to the states at the Fermi level, has been introduced in the previous section. Also there, the superconducting states with different symmetries of the order parameter, as they follow from the random phase approximation for spin susceptibility and for the pairing interaction, have been discussed. In view of complexity of the problem and the sensitivity of numerical results with respect to parameters and details of a DFT calculation of the compounds' band structure, a more profound analysis of the model and a more detailed study of the superconducting order parameter seem to be important. In this section, we outline the results of [430], in which the five-orbital model, albeit in combination with a different calculation of the electronic structure, has been used. To provide a more self-contained outline, we re-iterate a mathematical formalism of the RPA in relation to the calculation of the spin and charge susceptibility and pairing interaction. This might be in part redundant, but now the detailed derivation will be given, as it is done in [430]. We start with the Hamiltonian.

The Hamiltonian of the five-orbital model H consists of two terms: H_0 describes the kinetic movement of electrons over the lattice, with transitions from one orbital to another; the H_{int} term describes the interaction of electrons. Let us choose H_0 and H_{int} in the following form (retaining all notations of [430]):

$$H_0 = \sum_{i\sigma}\sum_{s} \varepsilon_s n_{is\sigma} + \sum_{ij\sigma}\sum_{st} t_{ij}^{st}\, c_{is\sigma}^{+}\, c_{jt\sigma}, \tag{5.67}$$

$$\begin{aligned} H_{\text{int}} &= U \sum_{is} n_{is\uparrow} n_{is\downarrow} + \frac{V}{2} \sum_{ist} n_{is} n_{it} \\ &\quad - \frac{J}{2} \sum_{i,s\neq t} \mathbf{S}_{is}\, \mathbf{S}_{it} + \frac{J'}{2} \sum_{i,s\neq t} \sum_{\sigma} c_{is\sigma}^{+}\, c_{is\bar{\sigma}}^{+}\, c_{it\bar{\sigma}}\, c_{it\sigma}. \end{aligned} \tag{5.68}$$

Here, $c^+_{i\,s\,\sigma}$ ($c_{i\,s\,\sigma}$) are creation (annihilation) operators of electrons at site i, in the orbital s of spin σ; $n_{i\,s\,\sigma}$ is the number of electrons in the $|i\;s\;\sigma\rangle$ state; $n_{i\,s} = n_{i\,s\uparrow} + n_{i\,s\downarrow}$ is the full number of electrons on the site i, in the orbital s; $\mathbf{S}_{i\,s}$ is the electron spin operator at site i in the orbital s. The operators of electron number and spin, at a given site, are expressed via creation and annihilation operators via the following relations:

$$n_{is\sigma} = c^+_{is\sigma}\, c_{is\sigma}, \tag{5.69}$$

$$\mathbf{S}_{is} = \frac{1}{2} \sum_{\sigma\sigma'} c^+_{is\sigma}\, \boldsymbol{\sigma}_{\sigma\sigma'}\, c_{is\sigma'}, \tag{5.70}$$

where $\boldsymbol{\sigma}$ is the vector of Pauli matrices.

In $H_{\rm int}$, only the interaction between electrons situated at neighbouring sites are included; they describe the Coulomb repulsion of electrons within the same orbital and between different orbitals, as well as the exchange (Hund's) interaction of electrons in different orbitals, and the pair hopping of electrons from one orbital onto the other one. The Hamiltonian $H = H_0 + H_{\rm int}$ can be considered as a generalization of non-degenerate Hubbard model over the case of several orbitals per atom.

The authors of [430] used the Hamiltonian (5.67)–(5.68) for a construction of a model, which should be able to describe the FeAs-systems. In this model, all five orbitals of the Fe atoms have been included: d_{xz}, d_{yz}, $d_{x^2-y^2}$, d_{xy}, $d_{3z^2-r^2}$. To determine the parameters entering H_0, the results of DFT calculation for LaOFeAs [62] have been used. The Hamiltonian H_0 after the Fourier transformation can be written as follows:

$$H_0 = \sum_{\mathbf{k}\sigma} \sum_{st} \left(\xi_{st}(\mathbf{k}) + \varepsilon_s\, \delta_{st}\right) c^+_{s\sigma}(\mathbf{k})\, c_{t\sigma}(\mathbf{k}), \tag{5.71}$$

where the matrix elements $\xi_{st}(\mathbf{k})$ can be expressed via the hopping parameters for several next neighbours, t^{st}_{ij}. The corresponding expressions for $\xi_{st}(\mathbf{k})$ are derived, for the crystal structure of the LaOFeAs compound, in [430]. They contain a quite large number of unknown parameters to be gained from a comparison of calculated band spectra [62] with those resulting from the diagonalization of the tight-binding Hamiltonian (5.71). The latter diagonalization results are eigenvectors $E_\mu(\mathbf{k})$ of the $[\xi_{st}(\mathbf{k}) + \varepsilon_s\, \delta_{st}]$ matrix; the corresponding eigenvectors are $|\mu\,\mathbf{k}\rangle$. The $E_\mu(\mathbf{k})$ and $|\mu\,\mathbf{k}\rangle$ make a basis, over which, along the perturbation theory, the interaction H_{int} is taken into account. The thus calculated spectrum of the 5-orbital model is depicted in Fig. 5.33. Already from the dispersion curves, it is seen that the Fermi surface consists of two hole pockets (α_1 and α_2), centred at Γ, and two electron pockets (β_1 and β_2), centred near the $(0,\pi)$ and $(\pi,0)$ points of the extended Brillouin zone.

We bring our attention to [431] and [432], in which multiband tight-binding models were used for analysis of electron spectra in different FeAs- systems.

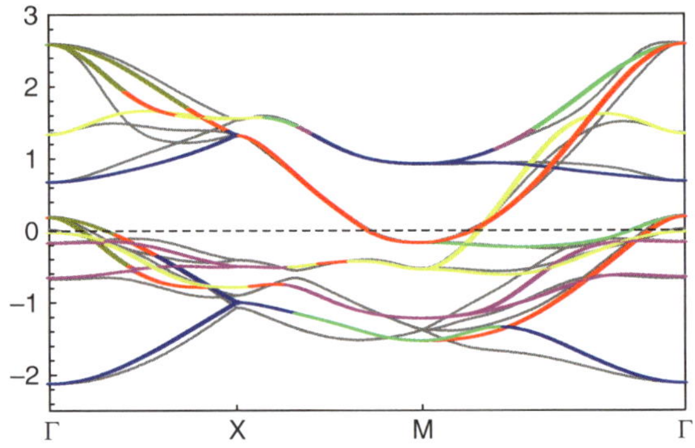

Fig. 5.33 Band spectrum of the five-orbital model for the LaOFeAs compound, from DFT calculations [62] and that obtained by diagonalization of the model Hamiltonian (5.71) [430]

5.5.2 *Spin and Charge Susceptibility*

Let us define the matrix elements of the spin χ_1 and the charge χ_0 Matsubara Green's functions:

$$\left(\chi_1\right)_t^s(\mathbf{q},\, i\,\omega) = \frac{1}{3}\int_0^\beta \mathrm{d}\tau\, e^{i\omega\tau}\,\langle\,\hat{T}_\tau\,\mathbf{S}_s(\mathbf{q},\,\tau)\,\mathbf{S}_t(-\mathbf{q},\,0)\rangle, \tag{5.72}$$

$$\left(\chi_0\right)_t^s(\mathbf{q},\, i\,\omega) = \int_0^\beta \mathrm{d}\tau\, e^{i\omega\tau}\,\langle\,\hat{T}_\tau\,n_s(\mathbf{q},\,\tau)\,n_t(-\mathbf{q},\,0)\rangle. \tag{5.73}$$

of the charge and spin operators, defined by (5.69) and (5.70), correspondingly. They make

$$\mathbf{S}_s(\mathbf{q}) = \frac{1}{2}\sum_{\mathbf{k}}\sum_{\sigma\sigma'} c^+_{s\sigma}(\mathbf{k}+\mathbf{q})\,\boldsymbol{\sigma}_{\sigma\sigma'}\,c_{s\sigma'}(\mathbf{k}), \tag{5.74}$$

$$n_s(\mathbf{q}) = \sum_{\mathbf{k}}\sum_{\sigma} c^+_{s\sigma}(\mathbf{k}+\mathbf{q})\,c_{s\sigma}(\mathbf{k}). \tag{5.75}$$

On calculating (5.72) and (5.73), by allowing an analytical continuation $i\,\omega \to \omega + i\,\delta$ we arrive at dynamical spin and charge susceptibilities.

In the course of constructing the equations for susceptibilities defined by the formulae (5.72) and (5.73), the matrix elements of more general form will be needed:

$$\chi^{pq}_{st}(\mathbf{q},\, i\,\omega) = \int_0^\beta \mathrm{d}\tau\, e^{i\omega\tau}\sum_{\mathbf{k}\,\mathbf{k}'}\sum_{\sigma\sigma'}\langle\hat{T}_\tau\,c^+_{p\sigma}(\mathbf{k},\,\tau)\,c_{q\sigma}(\mathbf{k}+\mathbf{q},\,\tau)\,c^+_{s\sigma'}(\mathbf{k}',\,0)\,c_{t\sigma'}(\mathbf{k}'-\mathbf{q},\,0)\rangle. \tag{5.76}$$

For non-interaction electrons (the H_0 Hamiltonian), the expression (5.76) can be immediately calculated,

$$\chi_{st}^{pq}(\mathbf{q}, i\,\omega) = -\frac{1}{N\,\beta} \sum_{\mathbf{k},\omega_n} G_{sp}(\mathbf{k}, i\,\omega_n)\, G_{qt}(\mathbf{k}+\mathbf{q}, i\,\omega_n + i\,\omega), \tag{5.77}$$

where $G_{sp}(\mathbf{k}, i\,\omega_n)$ is the Fourier image of the electron Green's function:

$$G_{sp}(i\,\tau, j\,0) = -\langle \hat{T}_c\, c_{is\sigma}(\tau)\, c_{jp\sigma}^{+}(0)\rangle\,, \tag{5.78}$$

for which the spectral representation holds:

$$G_{sp}(\mathbf{k}, i\,\omega_n) = \sum_{\mu} \frac{a_{\mu}^{s}(\mathbf{k})\, a_{\mu}^{p*}(\mathbf{k})}{i\,\omega_n - E_{\mu}(\mathbf{k})}. \tag{5.79}$$

Here, $a_{\mu}^{s}(\mathbf{k})$ is the eigenvector of the H_0 Hamiltonian, corresponding to the eigenvalue $E_{\mu}(\mathbf{k})$. On substituting this expression into (5.77) and performing summation over Matsubara frequencies ω_n, we arrive at an expression for the non-interacting susceptibility (for which we define in the identical way its analytical continuation, $i\,\omega_n \to \omega + i\,\delta$):

$$\chi_{st}^{pq}(\mathbf{q}, \omega) = -\frac{1}{N} \sum_{\mathbf{k}\mu\nu} \frac{a_{\mu}^{s}(\mathbf{k})\, a_{\mu}^{p*}(\mathbf{k})\, a_{\nu}^{q}(\mathbf{k}+\mathbf{q})\, a_{\nu}^{t\,*}(\mathbf{k}+\mathbf{q})}{\omega + E_{\nu}(\mathbf{k}+\mathbf{q}) - E_{\mu}(\mathbf{k}) + i\,\delta} \Big[f\big(E_{\nu}(\mathbf{k}+\mathbf{q})\big) - f\big(E_{\mu}(\mathbf{k})\big) \Big]. \tag{5.80}$$

Since the eigenvalues and eigenfunctions of the Hamiltonian of five-orbital model (5.71) are known (may be calculated), the dynamical susceptibilities in the zeroth approximation may as well be considered as known. The problem is how to take into account the interaction of electrons. In the RPA, for the charge susceptibility χ_0^{RPA} and the spin susceptibility χ_1^{RPA}, the following standard equations hold:

$$\left(\chi_0^{\mathrm{RPA}}\right)_{st}^{pq} = \chi_{st}^{pq} - \left(\chi_0^{\mathrm{RPA}}\right)_{uv}^{pq} \left(U^{c}\right)_{wz}^{uv} \chi_{st}^{wz}, \tag{5.81}$$

$$\left(\chi_1^{\mathrm{RPA}}\right)_{st}^{pq} = \chi_{st}^{qq} + \left(\chi_1^{\mathrm{RPA}}\right)_{uv}^{pq} \left(U^{s}\right)_{wz}^{uv} \chi_{st}^{wz} \tag{5.82}$$

(assuming summation over repeated indices). Therefore, in the zeroth approximation the charge and spin susceptibilities do coincide. We note that (5.81) and (5.82) are equivalent to (5.40) and (5.41) of the previous section; however, in the expressions for the matrix elements U^c and U^s the contributions from all those interactions are included which enter the Hamiltonian H_{int}.

Non-zero matrix elements for the charge and spin potential are (a, b numbering different orbitals):

$$\left(U^{c}\right)_{aa}^{aa} = U,\ \left(U^{c}\right)_{bb}^{aa} = 2V,\ \left(U^{c}\right)_{ab}^{ab} = \frac{3}{4}J - V,\ \left(U^{c}\right)_{ab}^{ba} = J'; \tag{5.83}$$

$$\left(U^{s}\right)_{aa}^{aa} = U,\ \left(U^{s}\right)_{bb}^{aa} = \frac{1}{2}V,\ \left(U^{s}\right)_{ab}^{ab} = \frac{1}{4}J + V,\ \left(U^{s}\right)_{ab}^{ba} = J'. \tag{5.84}$$

Therefore, the problem of calculating the dynamical susceptibilities reduces to a solution of the systems of matrix equations (5.81) and (5.82) with the coefficients determined by the expressions (5.77) and (5.79). The static spin susceptibility is determined by a sum of contributions from different orbitals:

$$\chi_s(\mathbf{q}) = \frac{1}{2}\sum_{sp}(\chi_1)_{ss}^{pp}(\mathbf{q}, 0), \tag{5.85}$$

and the homogeneous static susceptibility, $\chi_0 \equiv 2\,\chi_s(\mathbf{q}=0)$, at $T = 0$, according to the relation found, equals

$$\chi_0 = 2\sum_{\nu} N_\nu(0). \tag{5.86}$$

In Fig. 5.34, an example of calculation results for static spin susceptibility, performed along the formula (5.85), is shown, along with the charge susceptibility. It is seen from this figure that the charge susceptibility is by an order of magnitude smaller than the spin one, and moreover does not have any singularities throughout the Brillouin zone. In contrast, the spin susceptibility has peaks, of which the most prominent one is situated in the vicinity of the point $\mathbf{Q} = (\pi, 0)$ – or, correspondingly, $(0, \pi)$; most precisely, the peak falls onto the wave vector $\mathbf{Q}^* = (\pi, 0.16)$. For the undoped compound, the peak is exactly at the wave vector $\mathbf{Q}$.

In Fig. 5.35, the $\chi_s(\mathbf{q}, 0)$ dependence on the magnitude of the $U = V$ parameter for two different wave vectors, $\mathbf{Q}$ and $\mathbf{Q}^*$, is shown. We see a moderate and almost

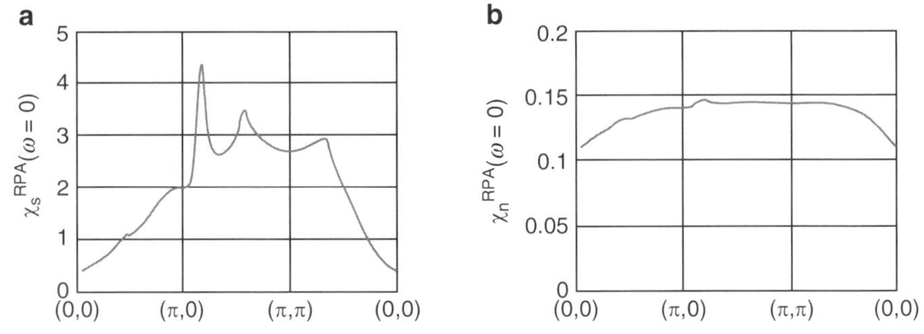

Fig. 5.34 Spin (**a**) and charge (**b**) susceptibilities, calculated for electron-doped ($x = 0.125$) LaOFeAs compound, at the parameter values: $U = V = 1.65\,\text{eV}$ and $J = 0$ for spin and charge susceptibilities [430]

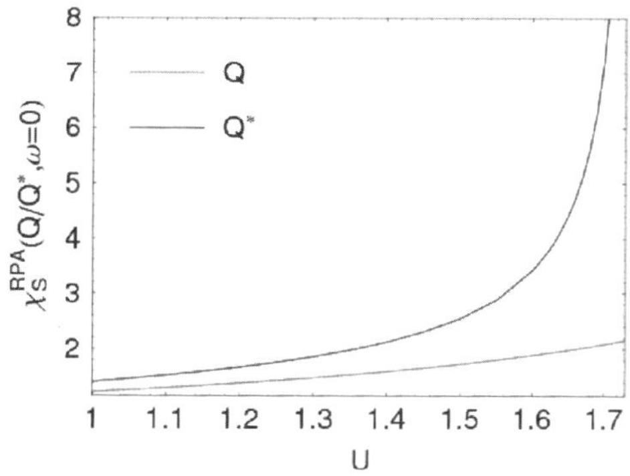

Fig. 5.35 Dependence of spin susceptibility $\chi_s(\mathbf{q}, 0)$ on the $U = V$ parameter for two wave vectors in the electron doped ($x = 0.125$) LaOFeAs compound [430]

uniform increase of $\chi_s(\mathbf{Q}, 0)$, whereas $\chi_s(\mathbf{Q}^*, 0)$ diverges, as a critical value of U is approached.

This calculated behaviour of the spin susceptibility agrees with experimental results. For instance, this relates to the temperature dependence of $\chi_s(\mathbf{q}, 0)$, which exhibits a quasi-linear augmentation with the rise of T within the range 100–500 K, according to direct measurements of magnetization and NMR studies of compounds [79, 98, 101, 433], in a qualitative agreement with the calculations [430]. With the same interaction parameters, the spin–lattice relaxation time T_1 has been calculated, and the $1/T_1 T$ property also turned out to be in a qualitative agreement with the NMR experimental data [79, 98, 101].

5.5.3 *Pairing of Electrons via Spin Fluctuations*

The above described calculations of susceptibilities in the paramagnetic phase of a metal demonstrate that the charge susceptibility is inferior to the spin one by an order of magnitude and does not contain peaks at any points of the Brillouin zone, whereas the spin susceptibility, at certain choices of interaction parameters, exhibits sharp peaks, in particular, at the $\mathbf{Q} = (0, \pi)$ vector, which is the wave vector of the SDW structure. By virtue of these observations, these are primarily the spin fluctuations which might be responsible for the electron pairing in the FeAs-systems.

In the multi-orbital model, the singlet pairing via spin and charge fluctuations is described by the vertex part [415] of the RPA:

$$\Gamma_{st}^{pq}(\mathbf{k}, \mathbf{k}', \omega) = \left[\frac{3}{4} U^s \chi_1^{\text{RPA}}(\mathbf{k}-\mathbf{k}', \omega) U^s + \frac{1}{2} U^s - \frac{1}{2} U^c \chi_0^{\text{RPA}}(\mathbf{k}-\mathbf{k}', \omega) U^c + \frac{1}{2} U^c\right]_{ps}^{tq}. \tag{5.87}$$

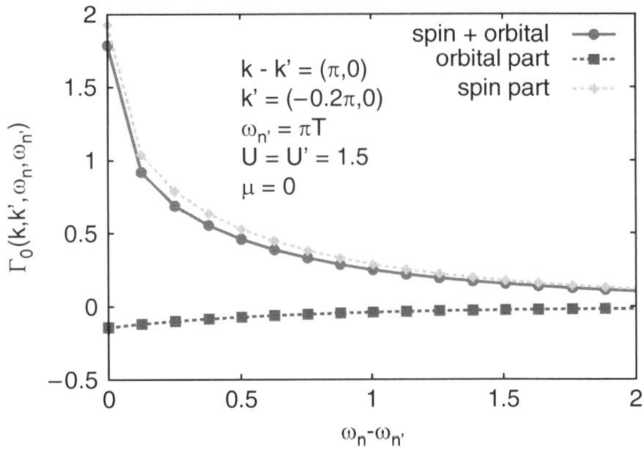

Fig. 5.36 Frequency dependence of the spin and charge contributions into the pairing interaction $\Gamma(\mathbf{k}, \mathbf{k}', \omega_n, \omega_{n'})$ [430]

This equation, up to the factor of $1/2$, is equivalent to (5.44), taking into account a different definition of matrix elements U^s and U^c, which now comprise all the interactions within the H_{int} Hamiltonian. In Fig. 5.36, the frequency dependence of the pair interaction is shown for the transfer wave vector $\mathbf{k} - \mathbf{k}' = (0, \pi)$ at typical magnitudes of the interaction parameters. We see that at small Matsubara frequencies, the dominating contribution comes from spin fluctuations in the vicinity of the wave vector $\mathbf{Q}$.

If limiting ourselves by wave vectors $\mathbf{k}$ and $\mathbf{k}'$ close to the Fermi surface, we can estimate the scattering amplitude of a Cooper pair $(\mathbf{k}, -\mathbf{k})$ on the C_i sheet of the Fermi surface into a state $(\mathbf{k}', -\mathbf{k}')$ on the sheet C_j as:

$$\Gamma_{ij}(\mathbf{k}, \mathbf{k}') = \sum_{stpq} a_{\nu_i}^{t*}(-\mathbf{k})\, a_{\nu_i}^{s*}(\mathbf{k})\, \mathrm{Re}\, \Gamma_{st}^{pq}(\mathbf{k}, \mathbf{k}', 0)\, a_{\nu_j}^{p}(\mathbf{k}')\, a_{\nu_j}^{q}(-\mathbf{k}'), \qquad (5.88)$$

where $\mathbf{k}$ and $\mathbf{k}'$ belong to the corresponding sheets of the Fermi surface: $\mathbf{k} \in C_i$, $\mathbf{k}' \in C_j$.

We introduce now an energy gap $\Delta(\mathbf{k})$, to be expressed as $\Delta(\mathbf{k}) = \Delta\, g(\mathbf{k})$, where $g(\mathbf{k})$ describes the symmetry of the superconducting order parameter. We can introduce a certain functional $\lambda\big[g(\mathbf{k})\big]$, which would define the strength of the pairing interaction:

$$\lambda\big[g(\mathbf{k})\big] = -\frac{\sum_{ij} \oint_{C_i} \frac{\mathrm{d}k_\parallel}{v_{\mathrm{F}}(\mathbf{k})} \oint_{C_j} \frac{\mathrm{d}k'_\parallel}{v_{\mathrm{F}}(\mathbf{k}')}\, g(\mathbf{k})\, \Gamma ij(\mathbf{k}, \mathbf{k}')\, g(\mathbf{k}')}{(2\pi)^2 \sum_i \oint_{C_i} \frac{\mathrm{d}k_\parallel}{v_{\mathrm{F}}(\mathbf{k})} \big[g(\mathbf{k})\big]^2}. \qquad (5.89)$$

Here, the integration is done over a closed contour which represents a sheet of the Fermi surface in the (k_x, k_y) plane; $v_{\mathrm{F}}(\mathbf{k})$ is the electron velocity on this sheet.

Proceeding from the stationarity condition, we arrive at the following eigenvalue problem for the $\Gamma_{ij}(\mathbf{k}, \mathbf{k}')$ amplitude averaged over the Fermi surface:

$$-\sum_j \oint_{C_j} \frac{dk'_\parallel}{2\pi} \frac{1}{2\pi v_F(k')} \Gamma_{ij}(\mathbf{k}, \mathbf{k}')\, g_\alpha(\mathbf{k}') = \lambda\, g_\alpha(\mathbf{k}). \qquad (5.90)$$

The largest eigenvalue λ_α will define the temperature of the superconducting transition, and its eigenfunction $g_\alpha(\mathbf{k})$ – the symmetry of the order parameter. This is the leading instability of the normal phase of a system. The following one (the next largest value of λ_α) would define the further instability, etc. We note in conclusion that (5.90) is equivalent to (5.43) of the previous section.

In an exhaustive work [434], the Eliashberg equations were solved within the 5-orbital model for the NdOFeAs compound, at different levels of the electron doping, when the number of electrons per Fe atom is $n > 6$. One of the main objectives of this work was to establish a relation between the strength of the electron pairing and the position of the As atoms relative to the Fe-planes. This position is characterized by the z_{As} coordinate, or by the "height" h_{As}. Depending of the h_{As} parameter, the topology of the Fermi surface changes. Along with the traditional α_1 and α_2 hole cylinders at the centre of the Brillouin zone and the electron β_1, β_2 cylinders at its corners, additional pockets γ may appear around the (π, π) point of the extended Brillouin zone. An additional β–γ nesting, along with the conventional α–β one, affects the structure of the spin susceptibility in the (k_x, k_y)-space, and through it the strength of the electron pairing.

In Fig. 5.37, one of the numerous results of [434] is shown, a dependence of the eigenvalue λ of the linearised Eliashberg equation (a measure of electron pairing) on the "height" of the As atom in a compound with electron concentration $n = 6.1$. We see that the $s^\pm$-pairing dominates over the d-pairing, and the difference between them increases as the h_{As} grows. As low h_{As} values, the s and d-pairings are in competition. The h_{As} value does, moreover, determine a possibility of the existence of zeros at the Fermi surface. At large h_{As}, the gap has no zeros, but they appear at small h_{As}. This tendency is maintained for other compounds of the 1111 type. Judging by the experimental h_{As} values, it was predicted that the (doped)

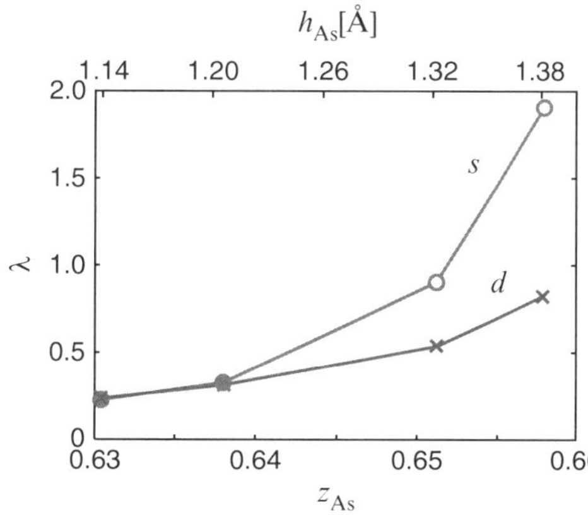

Fig. 5.37 Dependence of the parameter of electron–electron pairing in the NdOFeAs compound with $n = 6.1$ on the "height" of the As atom for two types of pairing, $s^\pm$ and d. For s-pairing, the *open circles* correspond to the states without zeros on the Fermi surface, and *shaded ones* – to those with zeros of the gap [434]

NdOFeAs compound with high T_c does not have gap zeros at the Fermi surface, whereas LaOFeAs with low T_c should have them, that is apparently in agreement with experiment.

We note, moreover, the attempts to describe the $s^{\pm}$ state in FeAs-compounds within the 3-orbital model [435–437].

5.5.4 Possible Symmetries of the Superconducting Order Parameter

From the solution of (5.90), we find the eigenfunctions $g_\alpha(\mathbf{k})$, which describe the $\mathbf{k}$-dependence of the gap on the Fermi surface, and the eigenvalues λ_α, yielding a dimensionless parameter of the corresponding pairing. Different symmetries of the order parameter are characterized by the following conditions on the $\lambda_\alpha(\mathbf{k})$ function.
s-symmetry:

$$\begin{aligned} g(-k_x, k_y) = g(k_x, -k_y) &= g(k_x, k_y), \\ g(k_y, k_x) &= g(k_x, k_y); \end{aligned} \tag{5.91}$$

$d_{x^2-y^2}$-symmetry:

$$\begin{aligned} g(-k_x, k_y) = g(k_x, -k_y) &= g(k_x, k_y), \\ g(k_y, k_x) &= -g(k_x, k_y); \end{aligned} \tag{5.92}$$

d_{xy}-symmetry:

$$\begin{aligned} g(-k_x, k_y) = g(k_x, -k_y) &= -g(k_x, k_y), \\ g(k_y, k_x) &= g(k_x, k_y); \end{aligned} \tag{5.93}$$

g-symmetry:

$$\begin{aligned} g(-k_x, k_y) = g(k_x, -k_y) &= -g(k_x, k_y), \\ g(k_y, k_x) &= -g(k_x, k_y). \end{aligned} \tag{5.94}$$

Within each symmetry, the states should be distinguished according to how the sign of the order parameter changes when coming from the α- to the β-sheet and from β_1 to β_2. If the signs of the order parameter are opposite on the α_1 and β_1 sheets, one speaks of the $s^{\pm}$-symmetry. As concerns the $d_{x^2-y^2}$-symmetry, it is called $d_{x^2-y^2}(1)$ when the signs of the order parameter are different on the α_1 and α_2 sheets, and $d_{x^2-y^2}(2)$ when these signs are equal.

In Fig. 5.38, the eigenvalues λ_α are depicted as functions of the U parameter for doped and undoped compounds. In both cases, the eigenvalues λ_α for all symmetries do sharply increase near the critical value $U \simeq 1.73$ for the doped case and $U \simeq 1.54$ for the undoped one. A remarkable fact is that the $\lambda_\alpha(U)$ curves for the

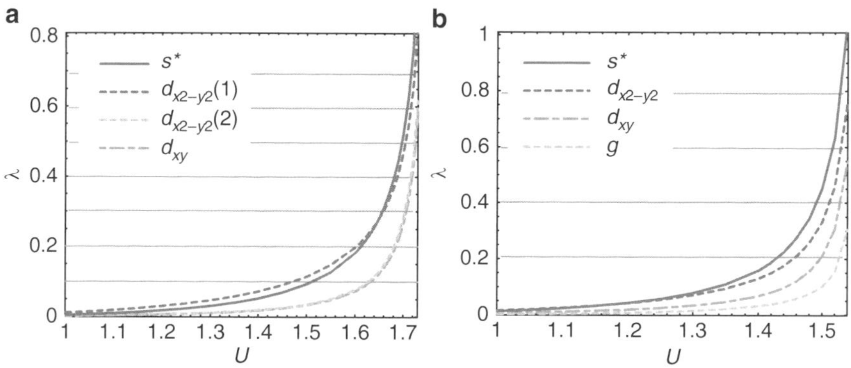

Fig. 5.38 Eigenvalues λ_α, calculated for (**a**) doped case ($x = 0.125$) and (**b**) the non-doped one, as functions of U at the following values of the other interaction parameters: $U = V, J = J' = 0$ [430]. The s^* symbol denotes an extended s-symmetry of the order parameter on all sheets of the Fermi surface

$s^\pm$- and $d_{x^2-y^2}$-symmetries pass substantially higher than those for all other symmetries. Therefore, the leading instabilities of the normal phase of LaOFeAs are the pairings of the $s^\pm$ and the $d_{x^2-y^2}$ types. The differences between $\lambda_\alpha(U)$ for these symmetries are very sensitive to the values of interaction parameters, so that the leading instability may turn to be either the $s^\pm$-, or the $d_{x^2-y^2}$-pairing.

Within a given symmetry of the order parameter, the behaviour of the eigenfunctions $g_\alpha(\mathbf{k})$ on different sheets of the Fermi surface may happen to be quite complex. As an example, we reproduce in Fig. 5.39 the results of a certain calculation. We see that on the inner α_1 sheet, the order parameter is anisotropic and does not have zeros. On the α_2 sheet, the zeros of the order parameter are present. On the β_1 and β_2 sheets, the order parameter is anisotropic and has different signs along different arcs of a contour. Similarly, complex distributions of the order parameter on different sheets of the Fermi surface take place in case of the $d_{x^2-y^2}$-symmetry [430]. Therefore, in case of the s^*-symmetry the order parameter changes sign between the hole and electron sheets of the Fermi surface, but also exhibits zeros on electron sheets, whereas the $d_{x^2-y^2}$ order parameter has zeros on the hole sheets. This distinguishes the five-orbital model from the minimal two-orbital model. It should be noted that the zeros at the Fermi surface itself, in case of the $s^\pm$-symmetry, are "accidental" and not enforced by symmetry. They may disappear under a different choice of the interaction parameters; however, the region where they exist is quite large.

The authors of [430] note that a closeness of magnitude of the pairing interaction for the $s^\pm$ and $d_{x^2-y^2}$ symmetries holds over a broad region of the interaction parameters' variation. A degeneracy of these states is typical for a situation when the radius of the Fermi surface for all four sheets is identical. Over different FeAs-systems, a certain scattering exists in these radii, that is why the $s^\pm$ and $d_{x^2-y^2}$-symmetries are competing. In view of these results, it does not seem

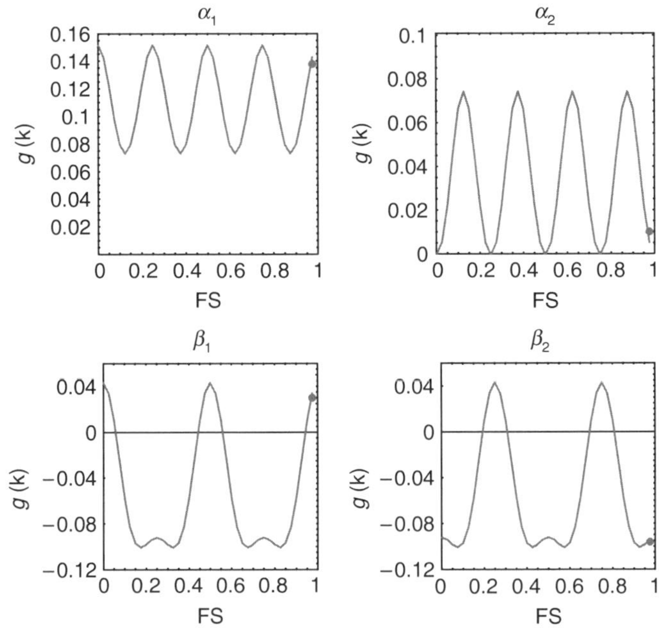

Fig. 5.39 The $g(\mathbf{k})$ function for the s^*-symmetry of the order parameter on four sheets of the Fermi surface, calculated for the following parameter values: $U = 1.5$, $J = J' = 0$, $x = 0$ [430]. Along the abscissa axis, the fraction of a full rotation along the contour of the Fermi surface, always staying on a single sheet chosen, is given [430]

surprising that experimental studies of the superconducting order parameter in different FeAs systems lead to conflicting conclusions about an existence or absence of zeros of the order parameter at the Fermi surface.

In conclusion, we discuss a relation of the results of the five-orbital model outlined here with other works done along this line. A suggestion about an existence in the FeAs-systems of the $s^{\pm}$-symmetry of the order parameter, with the change of its sign on passing from the hole to the electron sheet of the Fermi surface due to antiferromagnetic spin fluctuations, was first put forward by Mazin et al. [90]. Further on, Kuruoki et al. [117] addressed this issue in their study of the five-orbital model, parametrized on the basis of DFT calculations, within the RPA. In the parameter range close to those of [430], they found that the leading instability was that of the $s^{\pm}$-symmetry, with zeros at the electron sheets of the Fermi surface. They moreover established that the next instability would be that of the $d_{x^2-y^2}$-symmetry, and found the conditions under which this instability turns into the leading one. The difference between the results of [430] and [117] is in relative signs of the order parameter at hole and electron sheets in the case of the $s^{\pm}$-symmetry of the order parameter.

In another work [438], using the same parametrization of the 5-orbital model as Kuruoki et al. [117], it was shown that the leading instability is that of the

$s^{\pm}$-symmetry, and the next one – the $d_{x^2-y^2}$-symmetry of the order parameter. For their chosen values of the interaction parameters, the authors of [438] did not find zeros at the Fermi surface, even if large variations of the gap at the Fermi surface occured. They noticed that at a different choice of parameters, the order parameter at the electron sheet, in some points closest to Γ, may have the same sign as on the hole sheet.

All studies [117, 430, 438] of the five-orbital model agree in the main finding, namely that the spin fluctuations in the vicinity of the antiferromagnetic wave vector $\mathbf{Q}$ lead to an onset of superconducting pairing of either $s^{\pm}$- or $d_{x^2-y^2}$-symmetry. Which of these states would realize depends on the choice of interaction parameters and on the details of the band structure as calculated in the DFT. Similarly, sensitive to such technicalities are the details of the realization of different order parameter at different sheets of the Fermi surface. As yet, there is no unambiguous conclusion from the experiment concerning the symmetry of the order parameter. With all evidence, it must be of either $s^{\pm}$- or $d_{x^2-y^2}$-type.

Recently, a series of works has appeared [439–441], in which a possibility to determine coupled Andreev states in the vicinity of an impurity within the superconducting gap has been analyzed, with the aim to determine, on their basis, the symmetry of the superconducting order parameter. The Bogolyubov – De Gennes equations have been solved for a superconductor with the $d_{x^2y^2}$ and $d_{x^2-y^2}$-symmetry of the superconducting order parameter, in the presence of an isolated impurity. The data obtained in [439–441] are in agreement, therefore we discuss the results obtained in [441].

In [441], the states in the gap of a superconductor with the $s^{\pm}$ and s-symmetries of the order parameter, in the presence of a non-magnetic impurity, have been studied. In case of $s^{\pm}$ symmetry, two bound states, symmetrically situated relative to the Fermi level, are formed in the gap. In the case of conventional s symmetry of the order parameter, such bound states do not appear. For a magnetic impurity, only one bound state, characterized by a certain spin polarization, is formed. On increasing a magnitude of the electron scattering on the impurity, the system undergoes a phase transition from a non-spin-polarized ground state into a spin-polarized one. However, the results are too qualitatively close for $s^{\pm}$ and s-superconductors to justify a reliable judgment about the symmetry of the superconducting order parameter of the initial superconductor from the measurements of the impurity-related bound states.

Close to these works the [442] falls, according to which the bound states in the gap appear inside the fluxoides in the mixed phase of a superconductor. In case of a homogeneous superconductor without impurities, the $d_{x^2-y^2}$ state is more energetically favourable for pairing between electrons at the nearest-neighbouring Fe atoms than the $d_{x^2y^2}$ state is. This is in agreement with the earlier results [443]. Inside the fluxoid core of a $d_{x^2-y^2}$ superconductor, a bound state at the Fermi level is formed, whereas in the $d_{x^2y^2}$-superconductor, only a resonance state takes place. Therefore, there are differences in the character of bound states at the impurity or inside the fluxoid core, depending on $s^{\pm}$ or other symmetry of the superconductor, but it is difficult to identify the symmetry of the order parameter judging by them only. These studies need to be continued along both theoretical and experimental lines.

We mention moreover a fundamental work [444], in which, by means of the renormalization group functional method, a relation between the superconducting, antiferromagnetic and orbital order parameters in FeAs-compounds is analyzed, in comparison of the latter with cuprates. A comparison of the physics of the FeAs-systems and cuprates has been done in a fundamental [445] using the group theory methods. An analysis of similarities and differences between these two classes of high-temperature superconductors led the authors of [445] to a conclusion that they can be described from the point of view of symmetry, within an unified theory on the basis of the $SU(4)$ Lie group. The operators describing possible order parameters in the system with $SU(4)$ symmetry of interactions, form a non-Abelian algebra, whose structure determines the relation between the magnetic state of a system and a superconductivity. It is possible that this approach will get further development in the description of the FeAs systems and would be able to predict those symmetries of superconducting order parameter which are compatible with an onset of magnetic ordering in these systems.

5.6 Limit of Weak Coulomb Interaction

5.6.1 Renormalization Group Analysis

In two previous sections, we described the 5-orbital model and outlined the results of the study of spin susceptibility and pairing interaction in the FeAs-systems. The interaction between electrons was hereby taken into account by perturbation theory within the RPA. It is of interest to account for interactions in a more accurate way, namely, with the help of the renormalization group. To avoid complicating the analysis by the details of the electronic structure of FeAs compounds, as obtained within DFT calculations, and to concentrate instead on the electron interaction effects, it makes sense to consider a minimal two-band model, which comprises an existence of a hole sheet of the Fermi surface around Γ and an electron sheet near the M point of the Brillouin zone. Therefore, the problem becomes that of qualitatively studying the Coulomb interaction effects in an itinerant two-band model. This task was formulated by Chubukov et al. [423] and addressed by the renormalization group method.

On a phenomenological level, the Hamiltonian of the model can be written down as $H = H_0 + H_{int}$, where

$$H_0 = \sum_{\mathbf{k}\sigma} \Big[\varepsilon_{1\mathbf{k}} c^+_{1\mathbf{k}\sigma} c_{1\mathbf{k}\sigma} + \varepsilon_{2\mathbf{k}} c^+_{2\mathbf{k}\sigma} c_{2\mathbf{k}\sigma} + \Gamma_{\mathbf{k}} \big(c^+_{1\mathbf{k}\sigma} c_{1\mathbf{k}\sigma} + c^+_{2\mathbf{k}\sigma} c_{2\mathbf{k}\sigma} \big) \Big], \tag{5.95}$$

$$H_{\rm int} = \frac{U_{11}}{2} \sum_{\mathbf{k}_1 \dots \mathbf{k}_4 \sigma \sigma'} \left(c^+_{1\mathbf{k}_1\sigma} c^+_{1\mathbf{k}_2\sigma'} c_{1\mathbf{k}_3\sigma'} c_{1\mathbf{k}_4\sigma} + c^+_{2\mathbf{k}_1\sigma} c^+_{2\mathbf{k}_2\sigma'} c_{2\mathbf{k}_3\sigma'} c_{2\mathbf{k}_4\sigma} \right)$$
$$+ U_{12} \sum_{\mathbf{k}_1 \dots \mathbf{k}_4 \sigma \sigma'} c^+_{1\mathbf{k}_1\sigma} c^+_{2\mathbf{k}_2\sigma'} c_{2\mathbf{k}_3\sigma'} c_{1\mathbf{k}_4\sigma}. \tag{5.96}$$

Here, $c^+_{1\mathbf{k}\sigma}$ and $c^+_{2\mathbf{k}\sigma}$ are Fourier components of the creation operators for an electron in orbitals 1 and 2, $\varepsilon_{1\mathbf{k}}$ and $\varepsilon_{2\mathbf{k}}$ are Fourier components of transfer matrix elements on the lattice, $\Gamma_\mathbf{k}$ is the orbitals' hybridization parameter. $H_{\rm int}$ includes two parameters: the Coulomb interaction of electrons within the same orbital, U_{11}, and between two different orbitals –U_{12}. Summing up over the wave vectors in (5.96) presumes the condition $\mathbf{k}_1 + \mathbf{k}_2 = \mathbf{k}_3 + \mathbf{k}_4$.

The quadratic form H_0 can be diagonalized by a linear transformation from the initial operators to fermionic operators $c_{\mathbf{k}\sigma}$ and $f_{\mathbf{k}\sigma}$, corresponding to holes and electrons of the minimal model:

$$\begin{aligned} c_{1\mathbf{k}\sigma} &= \cos\theta_\mathbf{k}\, c_{\mathbf{k}\sigma} + \sin\theta_\mathbf{k}\, f_{\mathbf{k}\sigma}, \\ c_{2\mathbf{k}\sigma} &= \cos\theta_\mathbf{k}\, c_{\mathbf{k}\sigma} - \sin\theta_\mathbf{k}\, f_{\mathbf{k}\sigma}, \end{aligned} \tag{5.97}$$

where the $\theta_\mathbf{k}$ value is defined by the condition

$$\tanh\theta_\mathbf{k} = \frac{2\,\Gamma_\mathbf{k}}{\varepsilon_{2\mathbf{k}} - \varepsilon_{1\mathbf{k}}}. \tag{5.98}$$

After the transformation (5.97), the H_0 becomes diagonal:

$$H_0 = \sum_{\mathbf{k}\sigma} \varepsilon^c_\mathbf{k}\, c^+_{\mathbf{k}\sigma} c_{\mathbf{k}\sigma} + \sum_{\mathbf{k}\sigma} \varepsilon^f_\mathbf{k}\, f^+_{\mathbf{k}\sigma} f_{\mathbf{k}\sigma}, \tag{5.99}$$

where

$$\varepsilon^{c,f}_\mathbf{k} = \frac{\varepsilon_{1\mathbf{k}} + \varepsilon_{2\mathbf{k}}}{2} \pm \frac{1}{2}\sqrt{(\varepsilon_{1\mathbf{k}} - \varepsilon_{2\mathbf{k}})^2 + 4\,\Gamma^2_\mathbf{k}} \tag{5.100}$$

is the energy of fermion excitations.

Depending on the initial $\varepsilon_{1\mathbf{k}}$, $\varepsilon_{2\mathbf{k}}$, $\Gamma_\mathbf{k}$ parameters and chemical potential μ, the formula (5.100) describes different situations. It is necessary to select such parameter values that the two sheets of the Fermi surface, $\varepsilon^c_\mathbf{k} = \mu$ and $\varepsilon^f_\mathbf{k} = \mu$, would make small circles around the Γ and M points, corresponding to the hole and electron quasiparticles of the model describing the FeAs-compounds, and moreover that the radii of these two circles would be nearly equal. In this situation, a nesting would take place between the hole and electron sheets on the wave vector $\mathbf{Q} = (\pi, \pi)$ in the reduced Brillouin zone.

An analysis shows that this happens when the hybridization term dominates and has a property $\Gamma_\mathbf{k} = \Gamma_{\mathbf{k+Q}}$. Then $\varepsilon^c_\mathbf{k} \simeq \Gamma_\mathbf{k} \simeq -\varepsilon^f_{\mathbf{k+Q}}$ and $\theta_0 \simeq \theta_Q \simeq \pi/4$. In this case, $\varepsilon^c_\mathbf{k}$ describes holes with the maximal energy at $\Gamma = (0, 0)$, and $\varepsilon^f_\mathbf{k}$ – the electrons with the minimum of energy at $M = (\pi, \pi)$. In other words, in the first term of (5.99) the summation runs over small **k** vectors, and in the second one – over a small vicinity of the $\mathbf{k} = \mathbf{Q}$ point.

The Hamiltonian H_{int} after the linear transformation (5.97) acquires the following form:

$$\begin{aligned} H_{\text{int}} = & \, U_1^{(0)} \sum c_{\mathbf{k}_3\sigma}^+ f_{\mathbf{k}_4\sigma'}^+ f_{\mathbf{k}_2\sigma'} c_{\mathbf{k}_1\sigma} + U_2^{(0)} \sum f_{\mathbf{k}_3\sigma}^+ c_{\mathbf{k}_4\sigma'}^+ f_{\mathbf{k}_2\sigma'} c_{\mathbf{k}_1\sigma} \\ & + U_3^{(0)}/2 \sum \left[f_{\mathbf{k}_3\sigma}^+ f_{\mathbf{k}_4\sigma'}^+ c_{\mathbf{k}_2\sigma'} c_{\mathbf{k}_1\sigma} + \text{H.c.} \right] \\ & + U_4^{(0)}/2 \sum f_{\mathbf{k}_3\sigma}^+ f_{\mathbf{k}_4\sigma'}^+ f_{\mathbf{k}_2\sigma'} f_{\mathbf{k}_1\sigma} + U_5^{(0)}/2 \sum c_{\mathbf{k}_3\sigma}^+ c_{\mathbf{k}_4\sigma'}^+ c_{\mathbf{k}_2\sigma'} c_{\mathbf{k}_1\sigma}. \end{aligned} \tag{5.101}$$

The $U_i^{(0)}$ $(i = 1, \ldots 5)$ values are linear combinations of the Coulomb parameters U_{11} and U_{12} and of the properties depending on $\theta_{\mathbf{k}}$ in two points: $\mathbf{k} = 0$ and $\mathbf{k} = \mathbf{Q}$. We write them done for the case specified above, when $\Gamma_{\mathbf{k}}$ dominates, and $\theta_0 = \theta_Q = \pi/4$. Then

$$\begin{aligned} U_1^{(0)} &= U_4^{(0)} = U_5^{(0)} = \frac{U_{11} + U_{12}}{2}, \\ U_2^{(0)} &= U_3^{(0)} = \frac{U_{11} - U_{12}}{2}. \end{aligned} \tag{5.102}$$

The Coulomb parameters U_{11} and U_{12} are positive, therefore the values of $U_1^{(0)}$, $U_4^{(0)}$ and $U_5^{(0)}$, representing the hole–electron, electron–electron and hole–hole interactions, respectively, are positive as well. As regards the $U_2^{(0)}$ and $U_3^{(0)}$ parameters which define the exchange interaction and transfer, they can be of any sign. However, as the intra-orbital Coulomb interaction is likely to exceed the inter-orbital one, the $U_2^{(0)}$ and $U_3^{(0)}$ can be assumed to be positive. If in the FeAs-systems the electron correlations are weak (or, at least, moderate), one can assume that $u_i = U_i\, N(E_F) < 1$, and treat H_{int} by the perturbation theory.

The authors of [423], using the renormalization group method, have analyzed the perturbation series and evaluated the constants of effective interaction in the particle–hole channel, V^{sdw}, responsible for the onset of the magnetic order, and in the Cooper (particle–particle) channel V^{sc}, responsible for the superconducting order parameter.

The perturbation theory series contain loops constructed from the GF describing the propagation of c- and f-fermions (holes and electrons). To these loops, the polarization operators $\Pi_{\text{pp}}(\mathbf{q}, \Omega)$ do correspond in the particle–particle channel (with parallel electron lines in the loop) and $\Pi_{\text{ph}}(\mathbf{q}, \Omega)$ in the particle–hole channel (with antiparallel lines). They both logarithmically diverge for $\mathbf{q},\ \Omega \to 0$:

$$
\begin{aligned}
\Pi^{cc}_{\rm pp}(\mathbf{q},\Omega) &= \sum_{\mathbf{k}} \frac{1}{2\pi}\int \mathrm{d}\omega G^c_0(\mathbf{k},\omega)G^c_0(-\mathbf{k}+\mathbf{q},-\omega+\Omega)\\
= \Pi^{ff}_{\rm pp}(\mathbf{q},\Omega) &= \sum_{\mathbf{k}} \frac{1}{2\pi}\int \mathrm{d}\omega G^f_0(\mathbf{k},\omega)G^f_0(-\mathbf{k}+\mathbf{q},-\omega+\Omega) \qquad (5.103)\\
&= N(E_{\rm F})\ln\frac{\Lambda}{\max(\Omega,v_{\rm F}q)},
\end{aligned}
$$

where $G^c_0(\mathbf{k},\omega) = \left(\omega-\varepsilon^c_{\mathbf{k}}\right)^{-1}$, $G^f_0(\mathbf{k},\omega) = \left(\omega-\varepsilon^f_{\mathbf{k}}\right)^{-1}$ are Green's functions of free fermions, and Λ – the cutoff parameter (of the order of the bandwidth W of the whole electron spectrum).

In the particle–hole channel, the same divergence takes place:

$$
\begin{aligned}
\Pi^{cf}_{\rm ph}(\mathbf{q}+\mathbf{Q},\Omega) &= \sum_{\mathbf{k}} \frac{1}{2\pi}\int \mathrm{d}\omega G^c_0(\mathbf{k},\omega)G^f_0(\mathbf{k}+\mathbf{Q},\omega+\Omega)\\
&= -N(E_{\rm F})\ln\frac{\Lambda}{\max(\Omega,v_{\rm F}q)}.
\end{aligned} \qquad (5.104)
$$

In the latter expression, it is taken into account that the Fermi surface for electrons is displaced in the **k**-space by the vector **Q** relative to the hole surface.

The divergence of the loops of both types, $\Pi_{\rm pp}$ and $\Pi_{\rm ph}$, makes it necessary to take them into account within the perturbation theory simultaneously, in what concerns effective interactions in the particle–particle and particle–hole channels. Differently from the standard RPA in which the loops in only one channel are summed up, in the present situation all diagrams representing the combinations of loops of both kinds should be taken into consideration. This corresponds to summing up the parquet diagrams. Such type of perturbation theory was earlier used for cuprates [446]; with respect to the problem of FeAs-compounds with their two-sheet Fermi surface it was developed in [447,448]. We will follow the exposition of [448].

Summing up the parquet diagrams leads to the renormalization group equations for effective interaction parameters of the Hamiltonian, by integrating out a contribution of the states with high energies. The fixed point for the renormalization group determines the true values of renormalized interaction parameters which describe the low-energy physics of the system, in this case – the physics of the states near the Fermi energy.

In the course of the renormalization group procedure (coming up from the higher energies to the lower ones), the initial interaction parameters of the Hamiltonian u^0_1, u^0_2, u^0_3, u^0_4, u^0_5 are renormalized into the parameters u_1, u_2, u_3, u_4, u_5, which do now depend on energy E from the $E_{\rm F} < E < W$ range via the $\ln\frac{W}{E}$ function, which increases on lowering the E towards the Fermi energy $E_{\rm F}$.

Out of five u_i parameters, the linear combinations may be constructed, which will represent the effective interaction parameters (coupling constants) in different channels, giving rise to certain order parameters. Those in the problem under consideration are six and define the waves of spin and charge polarization (with real and imaginary order parameters), and superconducting pairing of the conventional s-type and of the $s^{\pm}$-type. The corresponding combinations are:

$$\begin{aligned} \Gamma^r_{\text{sdw}} &= u_1 + u_3, \quad \Gamma^i_{\text{sdw}} = u_1 - u_3, \\ \Gamma^r_{\text{cdw}} &= u_1 - u_3 - 2u_2, \quad \Gamma^i_{\text{cdw}} = u_1 + u_3 - 2u_2, \\ \Gamma^s_{\text{sc}} &= u_4 + u_3, \quad \Gamma^{s^{\pm}}_{\text{sc}} = u_4 - u_3. \end{aligned} \tag{5.105}$$

The above constants satisfy the renormalization group equations,

$$\frac{\mathrm{d}\Gamma_j}{\mathrm{d}L} = \Gamma_j^2, \tag{5.106}$$

where the j index runs over all six values represented in (5.106). The L parameter in the differential equation (5.106) is $L = \ln \frac{E_{\text{F}}}{E}$. Equation (5.106) has a simple solution:

$$\Gamma_j = \frac{\overline{\Gamma}_j}{1 - \overline{\Gamma}_j \ln \frac{E_{\text{F}}}{E}}, \tag{5.107}$$

where $\overline{\Gamma}_j$ is Γ_j at $E \simeq E_{\text{F}}$.

The coupling constant Γ_j shapes the possible order parameters of the system: Δ_{sdw}, Δ_{cdw}, Δ^c_{sc} and Δ^f_{sc}. They are determined by an insertion into the Hamiltonian of additional infinitesimal trial terms

$$\begin{aligned} &\Delta_{\text{sdw}} \sum_{\mathbf{k}} c^+_{\mathbf{k}\alpha} \sigma^z_{\alpha\beta} c_{\mathbf{k}+\mathbf{Q}\beta}, \\ &\Delta_{\text{cdw}} \sum_{\mathbf{k}} c^+_{\mathbf{k}\alpha} \delta_{\alpha\beta} f_{\mathbf{k}+\mathbf{Q}\beta}, \\ \Delta^c_{\text{sc}} \sum_{\mathbf{k}} c_{\mathbf{k}\alpha} \sigma^y_{\alpha\beta} c_{-\mathbf{k}\beta} &+ \Delta^f_{sc} \sum_{\mathbf{k}} f^+_{\mathbf{k}+\mathbf{Q}\alpha} \sigma^y_{\alpha\beta} f_{-\mathbf{k}-\mathbf{Q}\beta}. \end{aligned} \tag{5.108}$$

(α and β number the spin indices of a fermion).

It can be shown [448] how do these terms get renormalized owing to a renormalization of bare interaction constants u_i. The following formula connects the initial property Δ_j with its renormalized one $\widetilde{\Delta}_j$ [448]:

$$\widetilde{\Delta}_j = \Delta_j \left(1 + \Gamma_j \ln \frac{E_{\text{F}}}{E}\right), \tag{5.109}$$

where Γ_j are the combinations of the interaction constants u_i, shown in the expressions (5.105).

The condition under which a Γ_j diverges sets a non-vanished order parameter $\widetilde{\Delta}_j$. According to (5.107), the divergence condition for Γ_j is given by

$$\ln \frac{E_{\text{F}}}{E} = \frac{1}{\overline{\Gamma}_j}, \tag{5.110}$$

from which the equation on the phase transition temperature for an onset of the corresponding order parameter follows:

$$T_j = E_F e^{-1/\Gamma_j}. \tag{5.111}$$

For the three types of instability – the spin, orbital and superconducting ones – the critical temperatures are given by the following relations:

$$T^r_{\text{sdw}} = E_F \exp\left[-\frac{1}{\bar{u}_1 + \bar{u}_3}\right], \tag{5.112}$$

$$T^i_{\text{cdw}} = E_F \exp\left[-\frac{1}{\bar{u}_1 + \bar{u}_3 - 2\bar{u}_2}\right], \tag{5.113}$$

$$T^{s^\pm}_{\text{sc}} = E_F \exp\left[-\frac{1}{\bar{u}_3 - \bar{u}_4}\right], \tag{5.114}$$

under the condition $\bar{u}_3 > \bar{u}_4$.

Into these equations, the values of the $\bar{u}_i$ parameters in the fixed point have to be inserted. An analysis of equations for the fixed point reveals that $\bar{u}_3 > \bar{u}_4$, even if the bare constants obeyed the inverse relation $u^0_3 < u^0_4$, because u^0_3 is the exchange interaction and u^0_4 – the Coulomb one, which is always superior to the exchange. It is important to underline once more that the temperatures of different instabilities are not determined by bare interactions, but by the renormalized ones. The analysis given relates to the case when the hole and the electron Fermi surfaces are of the same size, and the $\mathbf{Q} = (\pi, \pi)$ vector is that of a perfect nesting. In such a situation, the highest temperature is that of the instability with respect to the formation of the SDW phase. As the nesting gets detuned out of its perfect value, the relation between different instabilities may change. Logarithmic singularities in the particle–particle channel (5.103) disappear, but persist in the particle–hole one (5.104), so that $T^r_{\text{sdw}}(\delta)$ decreases on an increase of doping δ, and the instability of Fermi particles against the formation of the $s^\pm$ superconducting state becomes the dominating one. As concerns the other superconducting state of the s-symmetry, it may appear only under an unprobable condition $\bar{u}_3 + \bar{u}_4 < 0$.

Therefore, the renormalization-group analysis of the two-band model in the weak coupling approximation shows that the u_3 parameter, responsible for the pair hopping of electrons from the hole sheet onto the electron one (and back), gets enhanced, as the contributions to the effective Hamiltonian from the states with high energies get increased. In this process, even if the bare interaction u^0_3 was repulsive, in the pairing channel of the $s^\pm$-symmetry it turns attractive, and assures the superconducting order parameter of variable sign, consistent with the $s^\pm$-symmetry. It turns out that in the weak coupling approximation, there are not the spin (or orbital) fluctuations which are responsible for the pairing of the $s^\pm$-type, as it followed from the RPA or FLEX approximation, but, instead, the pair transfer of electrons from the hole onto the electron sheet of the Fermi surface.

The spin fluctuation mechanism of pairing may become efficient under moderate or strong Coulomb interaction, at $u_i^{(0)} \gtrsim 1$, i.e. beyond the applicability of the present perturbation theory. In the previous sections, we have seen that the $s^{\pm}$-state appears from the solutions of the Eliashberg equations, where, for a coupling interaction, the mechanism of exchange via spin fluctuations has been taken. For this, sufficiently large magnitudes of the Coulomb interaction are required, at which the RPA is not sufficiently justified. Consequently, the question of whether the superconductivity in the FeAs-compounds results from an exchange by spin fluctuations cannot yet be at present considered as finally settled.

The results outlined in this section have been recently confirmed by another study [449], where the method of functional renormalization group has been used. Differently from [448], the authors of [449] proceeded from a model Hamiltonian with two hole pockets and two electron pockets of the Fermi surface. The Hamiltonian was of the similar structure, where the constants of the bare interaction, denoted g_1, g_2, g_3 and g_4, had the same meaning as the u_1, u_2, u_3 and u_4 of the two-band model [448]. In particular, g_3 described the interaction strength under pair transfer of electrons from the hole onto the electron sheet, similarly to the u_3 constant in [448].

It turned out that this interaction, studied over a broad region in all parameters, is responsible for two leading instabilities: the SDW state and the $s^{\pm}$ superconducting state. The temperatures of these instabilities, as functions of doping, are shown in Fig. 5.40. Along the ordinate axis, the T_{N} or T_{c} are marked, and along the abscissa axis – the magnitude of the electron ($x > 0$) or hole ($x < 0$) doping. In the figure, the values of the bare interaction constants are given. We note that for the upper curve, $g_3 = 0.4$ eV is substantially larger than the value $g_3 = 0.08$ eV for the bottom curve, that demonstrates the leading role of a pair electron transition from one sheet of the Fermi surface to the other one. It is remarkable that no superconducting instability of the d-symmetry has been found.

Therefore, the both works [448, 449], in which the renormalization-group analysis of the perturbation theory series has been applied, indicate that independently on the details of the band model (either two-band or four-band one), the leading instabilities against a formation of either SDW-ordering or $s^{\pm}$-superconductivity is

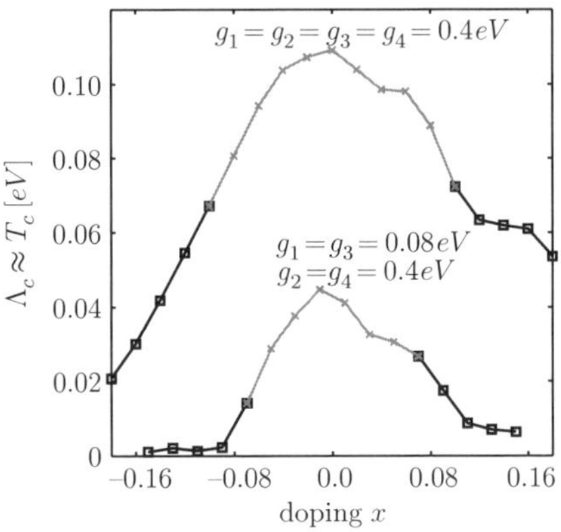

Fig. 5.40 Phase diagram on the (T, x) plane from calculations by the renormalization-group method for a four-bands model [449]. *Crosses* correspond to a transition from normal to the SDW-phase, *squares* – into the $S{\pm}SC$ state.

induced by the pair transfer of electrons from the hole onto the electron sheet of the Fermi surface.

The larger part of theory works in different models (whether 2-bands or 5-bands ones) predict the $s^{\pm}$-symmetry of superconducting order parameter in the FeAs-compounds [90,416,426,427,430,438,443,447,450]. It means that the signs of the gap at the hole and the electron sheets of the Fermi surface are opposite. However, there is no consensus so far concerning whether the superconducting gap may have zeros at the Fermi surface. The gaps without zeros have been found in both itinerant model [416,426,427,438,447] and the localized model [443]. However, Graser et al. [430] found an $s^{\pm}$-state with gap zeros within the 5-orbital Hubbard model.

To remove this ambiguity, Chubukov et al. [451] performed a special renormalization-group analysis of the two-band model, in which a competition of two interactions, the intra-band Coulomb repulsion u_4 and the inter-band pair transfer u_3, has been analyzed in detail. These interactions are schematically shown in Fig. 5.41a, where the Fermi surface of the two-band model in the extended Brillouin zone scheme is depicted. In Fig. 5.41b, the results of the renormalization-group analysis are shown, namely the flow lines of the u_4 and u_3 parameters, as functions of the scaling parameter $\ln \frac{W}{E_F}$.

As follows from Fig. 5.41a and the analysis of equations for superconducting gaps, in a situation when the renormalized constant of the pair hopping dominates over the Coulomb repulsion, $u_3 > u_4$ (the B line), the $s^{\pm}$-state with a zeroless gap on the Fermi surface is realized. In the opposite case $u_3 < u_4$ (the A line), the dependence of the pair hopping constant $u_3(\mathbf{q}-\mathbf{q}')$ on the momenta becomes important. Due to this dependence, in spite of a large magnitude of intra-band repulsion,

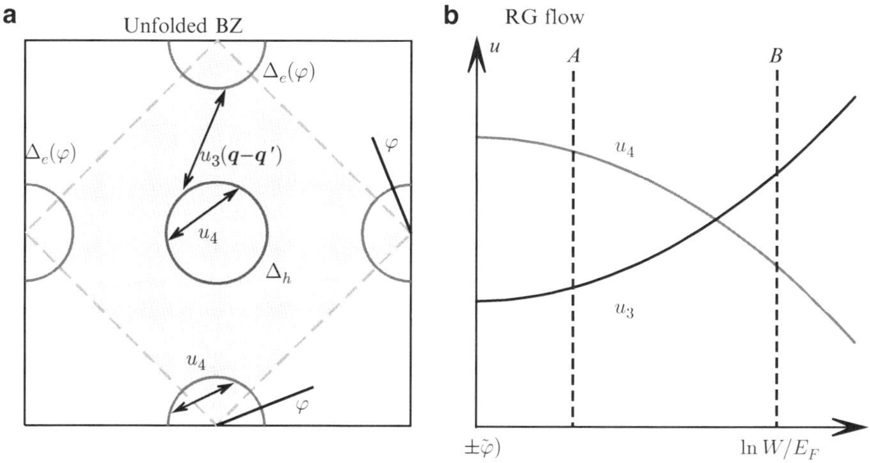

Fig. 5.41 (**a**) Hole and electron pockets of the Fermi surface in the two-band model; (**b**) the renormalization-group diagram of the flow of interaction constants u_4 and u_3 towards a low-energy effective Hamiltonian as a result of integrating out a contribution from the states of high-energy (of the order of W) electrons [451]

a formation of a paired state with zeros on the electron Fermi surface becomes possible, as a result of which a contribution of the repulsive interaction to the Cooper pair gets minimized (as in the case of the $d_{x^2-y^2}$ pairing in cuprates). Therefore, the $s^{\pm}$-state of the order parameter may emerge both without zeros at the electron sheet as with the zeros on it. Everything depends on the relation between the renormalized parameters of the intra-band repulsion and the pair transfer. An analysis of gap equations shows that if the tendencies versus magnetic ordering are stronger in a system, then the $s^{\pm}$-state without zeros gets realized. In the opposite situation, an $s^{\pm}$-state with zeros of gap at the Fermi surface is formed.

These results are consistent with the studies on the five-band Hubbard model in the RPA [430]. Results close to those of Chubukov et al. [451] have been simultaneously obtained on the basis of studying the 5-bands model [452, 453].

5.6.2 Equations for Superconducting and Magnetic Order Parameters

This question has been studied in detail in [454] on the basis of a two-band model, which included only one hole band centred at Γ and an electron one, centred at M. The Hamiltonian of the free carriers is given by (5.99). To maintain in the following the notations of [454], we rewrite this Hamiltonian as

$$H_0 = \sum_{\mathbf{k}\alpha} \left[\xi^c(\mathbf{k})\, c^+_{\mathbf{k}\alpha} c_{\mathbf{k}\alpha} + \xi^f(\mathbf{k})\, f^+_{\mathbf{k}\alpha} f_{\mathbf{k}\alpha} \right]. \tag{5.115}$$

Here, the hole operators are denoted $c^+_{\mathbf{k}\alpha}$ and $c_{\mathbf{k}\alpha}$, and electron ones – $f^+_{\mathbf{k}\alpha}$, $f_{\mathbf{k}\alpha}$; the spin components will be denoted by α and β. The summation over $\mathbf{k}$ runs over the wave vectors from the vicinity of the (0, 0) point in the first term and of the (π, π) in the second term. The reduced Brillouin zone, corresponding to the unit cell with two Fe atoms, is assumed.

We set for simplicity that the hole and electron sheets of the Fermi surface are circles of the same radius (for the undoped compound), and the dispersion relations for charge carriers, in the vicinity of the Fermi momentum $\mathbf{k}_F$, can be expressed as

$$\xi^{f,c}(\mathbf{k}) = \pm\varepsilon_{\mathbf{k}} + \delta, \quad \xi_{\mathbf{k}} = v_F(k - k_F). \tag{5.116}$$

The δ parameter represents the energy difference for both electrons and holes at the Fermi level in a doped metal, so that δ is immediately related to the dopant concentration.

We introduce the magnetic and superconducting order parameters, $m_{\mathbf{q}}$, Δ^c and Δ^f, to be defined by the following equations:

$$m_{\mathbf{q}} = V^{\text{sdw}} \sum_{\mathbf{k}} \sigma^z_{\alpha\beta} \langle f^+_{\mathbf{k}+\mathbf{q}\alpha} c_{\mathbf{k}\beta} \rangle, \tag{5.117}$$

$$\Delta^c = V^{sc} \sum_{\mathbf{k}} (-i\,\sigma^y)_{\alpha\beta}\, \langle c^+_{-\mathbf{k}\alpha}\, c_{\mathbf{k}\beta}\rangle, \tag{5.118}$$

$$\Delta^f = V^{sc} \sum_{\mathbf{k}} (-i\,\sigma^y)_{\alpha\beta}\, \langle f^+_{-\mathbf{k}\alpha}\, f_{\mathbf{k}\beta}\rangle. \tag{5.119}$$

Here, $\sigma^z_{\alpha\beta}$ and $\sigma^y_{\alpha\beta}$ are the Pauli matrices. The summation over $\mathbf{k}$ runs over a small vicinity of Γ. The magnetic order parameter $m_{\mathbf{q}}$ describes a spin density wave with the wave vector $\mathbf{Q}+\mathbf{q}$, whereas Δ^c and Δ^f take into account the Coulomb pairing on the hole and electron sheets. V^{sdw} and V^{sc} are coupling constants in the electron-hole SDW channel and in the superconducting particle–particle channel, correspondingly. The V^{sc} interaction, taken separately, leads to superconductivity of the $s^{\pm}$-type with the transition temperature T_c, while V^{sdw} – to the SDW ordering at the temperature T_s.

For an evaluation of correlators entering (5.117)–(5.119), it is necessary to know the Green's function

$$G_{\alpha\beta}(\mathbf{k},\tau) = -\langle \hat{T}_\tau\, \Psi_{\mathbf{k}\alpha}(\tau)\, \Psi^+_{\mathbf{k}\beta}(0)\rangle, \tag{5.120}$$

where $\Psi^+_{\mathbf{k}\alpha}$ will be chosen as a four-component spinor

$$\Psi^+_{\mathbf{k}\alpha} = \left(c^+_{\mathbf{k}\alpha},\; c_{-\mathbf{k}\alpha},\; f^+_{\mathbf{k}+\mathbf{q}\alpha},\; f_{-\mathbf{k}-\mathbf{q}\alpha}\right), \tag{5.121}$$

and $\Psi^+_{\mathbf{k}\alpha}$ makes an Hermitian conjugated four-component column.

The Hamiltonian of the system, in which the interaction term is taken in the mean field approximation, determined by the order parameters $m_{\mathbf{q}}$, Δ^c and Δ^f, can be written down as a quadratic form of the Ψ^+ and Ψ operators:

$$\hat{H} = \frac{1}{2} \sum_{\mathbf{k}\alpha\beta} \Psi^+_{\mathbf{k}\alpha}\, \hat{H}_{\alpha\beta}\, \Psi_{\mathbf{k}\beta}, \tag{5.122}$$

where $\hat{H}$ makes a 4×4 matrix:

$$\hat{H}_{\alpha\beta} = \begin{pmatrix} \xi^c(\mathbf{k}) & \Delta^c\, i\,\sigma^y_{\alpha\beta} & m_{\mathbf{q}}\,\sigma^z_{\alpha\beta} & 0 \\ -\Delta^{*c}\, i\,\sigma^y_{\alpha\beta} & -\xi^c(-\mathbf{k}) & 0 & -m_{\mathbf{q}}\,\sigma^z_{\alpha\beta} \\ m^*_{\mathbf{q}}\,\sigma^z_{\alpha\beta} & 0 & \xi^f(\mathbf{k}+\mathbf{q}) & \Delta^f\, i\,\sigma^y_{\alpha\beta} \\ 0 & -m^*_{\mathbf{q}}\,\sigma^z_{\alpha\beta} & -\Delta^{*f}\, i\,\sigma^y_{\alpha\beta} & -\xi^f(-\mathbf{k}-\mathbf{q}) \end{pmatrix}. \tag{5.123}$$

Two diagonal blocks of this matrix contain the $s^{\pm}$ superconducting order parameter $\Delta^c = -\Delta^f = \Delta$ for the two pockets of the Fermi surface. The off-diagonal blocks contain the SDW parameter $m_{\mathbf{q}}$. The band energies $\xi^c(\mathbf{k})$ and $\xi^f(\mathbf{k})$ are given by the expressions (5.116), where

$$\xi^f(\mathbf{k}+\mathbf{q}) = \varepsilon_{\mathbf{k}} + \delta + \mathbf{v}_{\mathrm{F}}\,\mathbf{q} \quad \text{for } q \ll k_{\mathrm{F}}. \tag{5.124}$$

The Green's function (5.120) in the Fourier representation over the τ parameter is determined as an inverse matrix,

$$G^{-1}(\mathbf{k},\, i\,\omega_n) = i\,\omega_n - \hat{H}\,; \quad \omega_n = (2n+1)\,\pi\,T. \tag{5.125}$$

On having calculated from this equation, the $G(\mathbf{k},\,\tau)$ matrix and, with its help, the correlators in (5.117)–(5.119) for the order parameters, one can get from them two equations on the order parameters $m_{\mathbf{q}}$ and Δ:

$$\ln\frac{T}{T_c} = 2\pi\,T\sum_{n>0}\mathrm{Re}\left(\frac{(E_n + i\,\delta_{\mathbf{q}})\,/\,E_n}{\sqrt{(E_n + i\,\delta_{\mathbf{q}})^2 + m_{\mathbf{q}}^2}} - \frac{1}{|\omega_n|}\right), \tag{5.126}$$

$$\ln\frac{T}{T_s} = 2\pi\,T\sum_{n>0}\mathrm{Re}\left(\frac{1}{\sqrt{(E_n + i\,\delta_{\mathbf{q}})^2 + m_{\mathbf{q}}^2}} - \frac{1}{|\omega_n|}\right), \tag{5.127}$$

where

$$E_n = \sqrt{\omega_n^2 + \Delta^2}\,; \quad \delta_{\mathbf{q}} = \delta + \frac{1}{2}\,\mathbf{v}_{\mathrm{F}}\,\mathbf{q}. \tag{5.128}$$

It is remarkable that in these equations, the interaction constants V^{sdw} and V^{sc} do not explicitly enter, thanks to an introduction of T_c and T_s for the magnetic and superconducting transitions, without taking into account, for each of them, the other order parameter. The values V^{sdw} and V^{sc} are hidden in the equations which determine T_c and T_s. They can be extracted from the linearized equations (5.117)–(5.119): for the SDW by setting $\Delta = 0$, $\delta = 0$, and for the superconductivity by setting $m_{\mathbf{q}} = 0$.

The magnetic order parameter $m_{\mathbf{q}}$ determines, for $\mathbf{q} = 0$, a commensurate SDW structure with the wave vector $\mathbf{Q} = (\pi,\,0)$ in the extended Brillouin zone. To find an incommensurate SDW structure (with $\mathbf{q}\neq 0$), one has to know the free energy of a system, so that, by minimizing it, to be able to find the value of $\mathbf{q}$. The energy $\Delta F(\Delta,\,m_{\mathbf{q}}) = F(\Delta,\,m_{\mathbf{q}}) - F(0,\,0)$ was calculated in [454]:

$$\begin{aligned}\frac{\Delta F(\Delta,\,m_{\mathbf{q}})}{4\,N_{\mathrm{F}}(0)} &= \frac{|\Delta|^2}{2}\ln\frac{T}{T_c} + \frac{m_{\mathbf{q}}^2}{2}\ln\frac{T}{T_s}\\ &-\pi\,T\sum_{\omega_n}\mathrm{Re}\left(\sqrt{(E + i\,\delta_{\mathbf{q}})^2 + m_{\mathbf{q}}^2} - |\omega_n| - \frac{|\Delta|^2}{2\,|\omega_n|} - \frac{m_{\mathbf{q}}^2}{2\,|\omega_n|}\right),\end{aligned} \tag{5.129}$$

where $N(E_{\mathrm{F}})$ is the density of states at the Fermi level, calculated per spin channel.

A strategy of search for the self-consistent solutions of (5.117)–(5.119) and (5.6.2) consists in the following: one has to find the solutions of (5.117)–(5.119) for Δ and $m_{\mathbf{q}}$ at a fixed δ (doping) and arbitrary $\mathbf{q}$, and then to choose a solution

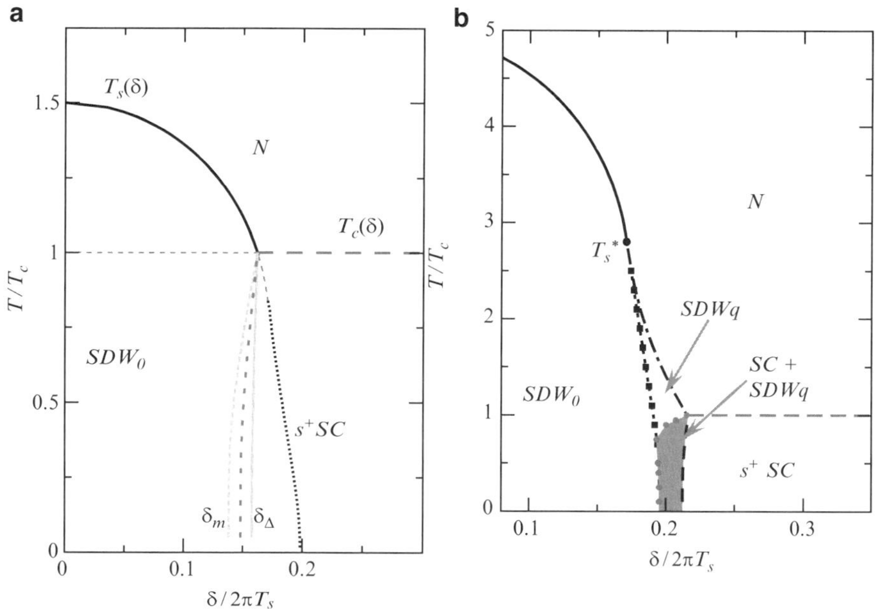

Fig. 5.42 Phase diagram in the $T - \delta$ plane for the two-band model, at two values of the $T_s/T_c parameter$: (**a**) $T_s/T_c = 1.5$ and (**b**) $T_s/T_c = 5$ [454]. The phases shown are: N – normal metal, $s^{\pm}$SC – superconductor with the $s^{\pm}$-symmetry of the order parameter, SDW_0 – a commensurate magnetic phase, $SDW_{\mathbf{q}}$ – non-commensurate magnetic phase with the modulation vector **q**

for the **q**, which minimizes the free energy (5.6.2). The solutions obtained permit to construct the phase diagram of the system.

5.6.3 *Phase Diagram of the Model*

The calculation results are depicted in Fig. 5.42. The system's behaviour depends on the $T_s/T_c > 1$ parameter, therefore the phase diagrams are shown in the figure for two values of this parameter. In both cases, the lines of the magnetic and superconducting phase transitions $T_s(\delta)$ and $T_c(\delta)$ from the paramagnetic normal phase are the lines of the second-kind transition (solid and dashed lines in the figures), whereby the superconducting transition temperature does not depend on doping. In Fig. 5.42a, below the tricritical point where $T_s(\delta) = T_c$, the phase transition between the states ($m \neq 0$, $\Delta = 0$) and ($\Delta \neq 0$, $m = 0$) is the transition of the first kind. Therefore, there is no such region where m and Δ would coexist.

The situation changes as T_s/T_c grows, and a broad region of doping emerges, in which $T_s(\delta) > T_c$ (Fig. 5.42b). In this situation, around the intersection of the $T_s(\delta)$ curve with $T = T_c$, an incommensurate $SDW_{\mathbf{q}}$ phase appears. In fact, a new phase is formed (the dashed one in Fig. 5.42b) in which superconductivity coexists with the $SDW_{\mathbf{q}}$ phase. The transition into the mixed $SC+SDW_{\mathbf{q}}$ phase from the superconducting state is of the second kind; from the SDW-phase – of the first kind. Therefore, the superconducting phase coexists only with the incommensurate $SDW_{\mathbf{q}}$ one.

The region of coexistence $SC+SDW_{\mathbf{q}}$ is maintained at intermediate values of the T_s/T_c parameter as well, as for instance for $T_s/T_c = 3$, but it shrinks in size and ascends towards the intersection point of the $T_s(\delta)$ and $T = T_c$ lines. Below it on the temperature scale, only a first-kind phase transition between $SDW_{\mathbf{q}}$ and the superconductivity takes place, as in the Fig. 5.42a. In other words, at intermediate values of T_s/T_c the phase diagram has an intermediate shape between those shown in Fig. 5.42a, b. At a further increase of the T_s/T_c parameter, the situation exhibited in Fig. 5.42b gets preserved. Summarizing, at low temperatures and small dopings the commensurate SDW-phase exists, at sufficiently high doping the superconducting phase appears, and in a narrow interval of doping a coexistence phase between superconductivity and an incommensurate SDW-phase takes place.

The results outlined refer to an assumption that the superconducting phase is homogeneous everywhere, including the coexistence area. It is pointed out that, in the $SDW_{\mathbf{q}}+SC$ region, an existence of a non-uniform superconducting phase, with a Cooper's pairing such that the summary momentum of the electrons in the pair be different from zero, is not ruled out. This may be a superconducting state of the Larkin – Ovchinnikov – Fulde – Ferrell (LOFF) type [455, 456].

The question of whether in genuine FeAs-systems the picture of coexistence of magnetic and superconducting order parameters, as obtained in a simplified two-band model, may really take place, can be resolved only by experiment. In this relation, it can be pointed out that in some compounds, an unhomogeneous SDW phase at finite dopings has been observed. It would make interest to analyze in more details, under which conditions does the superconductivity appear in these systems.

An experimental justification of the predicted picture of coexistence of superconductivity with collinear magnetic structure has been apparently obtained in [457], where on an $Fe_{1+y}Se_xTe_{1-x}$ system, a combined study involving magnetic measurements, μSR, and polarized neutron diffusion, has been done. In the phase diagram, three regions in the $0 < x < 0.5$ interval were identified: (1) with commensurate magnetic order at $x < 0.1$; (2) with a superconducting phase for $x \sim 0.5$; and (3) the intermediate phase in the range $0.25 < x < 0.45$, in which the superconductivity coexists with a static non-commensurate magnetic order. An evolution of the phase diagram depending on the Se concentration is shown in Fig. 5.43.

With the use of polarized neutrons, for an $Fe_{1.03}Se_{0.25}Te_{0.75}$ sample a magnetic peak (0.46, 0, 0.5) was detected at $T_N \sim 40$ K, revealing an incommensurate magnetic order. The magnetic order has been registered in samples of $x = 0.45$; 0.4; 0.25; 0.1 and 0.0 with the volume of magnetic fraction $\approx$75%; 98%; 98%; 95%, and 92% at $T = 0$. In each case, the sum of magnetic and superconducting fractions did not exceed 1, and in no sample was a macroscopic separation detected into the magnetic (M) and superconducting (SC) phases.

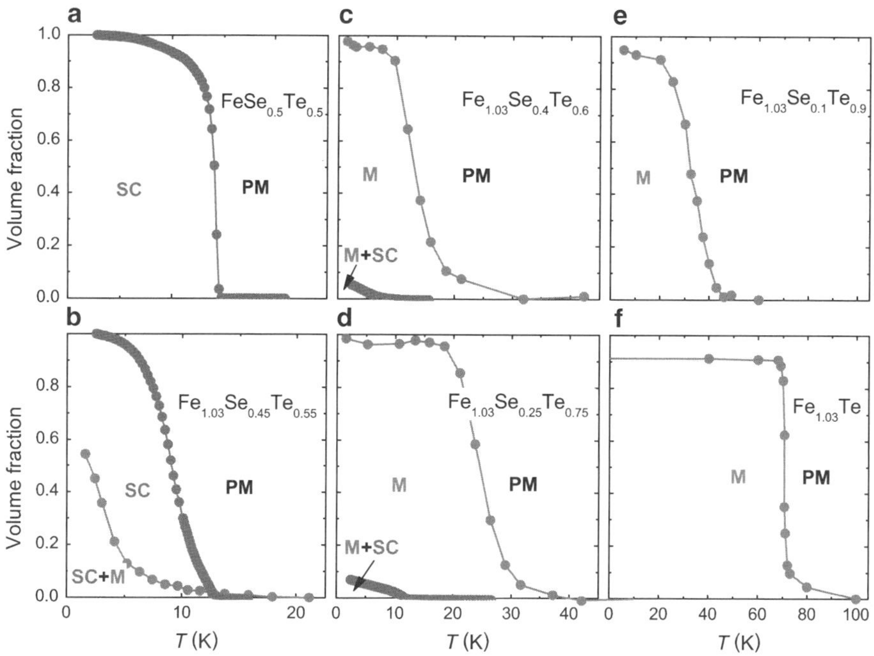

Fig. 5.43 Evolution of the phase diagram of the $Fe_{1.03}Se_xTe_{1-x}$ system on the plane (fraction volume – temperature) as function of the Se concentration x

In the SC+M coexistence region, the relation holds $T_N/T_c \sim 5$, that, according to Fig. 5.42c, corresponds to an appearance of the coexisting region between magnetic phase and superconductivity. On the other hand, for $x > 0.45$ $T_N/T_c \sim 1$, and no such coexistence phase appears, according to the theory of [454], see Fig. 5.42a.

The results shown in Fig. 5.43 are in qualitative agreement with another work [458], in which for the same system, using Mössbauer spectroscopy, a static magnetism – the spin glass (SG) state – has been detected in the region of existence of superconductivity, see Fig. 5.44.

5.6.4 Peculiarities of the $s^{\pm}$-Superconducting State

In this section, we address peculiar features of the $s^{\pm}$-superconducting state in what regards the effect on it of magnetic impurities. It turned out that if at two sheets of the Fermi surface – an electron and a hole one, separated by the $\mathbf{Q} = (\pi, \pi)$ vector, – the superconducting order parameter changes sign, the scattering on magnetic impurities with the transfer momentum $\mathbf{Q}$ results in a destruction of the Cooper pairs in a similar way as it happens under scattering on magnetic impurities in

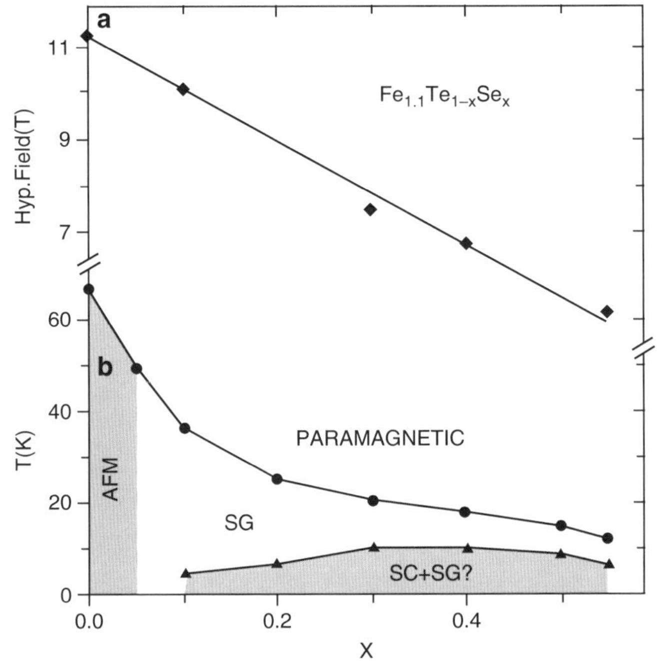

Fig. 5.44 Phase diagram of the $Fe_{1.1}Se_xTe_{1-x}$ system with a region of coexistence of superconductivity and spin glass, as established by Mössbauer spectroscopy [458]: (**a**) variation of the mean hyperfine field at the Fe nucleus with the Se concentration; (**b**) phase diagram in the (T, x) -plane

conventional superconductors. In Sect. 5.4, it was shown that the spin–lattice relaxation rate $1/T_1$, measurable by NMR, varies in this situation not exponentially with temperature as it would follow from the BCS theory, but along the power-law, as in those superconductors which have zeros at the Fermi surface. This conclusion permits to contest an interpretation of NMR experiments, in which a power-law behaviour of the $1/T_1$ was detected, in favour of a non-standard superconductivity.

In [459], a theoretical description of an effect of non-magnetic impurities on the $s^{\pm}$- has been extended. The superfluid density of states $\rho_s(T)$ with the $s^{\pm}$-symmetry of the order parameter in the presence of non-magnetic impurities has been calculated. It is related to the penetration depth by a relation:

$$\lambda(t) \sim \frac{1}{\sqrt{\rho_s(T)}}, \tag{5.130}$$

making use of which the calculated values of $\lambda(T)$ can be compared with experiment.

In a superconductor with impurities, the characteristics of the superconducting state depend on two parameters: Γ_0/T_{c0} and Γ_π/T_{c0}, where Γ_0 is the amplitude of the forward electron scattering, and Γ_π – that of the scattering along the nesting

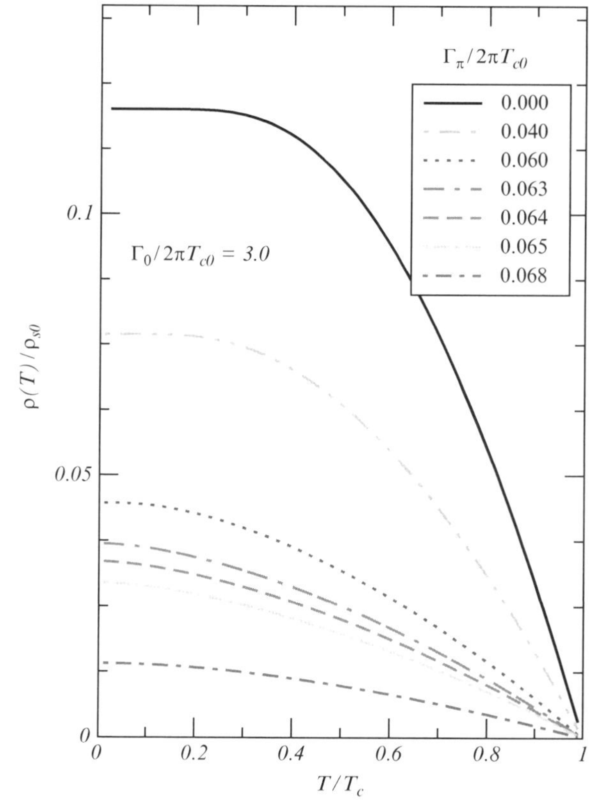

Fig. 5.45 Superfluid density $\rho_s(T)$ depending on the magnitude of the depairing parameter Γ_π/T_{c0} for a superconductor with the $s^{\pm}$-symmetry of the order parameter [459]

vector, $\mathbf{Q} = (\pi, \pi)$, i.e. of the scattering under which a fermion is hopped from one sheet of the Fermi surface onto the other one. Namely, this scattering destroys the Cooper pairs. In Fig. 5.45, the density $\rho_s(T)$ is shown depending on the Γ_π/T_{c0} parameter, at fixed Γ_0/T_{c0}, where T_{c0} is the superconducting transition temperature in the absence of impurities.

Taking notice of the logarithmic scale along the ordinate axis, we see that the region of exponential variation of $\rho_s(T)$ with temperature is dramatically narrowing as the parameter Γ_π/T_{c0} grows. As is shown in [459], with an exception of a very narrow temperature interval close to zero, the ρ_s follows the decrease of temperature according to the power-law $\sim T^2$.

Such behaviour is in good agreement with the data on the temperature dependence of the penetration depth, $\lambda(T) = \lambda(0) + \Delta\lambda(T)$. In Fig. 5.46, the $\Delta\lambda(T)$ variation is shown for the $Ba(Fe_{1-x}Co_x)_2As_2$ compound [460, 461], which does well fit the theory curves [459]. For an undoped sample, the chosen value of the penetration depth $\lambda(0) = 2{,}800$ nm is unrealistically large, and no agreement with theory is detected. For two other cases, the $\lambda(0)$ values are quite realistic.

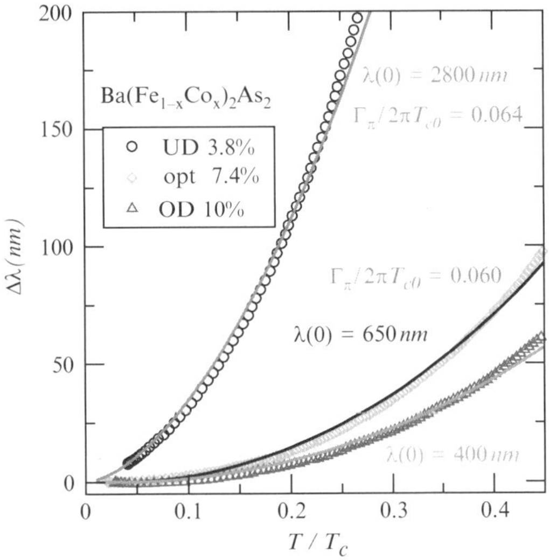

Fig. 5.46 Experimental data on the temperature dependence of the penetration depth $\Delta\lambda(T)$ for three samples of $Ba(Fe_{1-x}Co_x)_2As_2$: the optimally doped one (7.4%), the overdoped (10%) and the underdoped (3.8%). *Solid lines* indicate the theory [459] at the specified values of the Γ_π / T_{c0} parameter and $\lambda(0)$

In [459], it was reported that for the LaOFeP system, no satisfactory quantitative agreement could have been obtained.

Therefore, the experiments on the penetration depth in $Ba(Fe_{1-x}Co_x)_2As_2$, in which a power-law behaviour of $\Delta\,\lambda(T)$ has been found, do quite agree with a conclusion that this compound is an $s^{\pm}$-superconductor, under an assumption that the sample contains impurities.

5.7 The Limit of Strong Coulomb Interaction

5.7.1 The $t - J_1 - J_2$-Model

In Sect. 5.3–5.5, we considered the models with weak Coulomb interaction, which should be identified as itinerant models of FeAs-systems. At present, the relation between the U and W parameter values in these systems is not known. It seems somehow more probable that $U \lesssim W$. When considering the limit of weak Coulomb interaction ($U \ll W$), we must take, at the end of calculations, $U \simeq W$, to be able to attribute the results obtained to the FeAs-systems. It is possible, however, to approach the $U \simeq W$ case from the other side as well, departing from the limit $U \gg W$. It allows us to see in the electronic structure of FeAs-systems also some features of the localized model, which we will discuss below, and to analyze the appearance in these systems of the superconducting pairing with the same symmetries of the order parameter, which are possible in the weak Coulomb interaction limit, i.e. in the itinerant model.

Under the conditions of strong correlations, when the Coulomb repulsion parameters on a site (U within the same orbital and U' – between different orbitals) exceed the d-band width W, it is possible, by using a small parameter W/U or W/U', to pass to an effective Hamiltonian. This approach does fully correspond to a transition from the Hubbard model to the $t - J$-model. In the effective Hamiltonian, the exchange interaction of the antiferromagnetic sign appears, $J \sim W^2/U$ for electrons at neighbouring sites.

In relation to the FeAs-systems, the effective exchange interaction appears not only between the nearest-neighbouring Fe atoms, but also between those situated at more distant sites, because of the complex structure of hoppings (Fig. 5.21). To the Hamiltonian of the two-orbital model H_0 [404], two types of interactions have to be added: H_1 describes the exchange of electrons at different sites, and H_2 – the Hund's exchange at the same site. Let us write down the both expressions:

$$H_1 = \sum_{iab} \sum_{n} J_n^{ab} \left[\mathbf{S}_{ai} \, \mathbf{S}_{bi+\delta_n} - n_{ai} \, n_{bi+\delta_n} \right], \tag{5.131}$$

$$H_2 = - \sum_{ia} J \, \mathbf{S}_{\alpha i} \, \mathbf{S}_{\overline{\alpha} i}. \tag{5.132}$$

Here, $\mathbf{S}_{a\,i}$ is the operator of electron spin at site i and in the orbital a. It is expressed via the couple of Fermi operators $c_{i\,a\,\sigma}^{+}$ and $c_{i\,a\,\sigma}$ of creation and annihilation for an electron in this state, making use of the known formula:

$$\mathbf{S}_{\alpha i} = \sum_{\sigma\sigma'} c_{\alpha i \sigma}^{+} \, \boldsymbol{\sigma}_{\sigma\sigma'} \, c_{a i \sigma'}, \tag{5.133}$$

where $\boldsymbol{\sigma}$ is the vector composed out of the Pauli matrices. We remind that the $a = 1, 2$ index numbers the d_{xy} and d_{yz} orbitals included into the model. The number of particles $n_{a\,i}$ in the state $i\,a$ is also expressed in terms of the Fermi operators via the relation:

$$n_{ai} = \sum_{\sigma} c_{ai\sigma}^{+} \, c_{ai\sigma}. \tag{5.134}$$

Here, a runs over two values, 1 and 2, while $\boldsymbol{\delta}_1$ is a vector directed from the site i towards its nearest neighbour, and $\boldsymbol{\delta}_2$ – towards a next-nearest neighbour.

Similar to the $t - J$-model, the effective exchange parameter J_n^{ab} is defined by the following expression:

$$J_n^{ab} = 4 \left(t_{i,i+\delta_n}^{ab} \right)^2 / (U + 2J). \tag{5.135}$$

Using the notation of the hopping matrix elements as shown in Fig. 5.21, we see that the exchange between the nearest neighbours is determined by the electron hopping between identical orbitals, whereas to the exchange between next-nearest neighbours, the hoppings between identical as well as between different orbitals give contributions. Finally, the Hund's exchange H_2 is determined by electrons in the different orbitals. The Hamiltonian $H = H_0 + H_1 + H_2$ describes the so-called $t - J_1 - J_2$ model, which incorporates the antiferromagnetic exchange on the nearest and next-nearest neighbours, as well as the electron movement over the lattice. Further on, within the mean field approximation a possibility of a superconducting state to appear within this model, with different symmetries of the order parameter, will be analyzed.

5.7.2 Superconductivity with Different Order Parameters

To make the model analytically solvable, we will simplify it, assuming that the exchange interaction exists only between the electrons within the same orbital. Then the reduced Hamiltonian of the exchange interaction will be written down as [443]:

$$H_{\text{red}} = \sum_{\mathbf{k}\,\mathbf{k}'} \sum_{\alpha} V_{\mathbf{k}\,\mathbf{k}'}\, c^+_{\mathbf{k}\,\alpha\,\uparrow}\, c^+_{-\mathbf{k}\,\alpha\,\downarrow}\, c_{-\mathbf{k}'\,\alpha\,\downarrow}\, c_{\mathbf{k}'\,\alpha\,\uparrow}, \tag{5.136}$$

where

$$\begin{aligned} &V_{\mathbf{k}\,\mathbf{k}'} \\ &= -2\,J_1\Big[\big(\cos k_x + \cos k_y\big)\big(\cos k'_x + \cos k'_y\big) + \big(\cos k_x - \cos k_y\big)\big(\cos k'_x - \cos k'_y\big)\Big] \\ &\quad - 8J_2\,\big(\cos k_x \cos k_y \cos k'_x \cos k'_y + \sin k_x \sin k_y \sin k'_x \sin k'_y\big)\,. \end{aligned} \tag{5.137}$$

We note moreover that in H_{red}, the Hund's term is left out, and in the exchange term only the interaction of the Cooper pairs with opposite momenta and spins is retained. Thus, the reduced Hamiltonian corresponds to the BCS approximation.

We mark out in the Hamiltonian (5.136) the mean values for the operators of the Cooper pair:

$$\Delta_\alpha(\mathbf{k}') = \langle c_{\alpha\,-\mathbf{k}'\,\downarrow}\, c_{\alpha\,\mathbf{k}'\,\uparrow}\rangle\,; \tag{5.138}$$

we can write it down in the mean field approximation. In the Nambu representation with the four-component field operators

$$\Psi_{\mathbf{k}} = \Big(c_{1\,\mathbf{k}\,\uparrow},\, c^+_{1\mathbf{k}\,\downarrow},\, c_{2\,\mathbf{k}\,\uparrow},\, c^+_{2-\mathbf{k}\,\downarrow}\Big), \tag{5.139}$$

the full model Hamiltonian, including the kinetic term H_0 and the reduced interaction operator H_{red}, taken in the mean field approximation, is written down as

$$H = \sum_{\mathbf{k}} \Psi^+(\mathbf{k})\, A(\mathbf{k})\, \Psi(\mathbf{k}), \tag{5.140}$$

where $A(\mathbf{k})$ is a 4×4 matrix:

$$A(\mathbf{k}) = \begin{pmatrix} \xi_x(\mathbf{k}) - \mu & \Delta_1(\mathbf{k}) & \varepsilon_{xy}x(\mathbf{k}) & 0 \\ \Delta_1^*(\mathbf{k}) & -\varepsilon_x(\mathbf{k}) + \mu & 0 & -\varepsilon_{xy}(\mathbf{k}) \\ \varepsilon_{xy}(\mathbf{k}) & 0 & \varepsilon_y(\mathbf{k}) - \mu & \Delta_2(\mathbf{k}) \\ 0 & -\varepsilon_{xy}(\mathbf{k}) & \Delta_2^*(\mathbf{k}) & -\varepsilon_y(\mathbf{k}) + \mu \end{pmatrix}. \tag{5.141}$$

Here, $\Delta_a(\mathbf{k})$, $a = 1, 2$, is pairing amplitude, made out of pair of particles belonging to either the hole or the electron sheet of the Fermi surface. It consists of five contributions, corresponding to different symmetries of the operators:

$$\begin{aligned} \Delta_a(\mathbf{k}) &= s_{0a} + s_{x^2+y^2\,a}(\mathbf{k}) + s_{x^2y^2\,a}(\mathbf{k}) \\ &\quad + d_{x^2-y^2\,a}(\mathbf{k}) + d_{xy\,a}(\mathbf{k}), \end{aligned} \tag{5.142}$$

where

$$\begin{aligned} s_{x^2+y^2\,a}(\mathbf{k}) &= \Delta^0_{x^2+y^2\,a} \left(\cos k_x + \cos k_y\right), \\ s_{x^2y^2\,a}(\mathbf{k}) &= \Delta^0_{x^2\,y^2\,a} \cos k_x \cos k_y, \end{aligned} \tag{5.143}$$

$$\begin{aligned} d_{x^2-y^2\,a}(\mathbf{k}) &= \Delta^0_{x^2-y^2\,a} \left(\cos k_x - \cos k_y\right), \\ d_{x\,y\,a}(\mathbf{k}) &= \Delta^0_{x\,y\,a} \sin k_x \sin k_y, \end{aligned}$$

and S_{0a} does not depend on the momentum $\mathbf{k}$.

Therefore, the pairing amplitude consists of three contributions of the s-symmetry and two of the d-symmetry. The amplitudes of the corresponding order parameters are defined by the expressions:

$$\begin{aligned} \Delta^0_{x^2\pm y^2\,a} &= -\frac{2J_1}{N} \sum_{\mathbf{k}'} \left(\cos k'_x \pm \cos k'_y\right) \Delta_a(\mathbf{k}'), \\ \Delta^0_{x^2y^2\,a} &= -\frac{8J_2}{N} \sum_{\mathbf{k}'} \cos k'_x \cos k'_y \,\Delta_a(\mathbf{k}'), \\ \Delta^0_{x\,y\,a} &= -\frac{8J_2}{N} \sum_{\mathbf{k}'} \sin k'_x \sin k'_y \,\Delta_a(\mathbf{k}'). \end{aligned} \tag{5.144}$$

For an evaluation of the $\Delta_a(\mathbf{k})$ property, one has to construct the equation of motion for the electron Green's function in the superconducting state:

$$G(\mathbf{k}, \tau) = -\langle \hat{T}_\tau \,\Psi_{\mathbf{k}}(\tau)\, \Psi^+_{\mathbf{k}}(0)\rangle. \tag{5.145}$$

The (5.141) matrix can be diagonalized by a unitary transformation $U^+(\mathbf{k})A(\mathbf{k})$ $U(\mathbf{k})$. Its four eigenvalues are given by:

$$E_1(\mathbf{k}) = -E_2(\mathbf{k}), \quad E_3(\mathbf{k}) = -E_4(\mathbf{k}), \tag{5.146}$$

$$E_{1,3}(\mathbf{k}) = \frac{1}{\sqrt{2}}\Bigg(\xi_x^2 + \xi_y^2 + 2\varepsilon_{xy}^2 + \Delta_1^2 + \Delta_2^2$$
$$\pm\sqrt{\left(\xi_x^2 - \xi_y^2 + \Delta_1^2 - \Delta_2^2\right)^2 + 4\,\varepsilon_{xy}^2\left[\left(\xi_x + \xi_y\right)^2 + (\Delta_1 - \Delta_2)^2\right]}\Bigg)^{1/2}, \tag{5.147}$$

where $\xi_x = \varepsilon_x - \mu$, $\xi_y = \varepsilon_y - \mu$. Thus, the self-consistency equations follow for the pairing amplitudes and the occupation numbers. We have:

$$\Delta_1(\mathbf{k}) = \sum_{\mathbf{k}'\,m} V_{\mathbf{k}\,\mathbf{k}'}\,U_{2m}^*(\mathbf{k}')\,U_{1\,m}(\mathbf{k}')\,f\left(E_m\left(\mathbf{k}'\right)\right),$$
$$\Delta_2(\mathbf{k}) = \sum_{\mathbf{k}'\,m} V_{\mathbf{k}\,\mathbf{k}'}\,U_{4m}^*(\mathbf{k}')\,U_{3\,m}(\mathbf{k}')\,f\left(E_m\left(\mathbf{k}'\right)\right), \tag{5.148}$$

where

$$n_1 = 2\sum_{\mathbf{k}'\,m} U_{1m}^*(\mathbf{k})\,U_{1\,m}(\mathbf{k}')\,f\left(E_m\left(\mathbf{k}'\right)\right),$$
$$n_2 = 2\sum_{\mathbf{k}'\,m} U_{3m}^*(\mathbf{k})\,U_{3m}(\mathbf{k}')\,f\left(E_m\left(\mathbf{k}'\right)\right).$$

Here, $f(E)$ is the Fermi function. In (5.148), the summation runs over the index $m = 1, 2, 3, 4$, which numbers the components of the superspinor (5.139).

The equation for the superconducting transition temperature follows by the way of linearizing the equations for the amplitudes $\Delta_1(\mathbf{k})$ and $\Delta_2(\mathbf{k})$. Let us write down such an equation for $\Delta_2(\mathbf{k})$:

$$\Delta_2(\mathbf{k}) = \sum_{\mathbf{k}'} V_{\mathbf{k}\,\mathbf{k}'}\left[W_3(\mathbf{k}') - W_1(\mathbf{k}')\right], \tag{5.149}$$

where

$$W_i = \frac{\left((\varepsilon_x - \mu)^2 - \widetilde{E}_i^2\right)\Delta_2 + \varepsilon_{xy}^2\,\Delta_1}{2\left|\varepsilon_x + \varepsilon_y - 2\,\mu\right|\,\widetilde{E}_i\,\sqrt{4\,\varepsilon_{xy}^2 + \left(\varepsilon_x - \varepsilon_y\right)^2}}\tanh\frac{\widetilde{E}_i}{2T}, \tag{5.150}$$

and $\widetilde{E}_i = E_i(\Delta_1 = \Delta_2 = 0)$.

The numerical solution of these equations with the parameters of the two-orbital model leads to a phase diagram shown in Fig. 5.47. In the upper left corner, where $J_2 > J_{2c} \simeq 1.2$, a pure phase of the s-symmetry, $s_{x^2\,y^2}$, is realized. In the right bottom corner, where $J_1 > J_{1c} \simeq 1.05$, a mixed phase of $d_{x^2-y^2}$ and $s_{x^2+y^2}$ takes place. The remaining larger part of the $(J_1,\ J_2)$ plane is occupied by another mixed phase, $d_{x^2-y^2} + s_{x^2\,y^2}$. In this mixed phase, the sign of the order parameter of the

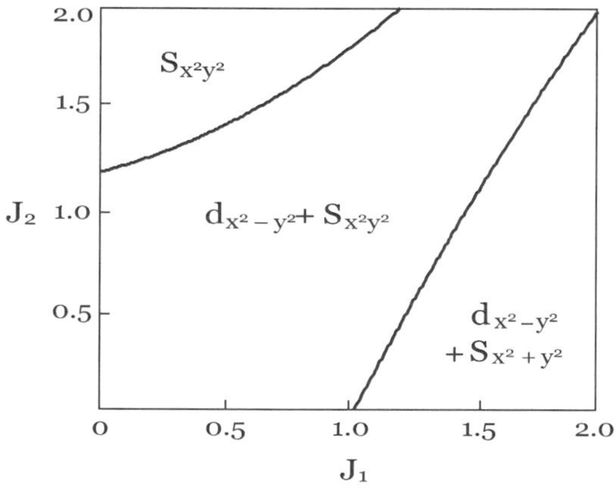

Fig. 5.47 Phase diagram in the $J_1 - J_2$ plane for superconducting states at different values of the order parameter, in the two-orbital model with electron doping $\delta = 0.18$ [443]

$d_{x^2-y^2}$ symmetry is different for the two orbitals. Thus, if $\Delta_1 = a\,\cos k_x\,\cos k_y + b\,(\cos k_x - \cos k_y)$, then $\Delta_2 = a\,\cos k_x\,\cos k_y - b\,(\cos k_x - \cos k_y)$. We note that no solution corresponding to the d_{xy}-symmetry of the order parameter has been found.

5.7.3 *Density of States and Differential Tunnel Conductivity*

The density of states in the electron spectrum of a superconductor has been calculated in another work [462], where the standard formulae have been used:

$$\begin{aligned}\rho(\omega) &= -\frac{1}{\pi}\sum_{\mathbf{k}} \operatorname{Im}\big\{G_{11}(\mathbf{k},\omega) + G_{33}(\mathbf{k},\omega)\big\}\\ &= -\frac{1}{\pi}\sum_{\mathbf{k}} \operatorname{Im}\big(\omega - A(\mathbf{k}) + i\,\delta\big)^{-1} \equiv \sum_{\mathbf{k}} \rho(\mathbf{k},\,\omega).\end{aligned} \tag{5.151}$$

An inversion of the 4×4 matrix $\omega - A(\mathbf{k})$ yields:

$$\begin{aligned}\rho(\mathbf{k},\omega) = {}& \frac{\varepsilon_{xy}^2\,(2\omega - \xi_x - \xi_y) - (\omega + \xi_y)(\omega^2 - \xi_x^2 - \Delta_1^2) - (\omega + \xi_x)(\omega^2 - \xi_y^2 - \Delta_2^2)}{E_1^2 - E_3^2}\\ &\times\left[\frac{1}{2E_3}\Big(\delta(E_3 - \omega) - \delta(E_3 + \omega)\Big) - \frac{1}{2E_1}\Big(\delta(E_1 - \omega) - \delta(E_1 + \omega)\Big)\right].\end{aligned} \tag{5.152}$$

A presence of the fourfold axis in crystalline samples of FeAs-compounds leads to the following relation between $\Delta_1(\mathbf{k})$ and $\Delta_2(\mathbf{k})$:

$$\left.\begin{aligned}\Delta_1(k_x,\,k_y) &= \Delta_2(k_y,\,k_x) \text{ for } s_{x^2+y^2},\ s_{x^2y^2},\ d_{xy}\\ \Delta_1(k_x,\,k_y) &= -\Delta_2(k_x,\,k_y) \text{ for } d_{x^2-y^2}\end{aligned}\right\}. \tag{5.153}$$

Thanks to the property mentioned, for all symmetries of the order parameter but the $d_{x^2-y^2}$, a simplified form of the expression (5.152) exists:

$$\rho(\mathbf{k},\omega) = \frac{\omega + E_-(\mathbf{k})}{2E_-^{\Delta}(\mathbf{k})}\left[\delta\left(E_-^{\Delta}(\mathbf{k}) - \omega\right) - \delta\left(E_-^{\Delta}(\mathbf{k}) + \omega\right)\right] + \frac{\omega + E_+(\mathbf{k})}{2E_+^{\Delta}(\mathbf{k})}\left[\delta\left(E_+^{\Delta}(\mathbf{k}) - \omega\right) - \delta\left(E_+^{\Delta}(\mathbf{k}) + \omega\right)\right], \quad (5.154)$$

where

$$E_{\pm}^{\Delta}(\mathbf{k}) = \sqrt{E_{\pm}^2(\mathbf{k}) + \Delta^2(\mathbf{k})}, \quad (5.155)$$

and we introduced the notation $\Delta_1(\mathbf{k}) = \Delta_2(\mathbf{k}) = \Delta(\mathbf{k})$.

The density of states $\rho(\omega)$ in a superconductor defines the differential tunnel conductivity, which can be experimentally measured. Its relation with $\rho(\omega)$ is given by

$$\frac{\mathrm{d}I}{\mathrm{d}V} \sim -\int_{-\infty}^{\infty} \rho(\omega)\, f'(\omega - e\,V)\,\mathrm{d}\omega, \quad (5.156)$$

where f' is the energy derivative of the Fermi function. It is obvious that in the limit $T \to 0$, $\mathrm{d}I/\mathrm{d}V \sim \rho(\omega)$, therefore the tunnel conductivity directly probes the density of states.

In Fig. 5.48, the calculated $\mathrm{d}I/\mathrm{d}V$ value is shown as function of the applied potential. For order parameters of the s_0- and the $s_{x^2+y^2}$-symmetry, a gap in the $\mathrm{d}I/\mathrm{d}V$ appears, as in the conventional BCS model. The d_{xy}, $d_{x^2-y^2}$ and $s_{x^2y^2}$

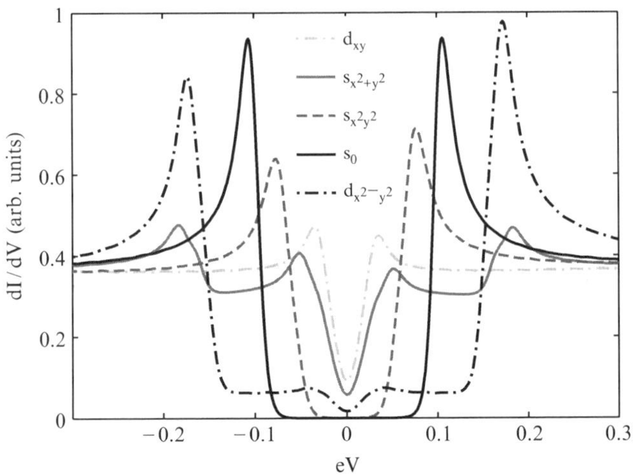

Fig. 5.48 Differential tunnel conductivity as function of the applied potential, calculated at the model parameters (5.29); $\mu = 1.6$, $T = 0.005$ and the depairing amplitude $\Delta^0 = 0.1$, for different symmetries of the order parameter [462]

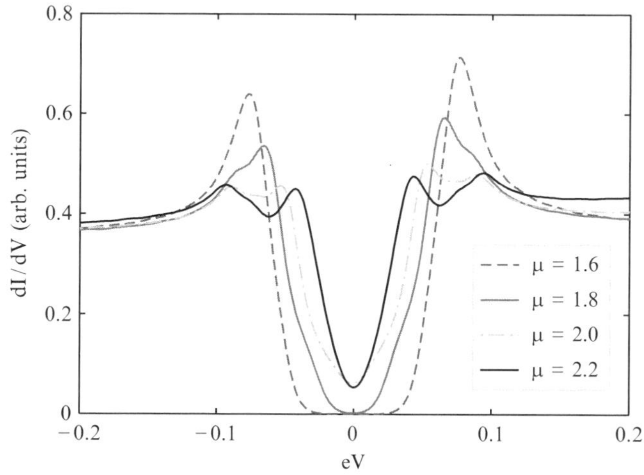

Fig. 5.49 Differential tunnel conductivity dI/dV as function of V for the order parameter $s_{x^2y^2}$, depending on doping, at the same parameter values as those in [462]

order parameters, which have zeros at the Fermi surface, yield non-zero density of states at $V = 0$.

The shape of the dI/dV curves depends on the magnitude of doping. The corresponding evolution in dependence on doping for the $s_{x^2y^2}$ order parameter is shown in Fig. 5.49.

A detailed analysis of possible symmetries of the superconducting order parameter in the $t-J_1-J_2$ model has been done in [463] within the two-orbital model [404] with the hopping parameters of (5.29). A solution of equations for the order parameter of different symmetry in the mean field approximation leads to the phase diagram shown in Fig. 5.50. Here, the superconducting order parameter is characterized by irreducible representations of the D_{4h} point group (see Sect. 5.3.4).

The solid lines indicates the line of the second-kind phase transition between the A_{1g} and $A_{1g} + iB_{1g}$ phases. The dashed line marks a crossover between the $s_{x^2y^2}$ and $s_{x^2+y^2}$ components, which dominate in the A_{1g} state. Therefore, at the hopping parameters chosen, the A_{1g} state dominates everywhere, but this state contains the $s_{x^2y^2}$ and $d_{x^2-y^2}$ components [463] – see (5.143). We note that a competition between the A_{1g} and B_{1g} states appears also in the models with weak coupling [404, 430].

5.7.4 *The Hubbard Model with the Hund's Exchange*

Above, the two-orbital $t - J_1 - J_2$ model has been considered. We address now another species of a two-orbital model which proceeds from the Hubbard model to which the Hund's exchange term is added, along with yet another term describing a

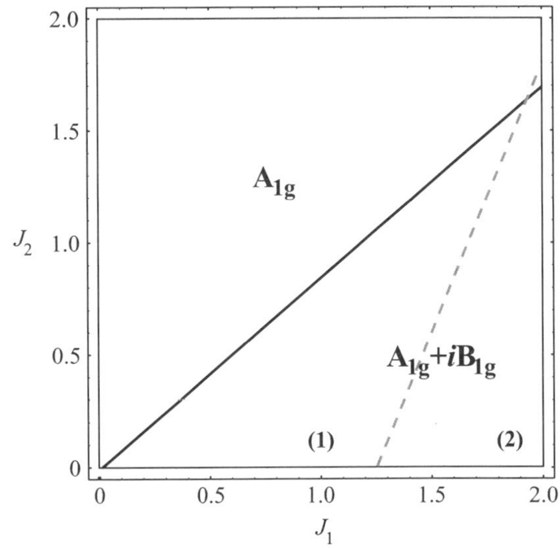

Fig. 5.50 Phase diagram of the superconducting state for $T = 0$ in the $J_1 - J_2$ plane [463]

pair transfer of electrons within the same site. We have already discussed a similar (three-orbitals) model with the Hamiltonian (5.57), making use of a certain version of the perturbation theory, the FLEX approximation. In this model we will treat the case according to the perturbation theory in an opposite small parameter, assuming that $U, J \gg t$ [464].

Under these conditions, one can transfer from the initial Hamiltonian towards an effective Hamiltonian of the $t - J$-model type. An interaction term in the effective Hamiltonian of singlet pairs can be obtained in the following form:

$$H_{\mathrm{eff}} = -\sum_{ij} \sum_{nmn'm'} A_{nm}^{m'n'}(ij)\, b_{nm}^{+}(ij)\, b_{n'm'}(ij). \tag{5.157}$$

Here,

$$b_{nm}(ij) = \frac{1}{\sqrt{2}} \left(c_{in\uparrow} c_{jm\downarrow} - c_{in\downarrow} c_{jm\uparrow} \right) \tag{5.158}$$

is the pair operator in the coordinate representation, and $A_{nm}^{m'n'}(ij)$ is the matrix element obtained in the first order over the small parameter:

$$A_{nm}^{m'n'}(ij) = \left[\frac{(-1)^{m+m'}}{U - J} + \frac{1}{U + J} \right] t_{ij}^{nm}\, t_{ji}^{m'n'} + \frac{t_{ij}^{n\overline{m}}\, t_{ji}^{\overline{m}'n'}}{U' + J} \tag{5.159}$$

(the orbitals are numbered by the indices n and m which may acquire two values, 1 and 2; $\overline{m}$ indicates an index complementary to m).

Let us introduce a superconducting order parameter:

$$\Delta_{nm}(\boldsymbol{\delta}) = \frac{1}{\sqrt{2}} \langle b_{nm}(i, i + \delta) \rangle . \tag{5.160}$$

In the mean field approximation, the Hamiltonian of the model $H_0 + H_{\text{int}}$ can be expressed as a quadratic form

$$H_{\text{MF}} = \sum_{\mathbf{k}} \psi_{\mathbf{k}}^{+} \begin{pmatrix} \xi_{\mathbf{k}} & V(\mathbf{k}) \\ V^{+}(\mathbf{k}) & -\xi_{\mathbf{k}} \end{pmatrix} \psi_{\mathbf{k}}, \tag{5.161}$$

where $\psi_{\mathbf{k}}^{+}$ is a four-component spinor, and $V(\mathbf{k})$ a two-component matrix

$$V_{\alpha\beta}(\mathbf{k}) = \sum_{n\,m\,n'\,m'\,\delta} A_{nm}^{m'n'}(\delta)\, \Delta_{nm}^{*}(\delta)\, e^{-i\mathbf{k}\delta}\, u_{m'\alpha}(\mathbf{k}) u_{n'\beta}(\mathbf{k}). \tag{5.162}$$

Here, $u_{m\alpha}(\mathbf{k})$ is the unitary transformation matrix, in the H_0 Hamiltonian, from the initial expression to the diagonal one:

$$H_0 = \sum_{\mathbf{k}nm\sigma} (\varepsilon_{\mathbf{k}}^{nm} - \mu)\, c_{\mathbf{k}n\sigma}^{+} c_{\mathbf{k}m\sigma} = \sum_{\mathbf{k}\alpha\sigma} \xi_{\mathbf{k}\alpha} c_{\mathbf{k}\alpha\sigma}^{+} c_{\mathbf{k}\alpha\sigma}. \tag{5.163}$$

The electron operators $c_{\mathbf{k}n\sigma}$ are related to the quasiparticle operators $c_{\mathbf{k}\alpha\sigma}$ $(\alpha = \pm)$ by an unitary transformation:

$$c_{\mathbf{k}n\sigma} = \sum_{\alpha} u_{n\alpha}(\mathbf{k}) c_{\mathbf{k}\alpha\sigma}. \tag{5.164}$$

The four-component spinor equals

$$\psi_{\mathbf{k}}^{+} = \left(c_{\mathbf{k}+\uparrow}^{+}, c_{\mathbf{k}-\uparrow}^{+}, c_{-\mathbf{k}+\downarrow}, c_{-\mathbf{k}-\downarrow} \right). \tag{5.165}$$

The 4×4 matrix (5.161) has two eigenvalues

$$E_{\pm}(\mathbf{k}) = \sqrt{w_{+}^{2} + V_{+-}^{2} \pm \sqrt{w_{-}^{2} + V_{+-}^{2}\left[(\delta\xi)^{2} + 4\overline{V}^{2}\right]}}, \tag{5.166}$$

where all properties depend on $\mathbf{k}$: $\delta\xi = \xi_{+} - \xi_{-}$, $\overline{V} = \frac{1}{2}(V_{++} + V_{--})$, $w_{\pm}^{2} = \frac{1}{2}\left[\xi_{+}^{2} + V_{++}^{2} \pm (\xi_{-}^{2} + V_{--}^{2})\right]$.

The total energy per site equals

$$E = -\frac{1}{N} \sum_{\mathbf{k}} \left[E_{+}(\mathbf{k}) + E_{-}(\mathbf{k}) \right]. \tag{5.167}$$

Therefore, two branches of the quasiparticle spectrum exist, with the energies $E_{+}(\mathbf{k})$ and $E_{-}(\mathbf{k})$ in the upper and lower bands. The equations for the superconducting order parameter (5.160) can easily be written in the standard form. Their numerical solution has been obtained for two values of the t/U parameter, equal

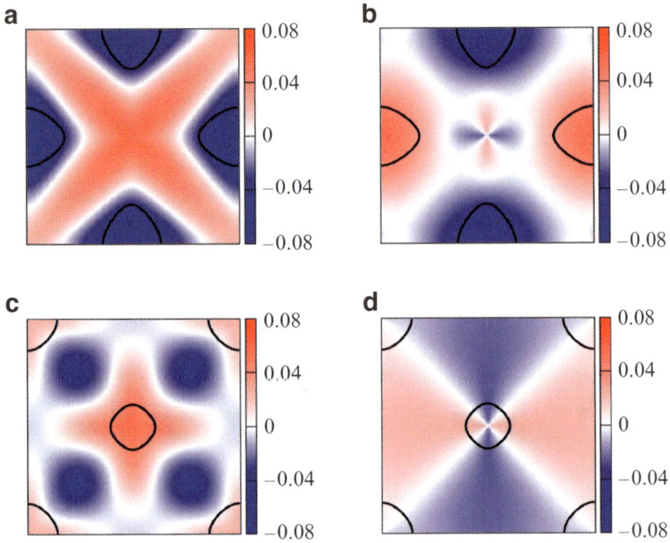

Fig. 5.51 Intraband pairing interaction V_{++} in the electron band, (**a**) and (**b**) and V_{--} – in the hole band, (**c**) and (**d**). Its signs are different near different sheets of the Fermi surface. *Left panels* are for the s-symmetry, *right* ones – for the d-symmetry [464]

to 0.1 and 0.2, and a free parameter J/U. The equations were being solved for the order parameter of the extended s-symmetry and the d-symmetry. The main result of the numerical solution is the following: the solution of the s-symmetry is energetically more favourable than that of the d-symmetry at all J values for $t/U = 0.2$. For $t/U = 0.1$, it stays more favourable at $J < J_c$ only, where $J_c/U \simeq 0.16$. At $J > J_c$, the ground state corresponds to the d-symmetry. A distribution of the pairing interaction $V_{\alpha\alpha}$ is depicted in Fig. 5.51.

As we see from the figure, in case of the s-symmetry the $V(\mathbf{k})$ is invariant with respect to a rotation by $\pi/2$, whereas for the d-symmetry the $V(\mathbf{k})$ changes its sign under such a rotation. For the s-symmetry, the line of zeros of the order parameter passes in the Brillouin zone out of the sheets of the Fermi surface, while for the d-symmetry the lines of zeros are situated at the diagonals of the square, i.e. beyond the electron sheets, but well passing through the hole sheet.

Remarkable is a fact of changing the sign of the order parameter on the electron and hole sheets. Therefore for yet another time we got a demonstration of energetical preferrability of the $s^{\pm}$-symmetry of the order parameter, now for the limit of strong Coulomb interaction.

We bring into attention one more work [465], in which, with the use of group theory analysis and sum rules, certain constraints have been established concerning a possibility of coexistence of different superconducting order parameters in the two-orbital model, with the exchange of the type H_{int} of (5.58). It is shown in this work that under the electron doping, a coexistence of the $d_{x^2-y^2}$ and s_{xy}

pairings is preferable, whereas under the hole doping a coexistence of the $s^{\pm}$ and s_{xy} pairings wins.

5.8 Magnetic Long-Range Order and Its Fluctuations

5.8.1 *Two Approaches to the Problem*

The duality in the behaviour of FeAs-systems, represented by the fact that they reveal localized and itinerant features at the same time, is primarily expressed in their magnetic properties. For this reason, from the very beginning the two alternative approaches have been used in the attempts to describe their magnetic properties. In one of them, the itinerant model is used, in which one attempt to relate the characteristics of long-range order and peculiarities of the spin fluctuation spectrum with the features of the Fermi surface of the compounds in question [60, 62, 63, 87, 90, 466–468]. In the other approach, the localized Heisenberg model is used with the exchange interaction J_1 and J_2 included between the nearest and between the next-to-nearest neighbours of the Fe atoms, and attempts are done to determine the conditions under which an observed magnetic order is formed [65, 280, 281, 469, 470].

Obviously, the two approaches represent just two limiting cases, when in the abovementioned dualism the one or the other side is dominating, whereas the reality does apparently comprise the both aspects on equal footing. Here, we face the same situation as in the description of strongly correlated systems, in which at $U \simeq W$ one must also consider both the itinerant and the localized nature of the electronic states. The theory which incorporates both is the DMFT. One can expect that an application of the DMFT model for a description of magnetic properties in the FeAs-systems would permit to take into account the both aspects, the itinerant and the localized one, on the same footing. So far such approach has not yet been realized, therefore we consider separately the theories which use either the localized or the itinerant approach.

As we have seen in Chap. 4, there are two approaches in the theoretical description of the FeAs-compounds, the "localized" and the "itinerant" one. In what regards the nature of magnetism of these compounds, in the "itinerant" approach it is assumed that the SDW antiferromagnetic ordering is promoted by the Fermi surface topology, namely the presence of hole and electron sheets which are related by the nesting vector. Indeed, in many compounds a (non-ideal) nesting takes place, such that the nesting vector coincides with the vector of the SDW structure. In the "localized" approach, the reason for the magnetic ordering is supposed to be in superexchange interaction between Fe atoms, mediated by their neighbouring As atoms. Driven by the superexchange mechanism, the antiferromagnetic exchange coupling is established between the nearest J_1 and the next-nearest J_2 neighbours over the Fe sublattice.

These two approaches are alternative ones. In explaining the magnetic structure and magnetic moments of FeAs-compounds, they do in fact declare different physics, which lays foundation for the nature of magnetic ordering. Recently, Johannes and Mazin suggested a novel concept [471], criticizing both "localized" and "itinerant" approaches and suggesting a certain third view on this problem.

The core of the new approach is the idea that it is not the Coulomb inter-electron interaction U which is responsible for magnetism in the FeAs-compounds, but the Hund's exchange J. As both spectroscopic data and numerical calculations show, the FeAs-compounds are moderately correlated systems, far from the Mott–Hubbard transition. The magnitude of the corresponding Coulomb repulsion U is of the order of 1 eV, and it cannot alone be responsible for the formation of local moments. At the same time, the Hund's exchange is of the same order of magnitude, and its existence in the many-orbital model does automatically guarantee a formation of localized magnetic moments at Fe sites. Therefore, according to the concept of [471], the Hund's exchange is responsible for an existence of localized magnetic moments in compounds, whereas for their magnetic ordering the structure of one-electron states is responsible, which is well described within the LDA. In this scenario, the nesting plays a certain, but not the principal, role.

To verify this hypothesis, the authors of [471] calculated the electron DOS for $BaFe_2As_2$ and FeTe in three magnetically ordered states: the AFM with chessboard arrangements of spins, the SDW (stripe) phase, and the double-stripe phase, as depicted in Fig. 5.52. In Fig. 5.53, the calculated DOS of $BaFe_2As_2$ for three magnetically ordered phases is shown, in comparison with the non-magnetic state. It is seen that all three magnetically ordered phases have a lower states density, as compared to the non-magnetic case, within a certain interval of energies below the Fermi level. This provides an energy gain due to the magnetic ordering. This gain is the largest for the SDW phase of $BaFe_2As_2$ (see Table 5.4), in agreement with experimental data. For the other compound, FeTe, the energy gain on magnetic ordering is particularly high, and the lowest energy is that of the double-stripe phase, consistently with the known neutron diffraction data. Therefore, the magnetic structures in the two compounds in question, $BaFe_2As_2$ and FeTe, are different, despite the fact that their Fermi surfaces are rather similar. This apparently suggests that the nesting does not play any important role in shaping the magnetic ordering in these substances.

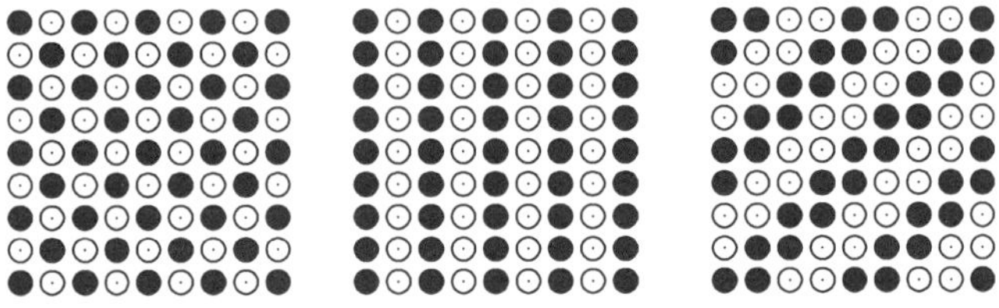

Fig. 5.52 Three types of magnetic ordering on a square Fe sublattice: AFM (checkerboard), stripe and double-stripe. *Black* and *white circles* correspond to opposite orientations of spins [471]

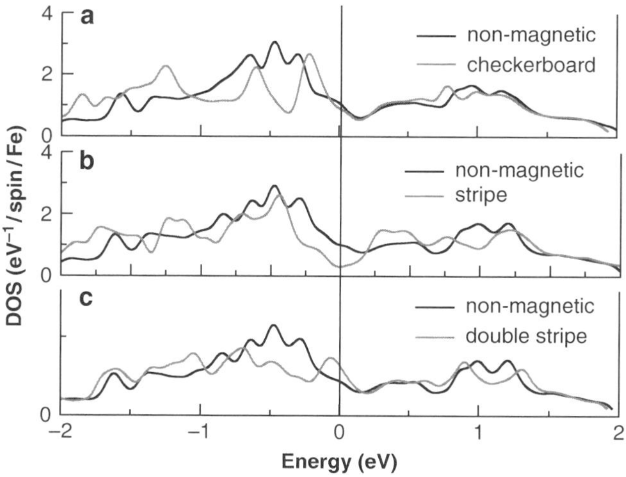

Fig. 5.53 Density of states of $BaFe_2As_2$ in three magnetically ordered phases (**a**–**c**), shown in Fig. 5.52, in comparison with the nonmagnetic state [471]

Table 5.4 Renormalized effective masses of quasiparticles in $BaFe_2As_2$ for different d-orbitals [393]

Orbitals:	d_{xy}	d_{yz}, d_{xz}	$d_{3z^2-r^2}$	$d_{x^2-y^2}$
m^*/m	2.06	2.07	2.05	1.83

The authors of [471] do also analyze the situation with superexchange in these compounds. It is seen from Fig. 5.52 that the J_1 magnetic ordering does not yield contribution to the crystal energy in stripe and double-stripe magnetic structures considered, because in each of them the number of spins set up or down on the nearest neighbours with respect to every centre is the same. For the realization of the stripe structure, the condition $J_2 > \frac{1}{2}J_1$ has to be satisfied. In what concerns the double-stripe structure, the positive and negative contributions exactly cancel down not only over the nearest neighbours, but over the next-nearest neighbours with their J_2 interactions as well, so that this structure can only be stabilized taking into account the interaction between even further neighbours, J_3. Therefore, for the realization of the stripe structure in $BaFe_2As_2$ it is necessary to have J_2 and J_1 of the same order, and for the realization of the double-stripe structure in FeTe one needs to have moreover J_3 and J_2 of the same order. These conditions are difficult to satisfy, because the corresponding distance in the Fe–As–Fe fragment are different for the Fe atoms which are neighbours of the different order.

A further argument against the "localized" model of magnetism in the FeAs-compounds is that it makes use of the $t - J_1 - J_2$ model, which combines in itself, additively, the Heisenberg model and the conventional band model. However, it is difficult to find a microscopic justification of such model. For example, departing from the Hubbard model under the conditions of strong Coulomb interaction $U \gg W$, we arrive at the known tJ-model, in which the Heisenberg exchange term appears; however, the electron part of the energy is described not by conventional band electrons, but by correlated electrons, with the consequence that the Hamiltonian of the tJ-model does not resemble that of the $t - J_1 - J_2$ model.

Therefore, in [471] a new approach to treating the magnetism in the FeAs-compounds has been proposed. It considers the electron system neither as fully localized nor as an entirely itinerant one. The magnetic ordering appears due to a shift of energy in the one-particle spectrum within the energy interval of about 1 eV from the Fermi level. The magnetic ordering is not directly related to peculiarities of the Fermi surface topology, e.g. nesting, but has, rather, to do with a gain due to the above shift of one-electron states. However, it may well happen that the structure of spin fluctuations be to a great extent determined by the Fermi surface topology and, in particular, by nesting.

5.8.2 *The Itinerant Model*

Out of many works [60,62,63,87,90,466–468] dedicated to the description of magnetic properties in the framework of itinerant models, we pick out [468], in which the dynamical spin susceptibility was calculated in the RPA:

$$\chi^{\mathrm{RPA}}(\mathbf{q}, i\,\omega_n) = \left[1 - V\,\chi_0(\mathbf{q}, i\,\omega_n)\right]^{-1} \chi_0(\mathbf{q}, i\,\omega_n), \tag{5.168}$$

where $\chi_0(\mathbf{q}, i\,\omega_n)$ is the susceptibility of the isotrope system without taking into account the interactions, and V is the interaction parameter.

In the Sect. 5.4, we have already discussed a calculation of the spin susceptibility in the RPA within an itinerant model [430]; however, the question about the long-range magnetic ordering and mean spin values at the Fe atom have not been addressed. Now we consider all these questions in full.

For a multi-orbital model, χ_0, V and χ_{RPA} are matrices, whose size is determined by the number of electron bands (or orbitals) included in the model. For FeAs-compounds, usually two hole bands (α_1, α_2) are included, with the sheets of the Fermi surface having shapes of two circles around Γ and two electron bands (β_1, β_2) with corresponding sheets near the M point. This corresponds to a Hamiltonian

$$H_0 = -\sum_{\mathbf{k}\nu\sigma} \varepsilon^{\nu}\, c^{+}_{\mathbf{k}\nu\sigma}\, c_{\mathbf{k}\nu\sigma} - \sum_{\mathbf{k}\nu\sigma} t^{\nu}_{\mathbf{k}}\, c^{+}_{\mathbf{k}\nu\sigma}\, c_{\mathbf{k}\nu\sigma}, \tag{5.169}$$

where $\nu = \alpha_1, \alpha_2, \beta_1, \beta_2$ numbers these bands, ε^{ν} are energies of the centre of each corresponding band, and $t^{\nu}_{\mathbf{k}}$ – the dispersion law. For hole and electron bands, $t^{\nu}_{\mathbf{k}}$ is expressed via the matrix elements of hopping t_1 and t_2 between the nearest and next-to-nearest neighbours to the Fe atoms:

$$\left.\begin{aligned} t^{\nu}_{\mathbf{k}} &= t^{\nu}_1\left(\cos k_x + \cos k_y\right) + t^{\nu}_2\,\cos k_x\,\cos k_y, \quad &\nu = \alpha_1, \alpha_2 \\ t^{\nu}_{\mathbf{k}} &= t^{\nu}_1\left(\cos k_x + \cos k_y\right) + t^{\nu}_2\,\cos\tfrac{k_x}{2}\,\cos\tfrac{k_y}{2}, \quad &\nu = \beta_1, \beta_2 \end{aligned}\right\}. \tag{5.170}$$

The placement of all four bands with respect to each other and the Fermi level depends on the parameters ε^ν, t_1^ν, t_2^ν. They are adjusted in such a way that the dispersion curves (5.170) would yield the sheets of the Fermi surface and the electron velocities on them consistently with how they follow from the LDA calculations. For the undoped case, the number of electrons per atom is 4 (since only four bands are taken into account; the fifth one is situated below the Fermi surface and is filed with two electrons). At the parameter values

$$\begin{aligned} &\alpha_1: \big(-0.60,\ 0.30,\ 0.24\big), \quad &\alpha_2: \big(-0.40,\ 0.20,\ 0.24\big), \\ &\beta_1: \big(1.70,\ 0.14,\ 0.74\big), \quad &\beta_2: \big(1.70,\ 0.14,\ -0.64\big) \end{aligned} \tag{5.171}$$

the dispersion curves along the principal directions are shown in Fig. 5.54. This spectrum agrees very well with that calculated in the five-orbital model (five d-orbitals and two Fe atoms in the unit cell) [117].

Now we come back to the formula (5.168) for the spin susceptibility. In the model outlined, all the properties appearing in (5.168) are 4×4 matrices. The χ_0 matrix is expressed by the known formula of the one-loop approximation in terms of the electron Green's function:

$$\chi_0^{\nu\mu}(\mathbf{q}, i\,\omega_m) = -\frac{T}{2N} \sum_{\mathbf{k}\,\omega_n} \mathrm{Tr}\big\{G^\nu(\mathbf{k}+\mathbf{q}, i\,\omega_n + i\,\omega_m)\, G^\mu(\mathbf{k}, i\,\omega_n)\big\}. \tag{5.172}$$

If considering only the Coulomb repulsion U between the electrons within the same orbital at the same site, and the Hund's exchange J, the V parameter in (5.168) is a matrix

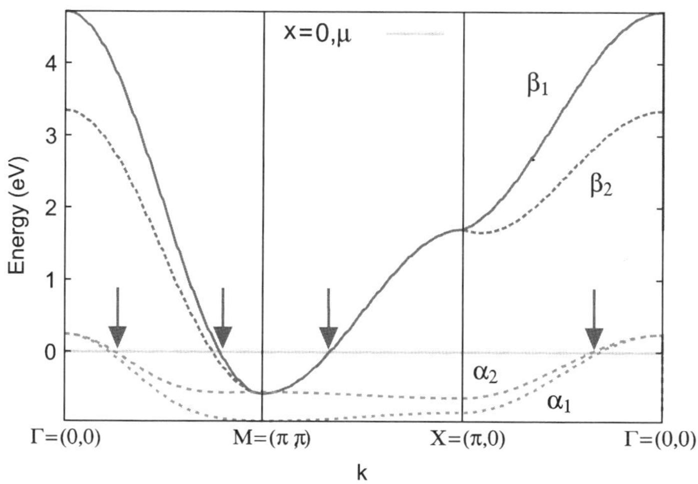

Fig. 5.54 Dispersion curves calculated for a non-doped case with the choice of parameters as in (5.171). The *arrows* indicate the points where the bands cross the Fermi level [468]

$$V = \begin{pmatrix} U & J/2 & J/2 & J/2 \\ J/2 & U & J/2 & J/2 \\ J/2 & J/2 & U & J/2 \\ J/2 & J/2 & J/2 & U \end{pmatrix}. \tag{5.173}$$

The U and J parameters have been chosen such as to obtain $T_N = 138\,K$, the Néel temperature for the undoped LaOFeAs. It turned out in this case that the mean Fe magnetic moment (the S^z-projection over a sublattice) equals $0.33\,\mu_B$. The parameters chosen,

$$U = 0.32\text{ eV}, \quad J = 0.07\text{ eV}, \tag{5.174}$$

are certainly very small; however, the U and J values calculated from the first principles turned out also to be much smaller than is usually under discussion for the FeAs-systems ($U \simeq 4\,\text{eV}$).

The spin susceptibility calculated along the formulae (5.168) and (5.172) is depicted in Fig. 5.55. In the M (π, π) point, a large maximum is seen. It appears on the antiferromagnetic wave vector $\mathbf{Q} = (\pi, \pi)$, which connects the hole and electron sheets of the Fermi surface and is thus the nesting vector:

$$\chi_0(\mathbf{q}, 0) = \sum_{\nu\mu} \chi_0^{\nu\mu}(\mathbf{q}, 0). \tag{5.175}$$

In Fig. 5.56, it is shown how does the imaginary part of the susceptibility vary with doping. Despite the fact that the peak height in Re $\chi(\mathbf{q}, 0)$ at $T = 0$ is hardly sensitive to doping, the imaginary part Im $\chi(\mathbf{q}, 0)/\omega$ for $\mathbf{q} = \mathbf{Q}$ exhibits a strong dependence on x. This results in a strong variation of T_N with doping (inset in

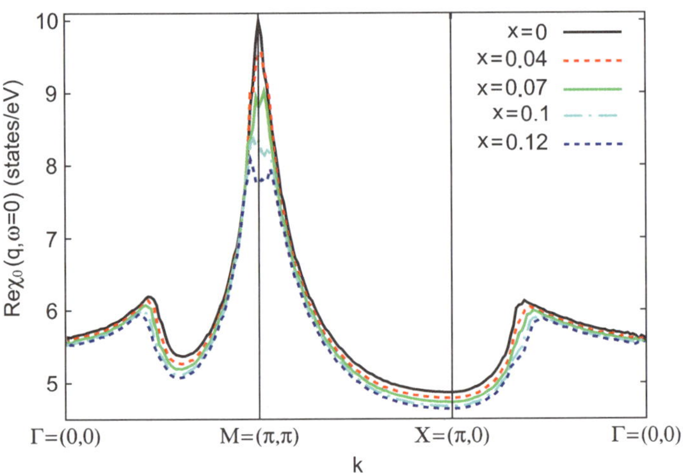

Fig. 5.55 Real part of the static spin susceptibility and its evolution on doping ($n = 4 + x$) [468]

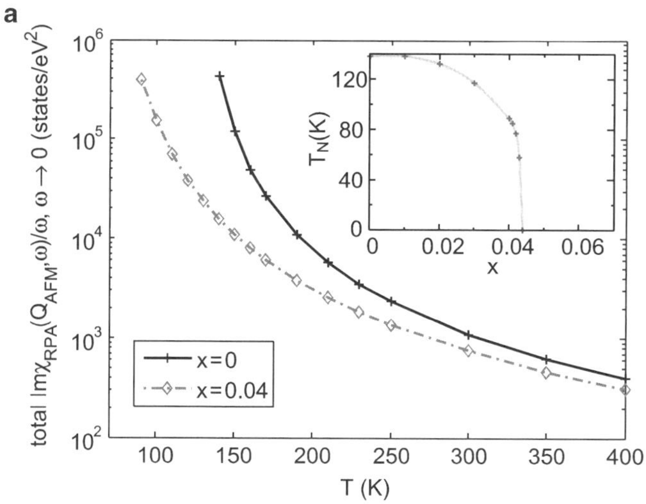

Fig. 5.56 Calculated value of Im $\chi_{RPA}(\mathbf{Q}, \omega)/\omega$ at $\omega \longrightarrow 0$, for non-doped and doped compounds. In the inset, the temperature dependence of the magnetic ordering on doping is shown [468]

Fig. 5.56), and already at $x \simeq 0.04$ the antiferromagnetic ordering breaks down. On further increase of x, the antiferromagnetic fluctuations are strongly suppressed.

A uniform static susceptibility Re $\chi(0, 0)$ exhibits an unusual temperature dependence. Above T_N it is neither Pauli-like nor Curie–Weiss-like. For an undoped compound, it slowly increases with temperature, achieving a maximum at $T \simeq 600$ K, and then decreases. On an increase of doping up to $x \simeq 0.1$, the susceptibility varies weakly up to $T \simeq 200$ K, and then rapidly decreases as the temperature rises further. This means that the short-wavelength spin fluctuations get suppressed by doping of $x \simeq 0.12$.

The main result of the discussed study is that the spin instability within the paramagnetic phase of the LaOFeAs system appears on the wave vector $\mathbf{Q} = (\pi, \pi)$, which is the wave vector of the SDW phase (stripe structure), obtained in experiment. It turns out therefore that the wave vector of the SDW ordering coincides with the nesting vector which connects the hole and electron sheets of the Fermi surface. The scattering of electrons on passing from the hole to the electron sheet of the Fermi surface (and back) forms the magnetic properties of the system: the spin fluctuations spectrum and the magnetic structure.

Even as the main features of the physics of FeAs-systems, relating the structure of the electron spectra (Fermi surface) with the details of the spin susceptibility, do follow from the itinerant model just outlined, a choice of the interaction parameter (5.174) remains unsatisfactory. The magnitude of the Coulomb interaction is too weak; it is by an order of magnitude smaller than various estimates under discussion in relation to the properties of the FeAs-systems. It would make interest not to extract U from the adjustment of the Néel temperature to its experimental value, but to calculate T_N departing from a more realistic estimate for U. It would be interesting to find out how would the features and the magnitude of spin susceptibility change on applying such a procedure.

5.8.3 The Localized Model: Spin Waves

We consider now a localized model of undoped LaOFeAs, the $J_1 - J_2$ Heisenberg model [472]:

$$H = J_1 \sum_{\langle ij \rangle} \mathbf{S}_i \, \mathbf{S}_j + J_2 \sum_{\langle\langle ij \rangle\rangle} \mathbf{S}_i \, \mathbf{S}_j . \tag{5.176}$$

Here, the first term takes into account the exchange interaction between the nearest-neighbouring Fe atoms, and the second one – between the next-nearest neighbours. A justification for a choice for such model is the fact that both interactions are mediated by the As atoms situated below or above the centres of the plackets made of Fe atoms. An analysis of overlaps of orbitals at Fe and As atoms gives that $J_2 > 0$ and $J_1 > 0$, i.e. both exchange interactions are antiferromagnetic, and their magnitudes are close to each other. At the same time, the first-principles calculations [469, 473] show the J_1 to be a ferromagnetic one. Due to an ambiguity in the data concerning the J_1 magnetic interaction, both signs for it should be admitted. Interactions J_1 and J_2 are marked in Fig. 5.57a, where also the experimentally determined magnetic structure of the SDW-type for LaOFeAs is shown.

The spin wave vector for such structures in the linear approximation (LSW) has been calculated in [472]. Using the known Holstein–Primakoff formulae relating the spin operators on a site with the Bose-operators of spin deviations, one can obtain, for the magnetic structure shown in Fig. 5.57a, the model Hamiltonian in terms of the Bose-operators $a_{\mathbf{k}}$ and $a^+_{\mathbf{k}}$ [472]:

$$H = E_0 + S \sum_{\mathbf{q}} \left[A_{\mathbf{q}} \, a^+_{\mathbf{q}} \, a_{\mathbf{q}} + \frac{1}{2} \, B_{\mathbf{q}} \, a^+_{\mathbf{q}} \, a^+_{-\mathbf{q}} + \frac{1}{2} \, B^*_{\mathbf{q}} \, a_{-\mathbf{q}} \, a_{\mathbf{q}} \right], \tag{5.177}$$

where $E_0 = -2 \, J_2 \, S^2 \, N$ is the ground state energy, and

$$\begin{aligned} A_{\mathbf{q}} &= 4J_2 + 2J_1 \cos q_x , \\ B_{\mathbf{q}} &= 2J_1 \cos k_y + 8J_2 \cos q_x \cos q_y . \end{aligned} \tag{5.178}$$

With the help of an unitary transformation to the new Bose operators

$$b_{\mathbf{q}} = \cosh \theta_{\mathbf{q}} \, a_{\mathbf{q}} - \sinh \theta_{\mathbf{q}} \, a^+_{-\mathbf{q}}$$

the Hamiltonian becomes diagonal,

$$H = \sum_{\mathbf{q}} \omega(\mathbf{q}) \, b^+_{\mathbf{q}} \, b_{\mathbf{q}}, \tag{5.179}$$

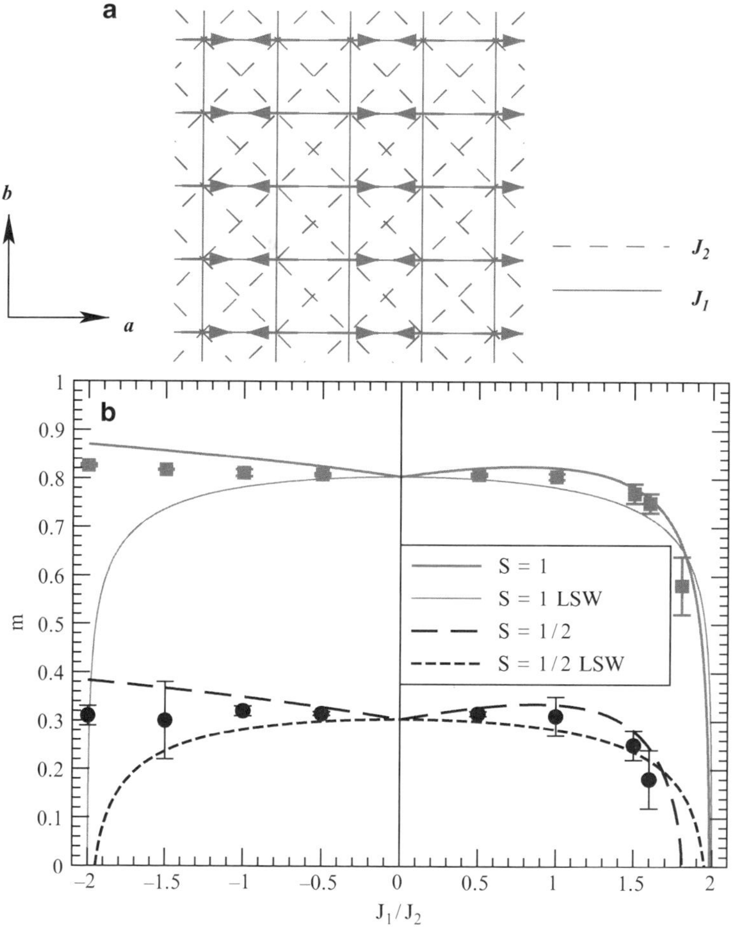

Fig. 5.57 (**a**) Magnetic structure of the LaOFeAs compound; (**b**) calculated values of the mean magnetic moment per Fe atom in the model (5.176), at two chosen spin values, $S = 1$ and $S = 1/2$ [473]

where the energies of spin waves are given by

$$\omega(\mathbf{q}) = S\sqrt{A_{\mathbf{q}}^2 - B_{\mathbf{q}}^2}$$
$$= 2S\sqrt{(2J_2 + J_1\cos q_x)^2 - \left(J_1\cos q_y + 2J_2\cos q_x\cos q_y\right)^2}. \quad (5.180)$$

From this, an expression for the vector of spin wave velocity can be obtained, defined from the dispersion law (5.8.3) at small $\mathbf{q}$, when $\omega(\mathbf{q}) \simeq \mathbf{v}\cdot\mathbf{q}$:

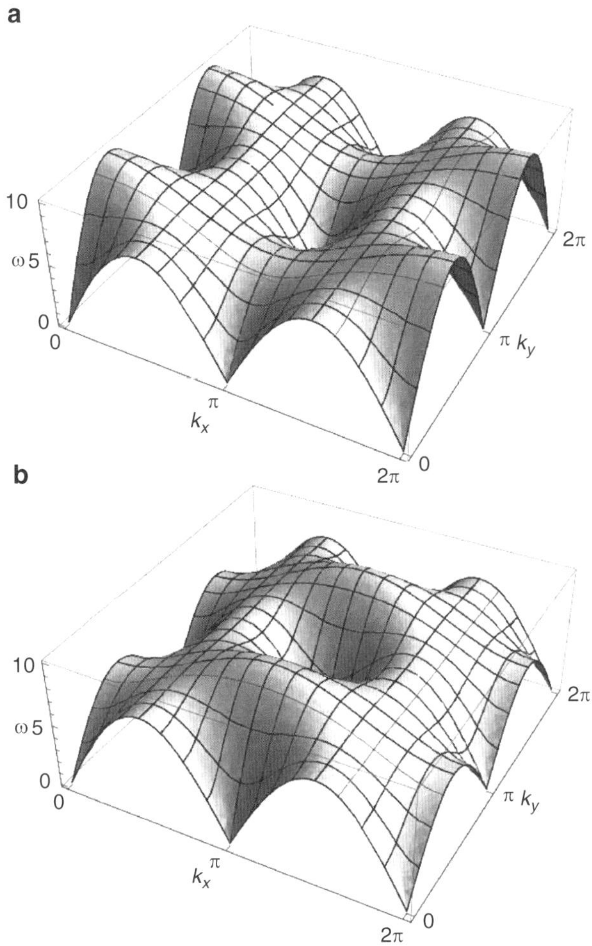

Fig. 5.58 Spin wave dispersions in two-sublattice magnetic structure of Fig. 5.1: (**a**) $J_1 = 1$, $J_2 = 2$; (**b**) $J_1 = -1$, $J_2 = 2$ [472]

$$\begin{aligned} v_x &= 2\,S\,\sqrt{-J_1^2 + 4J_2^2}, \\ v_y &= 2\,S\,|J_1 + 2J_2|. \end{aligned} \tag{5.181}$$

The dispersion law (5.8.3) is shown in Fig. 5.58 for two cases, with positive and negative J_1 values. In both cases, the spin wave energy falls down to zero at the wave vectors $\mathbf{q} = (\pi, \pi)$ and $\mathbf{q} = (\pi, 0)$.

The magnetization of a sublattice can be calculated using the relation

$$m = \langle S_i^z \rangle = S - \Delta m, \tag{5.182}$$

where

$$\Delta m = \langle a_i^+ a_i \rangle = \sum_{\mathbf{q}} \langle a_{\mathbf{q}}^+ a_{\mathbf{q}} \rangle$$
$$= \frac{1}{2N} \sum_{\mathbf{q}} \left[\frac{S A_{\mathbf{q}}}{\omega(\mathbf{q})} - 1 \right] + \frac{1}{N} \sum_{\mathbf{q}} \frac{S A_{\mathbf{q}}}{\omega(\mathbf{q})} \frac{1}{e^{\beta\, \omega(\mathbf{q})} - 1}. \quad (5.183)$$

The first term originates from quantum zero-point vibrations, the second one – from thermic fluctuations. The magnitude of the spin contraction Δm depends on the J_1/J_2 relation; at $J_1 \simeq J_2$, Δm makes about 10% of the S magnitude and cannot induce a noticeable decrease of $m = \langle S_i^z \rangle$ at Fe, observed in experiment. To clarify the situation, in [473] a calculation of spin waves beyond the linear approximation has been done, on the basis of self-consistent spin wave theory. The results of such calculations are depicted in Fig. 5.57, where the magnitude of mean spin is given as function of J_1/J_2, for positive and negative J_1. The dashed line shows the result of the linear (LSW) approximation. As is seen, for $J_1 > 0$ the results are qualitatively close between the linear and the self-consistent theories, whereas for $J_1 < 0$, as the $|J_1|/J_2$ grows, the deviation becomes enormous. For an antiferromagnetic exchange J_1, m dramatically drops down as J_1/J_2 approaches 2. The critical value of this parameter depends on the spin magnitude S and equals

$$\left.\frac{J_1}{J_2}\right|_{S=1/2} = 1.80, \quad \left.\frac{J_1}{J_2}\right|_{S=1} = 1.98.$$

Therefore, in the presence of a frustration in the spin system (both exchange interactions being antiferromagnetic), the mean value of the projection of the magnetic moment onto the magnetization axis can be very small. A peculiarity of the self-consistent spin wave spectrum is that $\omega(\mathbf{q})$ does not drop down to zero at the $\mathbf{q} = (0, \pi)$ wave vector (Fig. 5.59). We see that along the a axis (i.e. along the magnetic moments of the SDW structure), the both models, the linearized and the self-consistent ones, give consistent results. Along the b axis (perpendicularly to the magnetic moments), the solid and the dashed lines diverge. It should be specially pointed out that in neither of the models does the spin wave energy fall down to zero at $\mathbf{q} = (0, \pi)$. We underline moreover an anisotropy of the spin waves velocity along the a and b directions, seen from the figure. These both predictions of the theory allow, by making a comparison with experiment, to estimate the magnitude of exchange interactions. From the band calculations [65, 469], an estimate of $J_2 \simeq 33$ meV follows.

To make the situation more realistic, we should add to the Hamiltonian of the $J_1 - J_2$ model (5.176) the terms describing the single-ion anisotropy:

$$H_{\mathrm{A}} = \sum_i \left\{ K_c \left(S_i^z\right)^2 + K_{ab}\left[\left(S_i^x\right)^2 - \left(S_i^y\right)^2\right]\right\}. \quad (5.184)$$

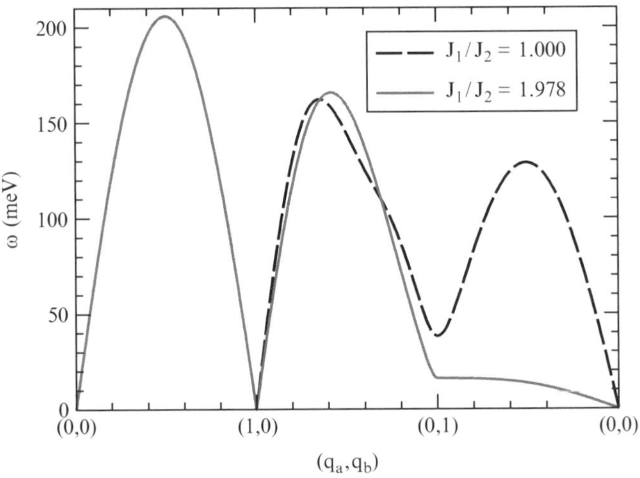

Fig. 5.59 Dispersion curves in the self-consistent spin wave theory (*solid curve*) with J_1/J_2 =1.978. For comparison, a spectrum is given as calculated in the itinerant model (*dashed line*) with $J_1/J_2 = 1$ [473]

where an anisotropy in the basal plane gives rise to an inequivalence of magnetic ordering along the x and y axes. The spin wave vector of the SDW structure with the wave vector $\mathbf{Q} = (0, \pi)$ and the full cross-section of the neutron scattering on them have been calculated in [474]. In the vicinity of the $\mathbf{Q}$ vector, the spin wave dispersion can be approximated by the following expression [475]:

$$\hbar\omega(\mathbf{q}) = \sqrt{\Delta^2 + v_{xy}^2\left(q_x^2 + q_y^2\right) + v_z^2 q_z^2}, \tag{5.185}$$

where the gap Δ and the spin wave velocities v_{xy}, v_z can be expressed via the model parameters: J_2, J_{1a}, J_{1b} and J_z – exchange interactions between the nearest and between the next-nearest neighbours, and the anisotropy constants K_{ab} and K_c.

Measurements of the spin wave spectrum by neutron spectroscopy have been done for a number of 122-compounds: $SrFe_2As_2$ [294], $CaFe_2As_2$ [302, 475] and $BaFe_2As_2$ [295, 474]. These experiments gave consistent results. Very sharp dispersion curves have been obtained, ascending from the $\mathbf{q} = \mathbf{Q}$ point with a substantial energy gap: 6.9 meV in $SrFe_2As_2$ and 9.8 meV in $BaFe_2As_2$. Low-energy measurements permitted to estimate the width of the spin wave spectrum along the formula for the dispersion curve [474]; it makes 175 meV.

From the comparison of the measured dispersion curves with the formula (5.189), the following estimates for the exchange interactions follow: in $SrFe_2As_2$, $J_{1a} + 2J_2 \simeq 100$ meV, $J_2 \simeq 5$ meV [294]; in $CaFe_2As_2$, $J_{1a} \simeq 41$ meV, $J_{1b} \simeq 10$ meV, $J_2 \simeq 2.1$ meV, $J_z \simeq 3$ meV. We note that the J_2 is sufficiently large to stabilize the SDW structure.

5.8.4 The Resonance Mode

The most important area of investigations is the study of spin fluctuations within the superconducting phase, because it is assumed, since the discovery of high-temperature superconductivity, that namely these fluctuations are responsible for the mechanism of electron pairing. By inelastic neutron scattering in cuprates, collective spin resonance modes have been found which are grouped around the wave vector $\mathbf{Q} = (1/2,\ 1/2)$ of non-magnetic structure in doped compounds, and whose energy falls within the superconducting gap (see [476] and references therein). The intensity of this mode changes with temperature as the superconducting order parameter does; note that no dispersion along the c axis has been found, which reveals that the mode observed is related to the dynamics of spins situated in the CuO_2-planes in cuprates. A discovery of the resonance mode in cuprates confirmed the $d_{x^2-y^2}$-symmetry of the order parameter and provided a strong argument in favour of the spin fluctuation mechanism of pairing.

A similar resonance mode is observed in the FeAs-compounds of the 122 type, because for them good-quality single crystals are available. With the help of inelastic neutron scattering, the resonance mode has been discovered in $Ba_{0.6}K_{0.4}Fe_2As_2$ ($T_c = 38$ K) [296, 477], $BaFe_{1.9}Ni_{0.1}As_2$ ($T_c = 20$ K) [478], $BaFe_{1.84}Co_{0.16}As_2$ ($T_c = 22$ K) [297, 479].

Let us consider in more detail the results of the study of $Ba_{0.6}K_{0.4}Fe_2As_2$ [477]. In Fig. 5.60, the intensities of the magnetic resonance of neutrons are shown over the field of variable transfer momenta (abscissa axis) and transfer energy (ordinate axis). In the (a) panel depicting the measurements at $T = 7$ K, a dark spot at $|\mathbf{Q}| = 1.15\,\text{Å}^{-1}$ and $\Delta E \simeq 15$ meV corresponds to a magnetic excitation localized in the momentum and energy spaces. No such spot is present in the panel (b) which corresponds to the normal phase ($T = 50$ K). The energy of this excitation ω_c falls within the superconducting gap, which, according to ARPES data, equals 12 meV [480], so that $\frac{\omega_c}{2\Delta} \sim 0.58$ [477].

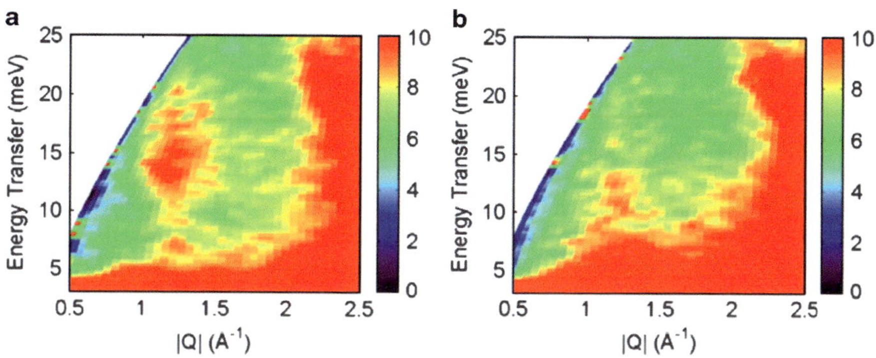

Fig. 5.60 Intensity of inelastic neutron scattering on $Ba_{0.6}K_{0.4}Fe_2As_2$ in the superconducting phase (**a**) and in the normal phase (**b**), after [477]

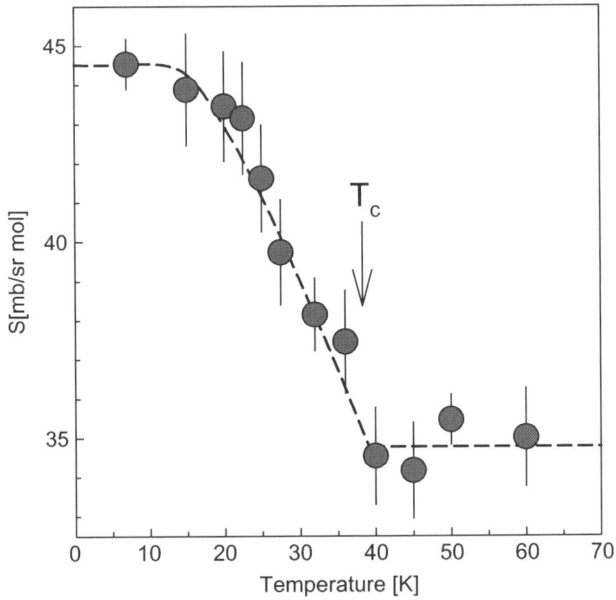

Fig. 5.61 Temperature dependence of the integral intensity of neutron scattering, corresponding to the region of maximum intensity of the resonance excitation (the *dark spot* area in Fig. 5.60a), after [477]

The intensity of the dark spot, integrated over an appropriate range of energies and momenta in the vicinity of the $\mathbf{Q}$ vector, varies with temperature in the same manner as the superconducting gap (Fig. 5.61) [481]. In the FeAs-compounds, it makes a strong argument in favour of the $s^{\pm}$-symmetry of the order parameter.

Indeed, let us turn to an expression for the coherence factor (Sect. 5.4), which describes the magnitude of the bare spin susceptibility in a superconductor $\chi_0(\omega, \mathbf{q})$. It comprises a product $\Delta_\mathbf{k}\, \Delta_{\mathbf{k}+\mathbf{q}}$. The integration in $\mathbf{k}$ runs over the whole Fermi surface. For the FeAs-compounds in case of $\mathbf{q} \simeq \mathbf{Q}$, the gaps $\Delta_\mathbf{k}$ and $\Delta_{\mathbf{k}+\mathbf{q}}$ belong to different sheets of the Fermi surface, a hole one and an electron one. In case of the $s^{\pm}$-symmetry of the order parameter they have opposite signs, therefore the coherence factor at the Fermi surface equals 2 (for a conventional s-symmetry of the order parameter, the coherence factor equals zero). In the RPA, the spin susceptibility is given by an expression:

$$\chi_{\mathrm{RPA}}(\omega, \mathbf{q}) = \frac{\chi_0(\omega, \mathbf{q})}{1 - U\chi_0(\omega, \mathbf{q})}. \tag{5.186}$$

Since for a $s^{\pm}$-superconductor, $\chi_0(\omega, \mathbf{q} = \mathbf{Q}) \neq 0$, the denominator may become zero at some frequency ω_0 that would mean an onset of a collective resonance mode. For the conventional s-symmetry, $\chi_0(\omega, \mathbf{q} = \mathbf{Q}) = 0$, therefore no resonance mode with a wave vector $\mathbf{q} = \mathbf{Q}$ appears.

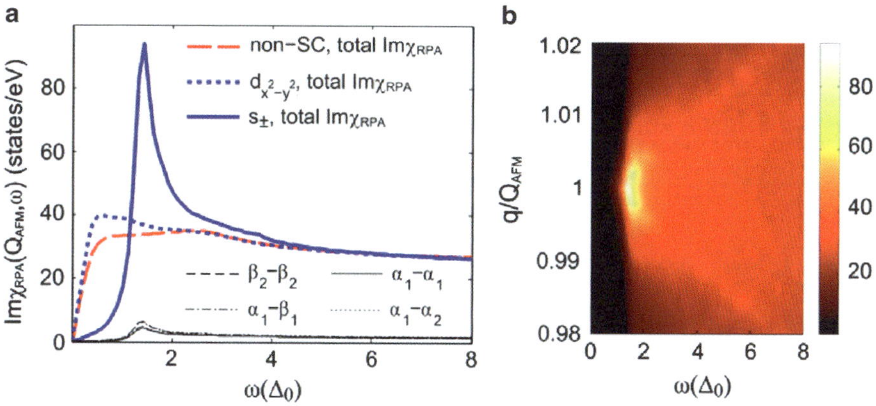

Fig. 5.62 (**a**) Calculated imaginary part of spin susceptibility in the RPA, at the wave vector $\mathbf{Q} = (\boldsymbol{\pi}, \mathbf{0})$, as function of frequency in the normal and superconducting state, for $s^{\pm}$ and $d_{x^2-y^2}$-symmetries of the order parameter (**b**) Calculated imaginary part of spin susceptibility for the $s^{\pm}$-symmetry of the order parameter, as function of frequency and momentum [477]

These simple arguments find support in numerical calculations of the spin susceptibility (Fig. 5.62). In the (b) panel, a light spot indicates a calculated intensity of spin fluctuations in the vicinity of the quasi-momentum **Q**. The energy of the resulting resonance mode falls into the superconducting gap. Therefore, a discovery of this resonance mode in the superconducting FeAs-compounds is consistent with a multi-sheet nature of the Fermi surface in these compounds and with an assumption about the $s^{\pm}$-symmetry of the superconducting order parameter.

In [478], an observation of three-dimensional resonance in the $BaFe_{1.9}Ni_{0.1}As_2$ compound has been reported. Resonance peaks have been found in the vicinity of wave vectors (101) and $(10\bar{1})$, see Fig. 5.63. Remarkable is a difference in the positions of resonances for the wave vectors (100) in the basal plane and $(10\bar{1})$, protruding out of it. In the first case, the resonance occurs at the energy $\hbar\omega = 9.1 \pm 0.4$ meV, in the second one – at $\hbar\omega = 7.0 \pm 0.5$ meV. In Fig. 5.63c, the temperature dependence of the intensity of the $\hbar\omega = 7$ meV peak, that for the at the wave vector $(10\bar{1})$, is shown.

If a resonance is a measure of pairing interaction between the electrons, an observation of a three-dimensional resonance is a manifestation of a fact that the superconducting order parameter does vary with z; in other words, the resonance mode exhibits a dispersion in the z direction. From simple assumptions, it can be estimated how does it depend on the wave vector q_z:

$$\hbar\omega(q_z) \simeq \Delta_0 - 2\delta \left|\sin \frac{q_z}{2}\right| . \tag{5.187}$$

From a comparison with experimental data, we get:

$$\delta/\Delta_0 = \left[\omega(100) - \omega(10\bar{1})\right]/\omega(100) \simeq 0.26 \pm 0.07.$$

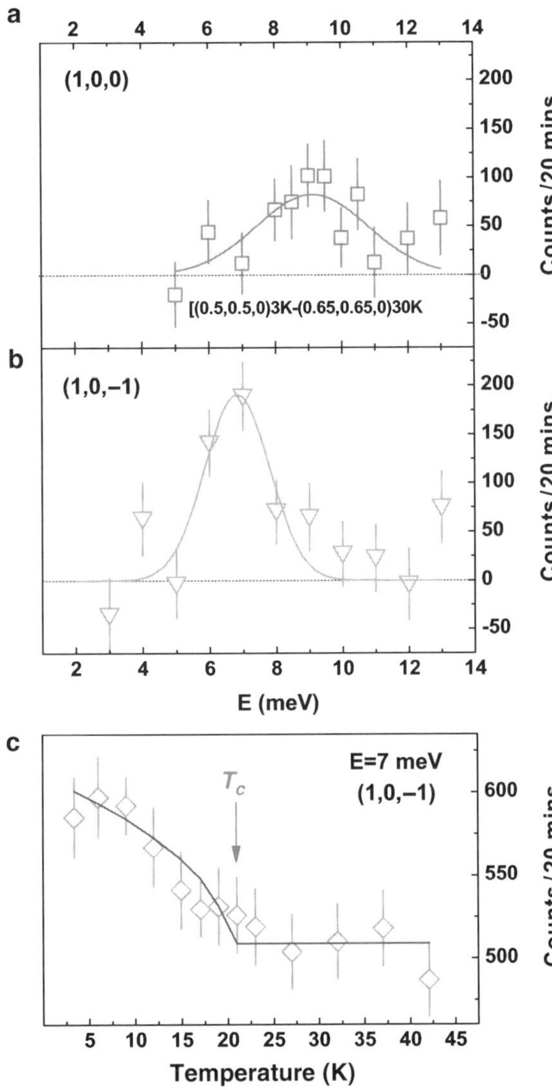

Fig. 5.63 Three-dimensional resonance in the $BaFe_{1.9}Ni_{0.1}As_2$ compound. (**a**, **b**): intensity peaks of the magnetic neutron scattering at the (100) and $(10\bar{1})$ wave vectors; (**c**): temperature dependence of the resonance peaks [478]

The Δ_0 and δ parameters turn out to be proportional to the exchange interactions $J_\perp$ and $J_\parallel$, correspondingly, therefore it is natural to assume that $\delta/\Delta_0 \sim J_\perp/J_\parallel$, where $J_\parallel$ is exchange interaction between Fe atoms in the basal plane, and $J_\perp$ – between the planes.

Thus, the resonance peaks of magnetic susceptibility within the superconducting gap make a common phenomenon in the FeAs-systems, along with their presence in cuprates. In case of the FeAs superconductors, a discovery of such peaks makes a strong argument for the $s^\pm$ symmetry of the order parameter.

A discovery of q_z-sensitivity of the resonance peak opens a possibility for an existence, in FeAs-systems, of a superconducting order parameter with zeros at the Fermi surface along the z axis, combined with an absence of zeros in the basal plane. A possibility of a formation of such superconducting state has been put forward in a work by Laad and Craco [482], based on a picture of FeAs-compounds like bad metals with non-coherent one-particle states, influenced by moderate (strong) electron correlations. In a number of papers [390, 482–485], these authors studied, by applying LDA+DMFT, a role of electron correlations in the shaping of the properties of dynamic spin fluctuations within multi-orbital models. Similar results concerning the dispersion of the resonance mode along the c axis were obtained in the localized model of a superconductor as well [480].

We reproduce here one of the results concerning the temperature dependence of the static susceptibility $\chi(T)$ for a compound based on La, with the doping level of $x = 0.1$ (Fig. 5.64). It is seen that at high temperatures $T > 200$ K, $\chi(T)$ does linearly depend on T, as it is usually the case in systems with strong correlations. At low temperatures, $\chi(T) \sim T^{1.8}$, as is well consistent with experimental data in a choice of the $U \simeq 4$ eV parmeter, which has been used in the works on LDA+DMFT calculations [18, 390, 483].

The resonance mode was also observed in the same $BaFe_2As_2$ system, in the optimally Ni-doped $BaFe_{1.9}Ni_{0.1}As_2$ compound with $T_c = 20$ K [486]. Measurements of the intensity of inelastic neutron scattering in the vicinity of the wave vector $(\frac{1}{2}\frac{1}{2}0)$ were done in an applied magnetic field of $H = 14.5$ T, and also at $H = 0$ T. In the zero field, the resonance have been detected at the energy $\hbar\omega = 8$ meV; an increase of the field suppresses superconductivity. The resonance

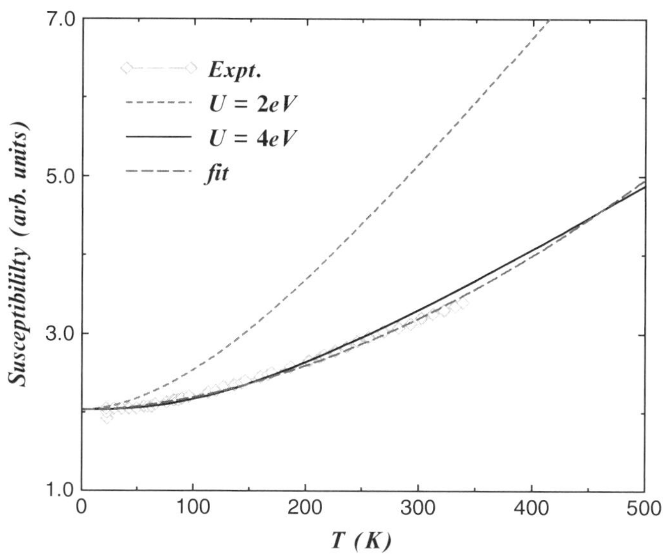

Fig. 5.64 Temperature dependence of χ(T), calculated along the LDA+DMFT method, in comparison with experimental data (*diamonds*) [485]

intensity decreases with the field; simultaneously the resonance energy shifts to 6.5 meV. This indicates that the resonance energy is related to the energy of electron pairing, and consequently the spin fluctuations do participate in the mechanism of the electron pairing.

We point out one more work [487], in which resonance conditions have been studied in the 5-orbital model within the LDA for the magnetic susceptibility. It turns out that both the $s^{\pm}$-symmetry and the extended s-symmetry do agree with the data available of neutron spectroscopy about a resonance on the wave vector $(\pi, 0)$. Measurements of resonance at other transfer wave vectors would help to resolve this ambiguity.

An observation of a resonance mode in superconducting $BaFe_2As_2$, doped with potassium [296], cobalt [297] and nickel [478] exhibited that this phenomenon is of universal character and reveals the dynamics of spin fluctuations within the superconducting phase. As a continuation of these works, spin dynamics has been studied in the $BaFe_{2-x}Co_xAs_2$ compound, in the region of coexistence of superconductivity and magnetic ordering [300]. In the range of doping $0.06 < x < 0.12$, this compound is both antiferromagnet and superconductor, although at presence stage it is impossible to judge whether this coexistence takes place at the microscopic scale, or in a form of strongly dispersed phase separation.

The neutron scattering experiment has been done on a sample with $x = 0.08$, in which $T_N = 58\,K$ and $T_c = 11\,K$. It was found that on cooling, the intensity of the magnetic Bragg peak reduces by 6% on passing through the T_c point. Above T_c, spin waves with a spin gap of 8 meV have been observed, and below T_c an inelastic scattering of neutrons with the energy $\sim$4.5 meV took place. The intensity of this resonance depends on the wave vector not in the basal plane only, as was the case of optimally doped superconductors without a long-rang magnetic order; it also depends on the component of momentum transfer along the c axis.

It is obvious that this observed spin resonance has been linked to the superconducting state; its intensity varied with temperature as the superconducting gap did, and dropped to zero at $T \to T_c$. A substantial difference of the phenomenon observed from the resonance in optimally doped superconductors consisted in the fact that an appearance of the spin resonance mode was accompanied by a reduction of intensity of the magnetic Bragg peak, i.e. a re-distribution of intensity of magnetic inelastic neutron scattering occurred between static magnetism and dynamic spin fluctuations. The phenomenon observed did clearly demonstrate that in the $BaFe_{1.92}Co_{0.08}As_2$ compound, a coexistence of superconductivity with long-range magnetic order may occur.

An important study of electron-doped (Co-doped) compound $BaFe_2As_2$ has been done with the use of inelastic neutron scattering [488]. Spin excitations in a single-crystal sample of $BaFe_{1.85}Co_{0.15}As_2$ ($T_c = 25\,K$) have been analyzed over a broad interval of temperatures up to 280 K and energies up to 32 meV. The measured dynamical structure factor $S(\mathbf{Q}, \omega)$ for the wave vector $\mathbf{Q} = \mathbf{Q}_{AFM} = (\frac{1}{2}\frac{1}{2}1)$, corresponding to the stripe magnetic structure of FeAs compounds, let the authors to determine the imaginary part of the magnetic susceptibility $\chi(\mathbf{Q}_{AFM}, \omega)$, see Fig. 5.65.

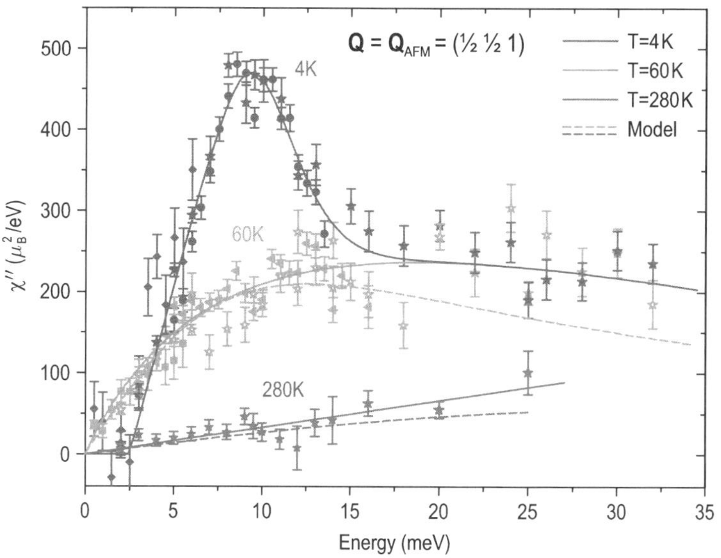

Fig. 5.65 Imaginary part of magnetic susceptibility in the superconducting ($T = 4$ K) and normal ($T = 60$ and 280 K) phases of $BaFe_{1.85}Co_{0.15}As_2$, reconstructed from the data on inelastic neutron scattering [488]

The curve corresponding to $T = 4$ K reveals a spin resonance at the energy $\varepsilon_{res} = 9.5$ meV, consistently with the measurements of [489]. At energies $\varepsilon < 3$ meV, a gap in the spectrum of spin excitations in the superconducting phase is observed, which compensates an increase of spin fluctuations in the resonance range. The measurement of the resonance energy at different temperatures has shown $\varepsilon_{res} \sim (1.6 \pm 0.3)\Delta(T)$, that is in good agreement with the prediction done for a superconductor with the $s^{\pm}$ symmetry of the order parameter [490, 491].

In the normal state at $T = 60$ K, in the spectrum of $\chi(\mathbf{Q}, \omega)$ the gapless fluctuations are observed with the maximum around 20 meV, and the linear ω-dependence at $\omega \to 0$. An increase of T to 280 K suppresses the intensity, but the low-frequency part of the spectrum maintains its linear behaviour with ω. We note that in the work cited, the spectrum of spin fluctuations is measured not only in the superconducting phase but in the normal one as well. The authors emphasize that in the FeAs-compounds, the spectral weight of spin fluctuations is of the same order as in cuprates, which provides an experimental foundation for the hypothesis on the electron pairing in the FeAs-superconductors via spin fluctuations.

A remarkable result of recent time is a discovery of resonance mode in another type of superconducting compounds, $FeSe_{0.4}Te_{0.6}$ with $T_c = 14$ K [335]. Within the superconducting phase, a spin resonance with the energy $\hbar\Omega_0 = 6.5$ meV $\approx 5.3\, kT_c$ and the width $\hbar\Gamma = 1.25$ meV was observed. The resonance mode has been detected at the wave vector $(\frac{1}{2}\,\frac{1}{2}\,L)$, which corresponds to the nesting vector for the Fermi surface. However, it differs from the wave vector $\mathbf{Q}_m = (\delta\, 0\, \frac{1}{2})$ of the observed magnetic structure of the $Fe_{1+y}Te$ compound, which is different from the SDW structure emerging in all undoped FeAs-compounds. The result obtained in [335] is very important, because it demonstrates a universal character of the resonance mode phenomenon in new high-temperature superconducting materials.

5.8.5 *Unified Models*

Itinerant and localized models considered above make limiting cases of a realistic situation in the FeAs-systems, which is characterized by an observation that the two principal parameters, U and W, are of the same order of magnitude. In such a situation, the features of both itinerant and localized electron states are manifested simultaneously. One of the simplified models, unifying the both aspects, is outlined in [492].

It is presumed that in the FeAs-systems, two groups of electronic states, formed by the Fe d-orbitals, can be singled out. One of them, prominent at the Fermi level, is itinerant one and forms hole and electron regions in the vicinity of the Γ and M points, that is well seen from LDA calculations and confirmed by many experiments. The other group of the d states, far from the Fermi level, forms localized magnetic moments. Such a situation might have take place at large enough U values, corresponding to the strong correlations regime, where a three-peaked structure of spectrum, with a central peak and two Hubbard bands, becomes pronounced (Fig. 5.66). If for these strongly correlated electrons the Mott gap is present, localized magnetic moments may get formed in the lower Hubbard band.

The next step consists of a choice of interaction between the itinerant and localized electrons. The simplest model of such type is given by three terms in the Hamiltonan, $H = H_0 + H_{J_1} + H_{J_2}$, where H_0 describes a contribution from itinerant electrons with their dispersion law $\varepsilon_{\mathbf{k}}$ (yielding a hole and an electron branches near the Γ and M points, respectively); H_{J_1} and H_{J_2} are the exchange interactions within the localized electrons, here

$$H_{J_2} = J_2 \sum_{\langle ij \rangle} \mathbf{M}_i \mathbf{M}_j, \tag{5.188}$$

where $\mathbf{M}_i$ is the magnetic moment at the i site, and the summation runs over neighbours beyond the nearest ones, with the antiferromagnetic exchange interaction J_2 (the J_1 interaction between the nearest neighbours is assumed small). A supposed coupling between the two groups of electrons is

$$H_{J_0} = J_0 \sum_{\langle ij \rangle} \mathbf{M}_i \mathbf{S}_j, \tag{5.189}$$

where $\mathbf{S}_i$ is an operator of spin of an itinerant electron at the site i.

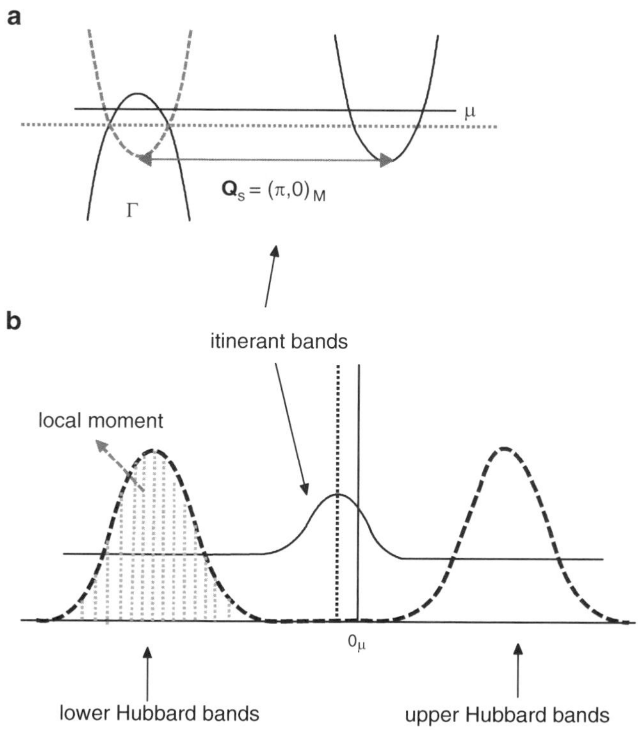

Fig. 5.66 Schematic structure of electron states in the unified model [492]. (**a**) Band spectrum of itinerant electrons in the vicinity of the Fermi surface; (**b**) density of states in the system of strongly localized electrons [492]

If in a field of localized electrons, due to antiferromagnetic exchange J_2, a two-sublattice ordering is formed, then in the system of itinerant electrons, due to the Hund's coupling, a splitting of the bare bands occurs, so that two bands are created, with the dispersion law

$$E_{\mathbf{q}}^{\pm} = \pm\sqrt{\varepsilon_{\mathbf{k}}^2 + \Delta_{\mathrm{SDW}}^2}, \tag{5.190}$$

separated by the gap $\Delta_{\mathrm{SDW}} = \frac{1}{2} J_0 M$, M being the mean magnitude of a moment over a sublattice of localized spins. This is a well-known fact, established already in the sd-model. Note that it is not necessary for this gap to fall onto the Fermi level; everything depends on the magnitude of doping. From now on, in the course of calculating the magnetic susceptibility or the strength of the Cooper pairing via the fluctuations of magnetic order, one should proceed not from the bare spectrum of itinerant electrons, but from the renormalized spectrum (5.190).

In Fig. 5.67, calculations of the temperature dependence of sublattice magnetization, gap in the electron spectrum Δ_{SDW} and gap in the spin wave spectrum

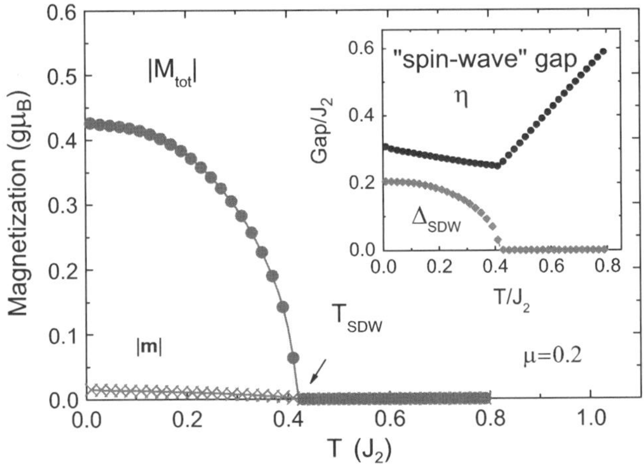

Fig. 5.67 Temperature dependence of sublattice magnetization in the unified model, at fixed chemical potential $\mu = 0.2\,\mathrm{J}$. Inset: the gap Δ_{SDW} in the electron spectrum and the gap η in the spin spectrum [492]

$\Omega_q = \sqrt{c^2 q^2 + \eta^2}$, ($c$ being the spin waves velocity) are shown. The calculation of the spin wave spectrum Ω_q of localized electrons was done within the two-dimensional σ-model, with the coupling constant $g_0 = 16\,J_2$. The magnitude of the Cooper pairing λ in the singlet channel is not varying from g_0 up to the critical value of the σ-model, g_c, and has a magnitude of the order of 1. At $g_0 > g_c$, λ is rapidly decreasing.

The authors of [492] report that the behaviour of magnetic susceptibility, magnetization, and the strength of the Cooper pairing in the unified model do well agree with experimental data for the FeAs-system. It remains unclarified how can in the 5-orbital model a clean separation be done between those states which should be considered as itinerant from those to be taken as localized ones. A DMFT calculation, as we have seen in Sect. 5.2.2, when done for realistic values of the Coulomb repulsion U, does not exhibit a presence, in these systems, of strong correlation effects, i.e. a formation of non-coherent Hubbard bands.

In relation to the issues under discussion, we point out an important work by Haule and Kotliar [493], in which within the LDA+DMFT approach the temperature dependencies of magnetic susceptibility and electrical resistivity at various magnitudes of the Hund's exchange J_{Hund} have been calculated. The experiments done on doped *Re*OFeAs compounds indicate that the temperature dependence of χ and ρ within the range of the normal phase contains anomalies: on a decrease of temperature, χ and ρ undergo a crossover from the localized type of behaviour to the itinerant one. This is manifested by a sharp increase of χ and decrease of ρ at some characteristic temperature T^*, which the authors of [493] do relate to the abovementioned crossover.

As suggested by the authors of [493], the observed effects are manifestations of strong electron correlations in the system, which are governed not only by the relative strength of the Coulomb interaction U/W, but by the single-atom Hund's exchange as well. To prove this suggestion, the authors calculate $\chi(T)$ and $\rho(T)$ within the LDA+DMFT, where the auxiliary single-impurity problem is solved by the quantum Monte-Carlo method with continuous time. The above properties are calculated as

$$\chi(T) = \frac{(g\mu_B)^2 N_A}{k_B} \int\limits_0^\beta d\tau \, \langle S^z(\tau) S^z(0) \rangle \,, \tag{5.191}$$

$$\frac{1}{\rho(\tau)} = \frac{\pi e^2}{V_0 \hbar} \int d\omega \left(-\frac{df}{d\omega} \right) \sum_{\mathbf{k}} \mathrm{Tr} \left[v_k(\omega) \rho_{\mathbf{k}}(\omega) v_k(\omega) \rho_{\mathbf{k}}(\omega) \right] . \tag{5.192}$$

In the first formula, S^z is the spin of an Fe atom, N_A is the Avogadro number; in the second formula, $v_{\mathbf{k}}(\omega)$ is the electron velocity, and $\rho_{\mathbf{k}}(\omega)$ – the electron spectral density for the bands formed by all five atom orbitals, so that these values are matrices; V_0 is the unit cell volume.

In the static limit, (5.191) transforms into the expression

$$\chi(T) = \frac{(g\mu_B)^2 S(S+1) N_A}{3kT} \,, \tag{5.193}$$

which yields the Curie–Weiss law for localized magnetic moments. In the opposite limit of itinerant magnetism, the spin susceptibility (5.191) leads to the Pauli susceptibility,

$$\chi(T) = \mu_B^2 N_A N(0), \tag{5.194}$$

where $N(0)$ is the density of states at the Fermi level. Between these two limits, the behaviour of $\chi(T)$ does strongly depend on the J_{Hund} parameter.

In Fig. 5.68, the calculated $\chi(T)$ and $\rho(T)$ are shown for different values of J_{Hund}. It is seen from the figure that, indeed, a certain temperature T^* exists for which a change of regime of temperature behaviour occurs for both susceptibility and resistivity. Moreover in [493], the γ coefficient was calculated which determines the temperature behaviour of heat capacity: $C_V \sim \gamma T$. It turned out that γ does rapidly increase with the Hund's exchange J_{Hund}. The behaviour of all three properties, $\chi(T)$, $\rho(T)$ and γ, is consistent between themselves and with the experiment, if the value $J_{\text{Hund}} = 0.35\,\text{eV}$ is chosen. This value of J_{Hund} does determine the crossover temperature, which for the compound under consideration makes about 100 K.

A rapid rise of γ with J_{Hund} means a drastic increase of the effective mass in dependence of this parameter, and an exponential suppression of the coherence parameter. This was first pointed out for the two-band Hubbard model in [494].

In [493] moreover the exchange integrals have been calculated between the nearest Fe neighbours J_1, and the next-nearest ones J_2. The relation $J_2 > J_1/2$, which assures, in *Re*OFeAs compounds, a realization of the experimentally observable SDW ordering, is exactly satisfied for the value $J_{\text{Hund}} = 0.35\,\text{eV}$ chosen from

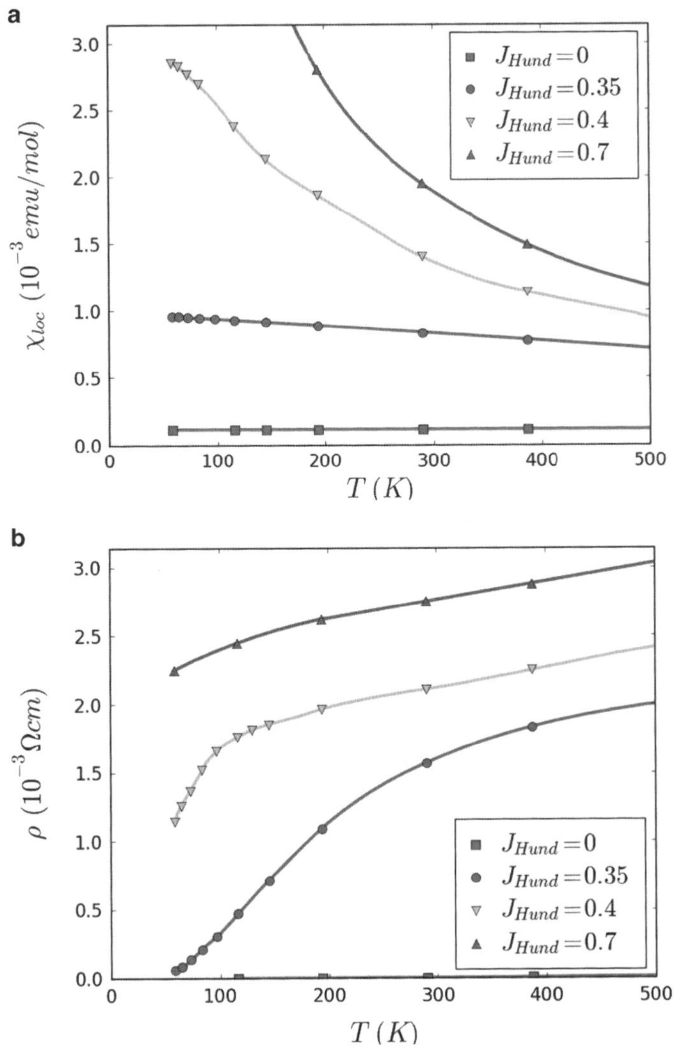

Fig. 5.68 Local magnetic susceptibility (**a**) and electrical resistivity (**b**) of $LaO_{0.9}F_{0.1}FeAs$ as functions of temperature, calculated within the LDA+DMFT at different values of the Hund's exchange (after [493])

the temperature dependencies of $\chi(T)$ and $\rho(T)$. This value is approximately three times smaller than the 1.2 eV value corresponding to a free Fe atom. It is two times smaller than the value 0.7 eV, calculated in [392] by the LDA+DMFT approach using the QMC algorithm by Hirsch and Fye.

Therefore, the study of transport properties of doped FeAs compounds shows that they belong to an intermediate case between those of itinerant and localized magnetic moments. At high temperatures, in their behaviour the trend typical for localized system gets more pronounced, at lower temperatures – that for itinerant ones gains. The reason for such duality is in electron correlations, which are controlled not only by the Coulomb interaction parameter, but by the intraatomic Hund's exchange as well.

5.8.6 FeAs-Compounds as Systems with Moderate Electron Correlations

Let us summarize the results of experimental and theoretical studies of the FeAs-systems in what regards the role of electron correlations in them. Transport properties reveal that these systems are bad metals. Indeed, the electrical resistivity at room temperatures is of the order of 5 mΩ·cm for polycrystals and 0.5 mΩ·cm for single crystals. This corresponds to the $k_F \ell$ parameter being of the order of one, whereas in good metals, e.g. in Cu, $k_F \ell \sim 10^2$. When the mean free path of carriers, ℓ, is of the order of mean distance between them, this is a manifestation of a bad metal.

Another signature of a bad metal is the magnitude of the Drude peak in these compounds, as well as an increase of a contribution to the optical conductivity due to high-energy transitions from coherent electronic states, which are forming the Drude peak, to non-coherent states with the energy 0.5–1.5 eV. An increase of effective masses m^*/m_0, according to the dHvA and ARPES measurements, makes a factor of 2–4, that reveals a substantial contribution into m^* of the electron–electron interaction.

A good agreement of measured Fermi surface with LDA calculations does not mean small role of electron correlations, because even in the case of weak one-site correlations the electron self-energy Σ does not depend on quasi-momentum and hence does not affect the shape of the Fermi surface.

The outlined peculiarities of electronic states of FeAs-systems witness a substantial role of electron correlations, even if spectroscopic experiments do not unambiguously reveal the features of strongly correlated systems, like an existence of manifested incoherent peaks substantially above and below the Fermi level. Nevertheless, judging by a bulk of evidence, one could have expected such peaks, i.e. FeAs-compounds should be attributed to systems with moderate electron correlations: not immediately close to the Mott transition, but yet feeling its possibility. The authors of [495] suggest that FeAs-systems resemble the known V_2O_3 system, which is intermediate between the localized and itinerant Mott antiferromagnets. They propose a certain phenomenological concept for a description of such intermediate case, which we do simply reproduce. A preliminary publication of this concept can be found in [496].

We suggest that, along with the quasiparticle peak at the Fermi level, an incoherent peak exists, and the integral intensity is distributed between them in the relation w to $1 - w$. We split the field operator d for an electron state into two components, $d = d^{\text{coh}} + d^{\text{inc}}$, related to the coherent (quasiparticle) and the incoherent peaks, respectively. Low-energy excitations (near the Fermi level) are described in terms of the d^{coh} operators, which are easy to normalize, introducing the $c = d^{\mathrm{coh}}/\sqrt{w}$ operators. The kinetic term of the Hamiltonian, $H_t = \sum t_{ij}^{ab} d_{ia\sigma}^{+} d_{jb\sigma}$, is then split into three parts:

$$H_{t1} = \sum t_{ij}^{ab} d_{ia\sigma}^{+\,\mathrm{coh}} d_{jb\sigma}^{\mathrm{coh}}, \tag{5.195}$$

$$H_{t2} = \sum t_{ij}^{ab} d_{ia\sigma}^{+\,\mathrm{inc}} d_{ja\sigma}^{\mathrm{inc}}, \tag{5.196}$$

$$H_{t3} = \sum t_{ij}^{ab} \left(d_{ia\sigma}^{+\,\mathrm{coh}} d_{ib\sigma}^{\mathrm{inc}} + \mathrm{H.c.} \right). \tag{5.197}$$

In terms of the c operators, the H_{t1} term transforms into

$$H_c = w \sum_{\mathbf{k}a\sigma} E_{\mathbf{k}a\sigma} c_{\mathbf{k}a\sigma}^{+} c_{\mathbf{k}a\sigma}. \tag{5.198}$$

The H_{t2} term can be considered as a perturbation, if the w parameter assumed to be small. Proceeding in the perturbation theory in the same manner as when deriving from the Hamiltonian of the Hubbard model the Hamiltonian of the $t - J$ model, we arrive at an effective Hamiltonian consisting of three terms:

$$H_{\mathrm{eff}} = H_J + H_c + H_m, \tag{5.199}$$

where H_J is the exchange Hamiltonian, including the J_1^{ab} exchange between the nearest neighbours, J_2^{ab} between the next-nearest neighbours, and the on-site Hund's exchange:

$$H_J = \sum_{\langle ij \rangle} J_1^{ab} \mathbf{S}_{ia} \mathbf{S}_{jb} + \sum_{\langle\langle ij \rangle\rangle} J_2^{ab} \mathbf{S}_{ia} \mathbf{S}_{jb} + \sum_{ia \neq b} J_{\mathrm{H}}^{ab} \mathbf{S}_{ia} \mathbf{S}_{ib}. \tag{5.200}$$

The H_m term describes the coupling between coherent carriers and local moments:

$$H_m = \frac{1}{2} w \sum_{\mathbf{kq}} \sum_{abc\sigma\sigma'} G_{\mathbf{kq}abc}\, c_{\mathbf{k+q}a\sigma}^{+}\, \boldsymbol{\sigma}_{\sigma\sigma'}\, c_{\mathbf{k}b\sigma'}\, \mathbf{S}_{\mathbf{q}c}. \tag{5.201}$$

In the expressions for H_J and H_m, $\mathbf{S}_{ia}$ are localized spin operators, which are expressed via the operators of noncoherent states d^{inc} via a standard relation:

$$\mathbf{S}_{ia} = \frac{1}{2} \sum_{\sigma\sigma'} d_{ia\sigma}^{+\,\mathrm{inc}}\, \boldsymbol{\sigma}_{\sigma\sigma'}\, d_{ia\sigma'}^{\mathrm{inc}}. \tag{5.202}$$

Therefore if one assumes that the intensity w of the coherent quasiparticle peak is small, the effective Hamiltonian consists of just three terms: the exchange term H_J of the Heisenberg type, due to incoherent states (including the Hund's term); the kinetic term of quasiparticle states H_c; and the term H_m, which describes the exchange coupling between localized and itinerant states.

Obviously, the description suggested is not but a phenomenological scheme. It would have been extremely useful to implement it in a microscopic model, having developed a mathematical toolkit which would take into account, in a formal way, a decomposition of electron field operators $d = d^{\text{coh}} + d^{\text{inc}}$. However, this is a difficult mathematical task. It can be hardly realized in terms of field operators, but, more probably, in term of the Green's functions, departing, e.g. from the DMFT equations, within which one could try to construct a perturbation theory over the w parameter of small intensity of the quasiparticle peak.

In all probability, FeAs-systems make an intermediary case between strong and weak electron correlations, or, in other words, between the localized and itinerant models. The intermediate case, as usual, is the most difficult one for the theory. At present, a strict theory of FeAs-systems is not yet constructed, and one is obliged to make use of either one or another of its opposite limits: that of tight binding (localized model), or weak binding (itinerant model), in an attempt to see the features of the both limits departing from only one of them. On this way, as the discussion in this chapter demonstrates, reasonable physical results have been obtained.

5.9 Orbital Ordering

5.9.1 *The Spin-Orbital Model*

The models of FeAs-compounds so far outlined neglected the orbital degrees of freedom in a sense that no assumption about orbital ordering has been done. Meanwhile, it is well known that in a number of other compounds based on transition metals, e.g. in manganites and ruthenates, different arrangements of the orbitals of transition metal atoms occur; hereby a strong spin-orbit coupling exists, due to which the magnetic and orbital orderings are closely related to each other [497–499].

In this section, we outline the work [500] in which the effects of spin-orbital coupling in FeAs-compounds, at their stoichiometric composition, are studied by means of theory. A model of this type of compounds in shown in Fig. 5.69, in which the state of ion Fe d^6, $S = 1$, is assumed. Under the conditions of tetrahedral coordination by As atoms, the Fe ion is occupied by six electrons, which fill all e_g-orbitals (four electrons) and two out of three t_{2g}-orbitals $|xy\rangle$, $|xz\rangle$ and $|yz\rangle$. The crystal field splits the t_{2g}-triplet into a $|xy\rangle$ singlet and $|xz\rangle$, $|yz\rangle$ doublet, and this splitting is inferior to the energy difference between the t_{2g} and e_g electrons. Under the conditions of moderate Hund's interaction, J_H on the $|xy\rangle$ orbitals and the degenerate $|xz\rangle$, $|yz\rangle$ orbitals are all occupied with one electron, so that on the Fe ion, a situation with the spin $S = 1$ is realized.

The effective Hubbard model for t_{2g}-electrons consists of the kinetic part H_t, the interaction with the crystal field H_{cf} and the one-site electron–electron interaction H_{int}:

$$H = H_t + H_{\text{cf}} + H_{\text{int}}. \tag{5.203}$$

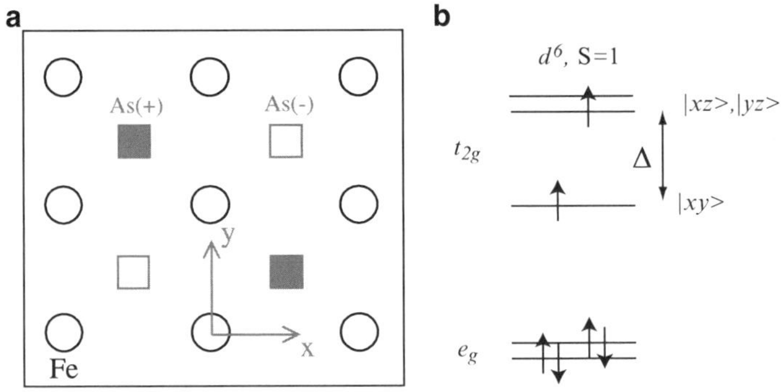

Fig. 5.69 Crystallochemistry of a FeAs-compound. (**a**) Arrangement of Fe and As atoms in the FeAs layer; (**b**) schematic drawing of the ground state of the d^6 Fe ion, corresponding to the spin $S = 1$ [500]

H_t includes the hopping of electrons between Fe ions, which occurs by way of both direct transitions and the indirect ones, via the As atoms:

$$H_t = -\sum_{ij}\sum_{\alpha\beta\sigma} t_{ij}^{\alpha\beta}\left(c_{i\alpha\sigma}^{+}c_{j\beta\sigma} + c_{j\beta\sigma}^{+}c_{i\alpha\sigma}\right), \tag{5.204}$$

where α numbers the t_{2g} orbitals $|xy\rangle$, $|xz\rangle$, $|yz\rangle$. The crystal field energy is

$$H_{\text{cf}} = \sum_{i\alpha} \varepsilon_\alpha n_{i\alpha}, \tag{5.205}$$

where $n_{i\,\alpha} = \sum_\sigma c_{i\,\alpha\,\sigma}^{+} c_{i\,\alpha\,\sigma}$. One can assume that $\varepsilon_\alpha = 0$ for the $|xy\rangle$-orbital and $\varepsilon_\alpha = \Delta$ for the $|xz\rangle$ and the $|yz\rangle$-orbitals.

The electron–electron interaction energy has the following form, including the Coulomb repulsion at the same orbital and between different orbitals, the pair transfer energy from one orbital to the other, and the Hund's exchange:

$$H_{\text{int}} = U\sum_{i\alpha} n_{i\alpha\uparrow}n_{i\alpha\downarrow} + \frac{1}{2}U'\sum_{i\,\alpha\neq\beta} n_{i\alpha}n_{i\beta} + J\sum_{i\,\alpha\neq\beta} c_{i\alpha\uparrow}^{+}c_{i\alpha\downarrow}^{+}c_{i\beta\downarrow}c_{i\beta\uparrow} - J_{\text{H}}\sum_{i\,\alpha\neq\beta}\mathbf{S}_{i\alpha}\mathbf{S}_{i\beta}. \tag{5.206}$$

According to [500], $U' = U - \frac{5}{2}J_{\text{H}}$, $J' = J_{\text{H}}$.

If assuming strong Coulomb interaction $U \gg t$, then in the second order of the perturbation theory, the Kugel–Khomsky effective Hamiltonian H_{KK} can be derived, which takes into account the spin–orbit interaction:

$$H_{\mathrm{KK}} = \sum_{ij} \left\{ -\frac{1}{3} \left(\mathbf{S}_i \, \mathbf{S}_j \, + 2 \right) Q^{(1)} \left(\mathbf{T}_i , \mathbf{T}_j \right) \right.$$
$$\left. + \frac{1}{3} \left(\mathbf{S}_i \, \mathbf{S}_j \, - 1 \right) Q^{(2)} \left(\mathbf{T}_i , \mathbf{T}_j \right) \right\} . \qquad (5.207)$$

In this Hamiltonian, along with the spin operators $\mathbf{S}_i$ related to a given site, the pseudospin operators $\mathbf{T}_i$ are introduced, of the magnitude $1/2$, which characterize the degenerate $|xz\rangle$ and $|yz\rangle$ orbitals at this site. The three components of the pseudospin operator are expressed in terms of spinless fermion operators a_i^+ (a_i) and b_i^+ (b_i), which create (annihilate) electron in the above orbitals, numbered by the a, b symbols, as follows:

$$T_i^z = \tfrac{1}{2} \left(n_{i\,a} - n_{i\,b} \right), \quad T_i^+ = b_i^+ a_i , \quad T_i^- = a_i^+ b_i , \qquad (5.208)$$

under a condition of homeopolarity: $n_{i\,a} + n_{i\,b} = 1$.

$Q^{(1)}$ and $Q^{(2)}$ in (5.207) are polynomials of the second order in pseudospin components:

$$Q^{[n]} \left(\mathbf{T}_i , \mathbf{T}_j \right) = f_{zz}^{[n]} T_i^z T_j^z + \frac{1}{2} f_{+-}^{[n]} \left(T_i^+ T_j^- + T_i^- T_j^+ \right) \qquad (5.209)$$
$$+ \frac{1}{2} f_{++}^{[n]} \left(T_i^+ T_j^+ + T_i^- T_j^- \right) + f_{xz}^{[n]} \left(T_i^z T_j^x + T_i^x T_j^z \right) \qquad (5.210)$$
$$+ f_z^{[n]} \left(T_i^z + T_j^z \right) + f_x^{[n]} \left(T_i^x + T_j^x \right) + f_0^{[n]} . \qquad (5.211)$$

Different $f^{[n]}$-coefficients here are expressed via the matrix elements of hoppings shown in Fig. 5.70. The matrix elements of indirect hopping via the As atoms depend on the director cosines l, m, n of the vector connecting the Fe and As atoms, and moreover on the $\gamma = (pd\pi)/(pd\sigma)$ relation. In the calculations, all the matrix elements $t_1, t_2, \ldots t_7$ are parametrized via the $\lambda = |n/l|$ property and γ. As a result, the $f^{[n]}$ coefficients of (5.211) can be expressed in terms of the $t_1 \ldots t_7$ matrix elements and the dimensionless parameters

$$\eta = J_H/U, \quad \delta = \Delta/U. \qquad (5.212)$$

In the FeAs-compounds, the λ parameter is close to 0.7. The γ parameter varies roughly within the range $-2 \leqslant \gamma \leqslant 2$; a realistic value for η would be $\eta < 0.3$; the δ parameter is small and has the order of magnitude 10^{-2}. In these conditions, the phase diagram of the FeAs-compounds is calculated, which includes both the spin and the orbital orderings.

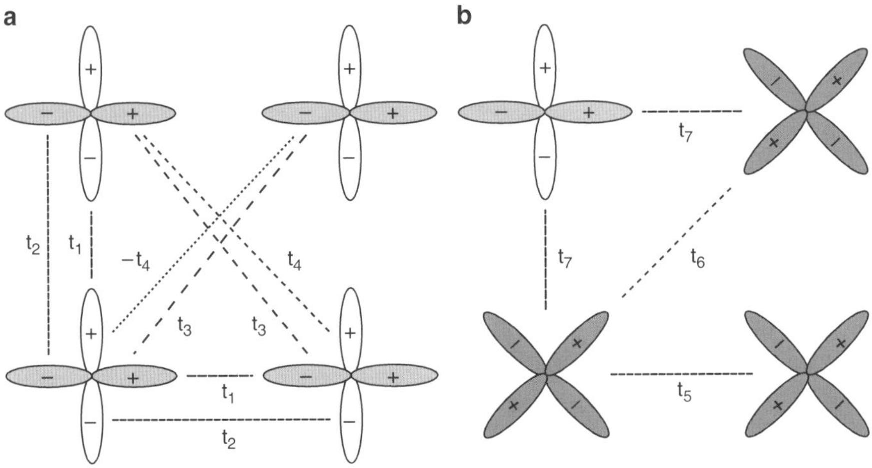

Fig. 5.70 Matrix elements of effective hopping parameters between (**a**) d_{xz} and d_{yz} orbitals; (**b**) between these orbitals and the $|xy\rangle$-orbitals [500]

Fig. 5.71 Three types of orbital orderings compatible with the spin ordering of the stripe-type: (**a**) orbital-ferro, (**b**) orbital-stripe, (**c**) orbital-antiferro [500]

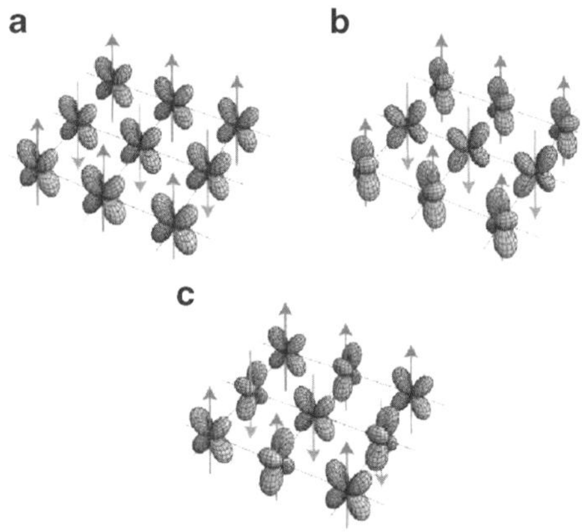

5.9.2 *Phase Diagrams with Spin and Orbital Orderings*

In Fig. 5.71, possible orbital orderings are depicted which are compatible with the spin ordering of the SDW type, with the wave vector $(\pi, 0)$. The phase diagram which takes these phases into account is constructed by calculating the energies of different configurations, which are specified in a Monte Carlo process, in the classical limit.

At $T = 0$, the SDW $(\pi, 0)$ state turns out to be stable within a broad range of parameters, whereas the orbital ordering does substantially depend on the η and γ, (Fig. 5.72). We see that at $\gamma \simeq 0.2$, the SDW state emerges which remains the ground state one under intermediate values of this parameter, but at $\gamma \simeq -0.2$ it is replaced by the antiferromagnetic (π, π) state. At larger values of η (i.e. large

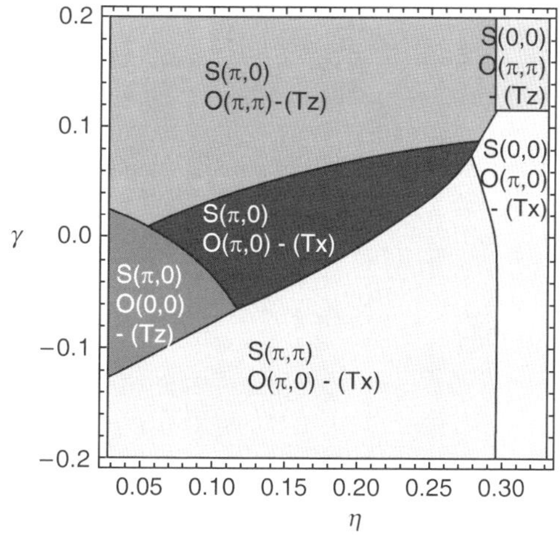

Fig. 5.72 $\gamma - \eta$ phase diagram at $T = 0$ for the parameter values $\lambda = 0.7$ and $\delta = 0.01$. The phases are marked by wave vectors of the spin (S) and orbital (O) ordering. Tz or Tx indicate the components of pseudospin which are saturated in the ordered state [500]

Hund's exchange J_H), a ferromagnetic state must emerge in the spin system. The phase diagram is strongly sensitive to the value of λ. Thus, already at $\lambda = 0.8$ the intermediate phases shown in Fig. 5.72 do disappear. However, the phase diagram is changing only weakly with variation of δ, i.e. in dependence on the value of splitting in the t_{2g} orbitals.

For low temperatures, the phase diagram is shown in Fig. 5.73. The SDW ordering exists throughout a continuous range of temperature at not so large η values, $\eta \lesssim 0.2$. As η grows, the type of ordering is replaced by the antiferromagnetic (π, π)-ordering, accompanied by the orbital $(\pi, 0)$-ordering.

We bring our attention to [501], in which an interaction of spin and orbital ordering was studied within the itinerant two-orbital model. Within the Hartree–Fock approximation, a competition of two phases has been studied, the SDW phase with the wave vector $\mathbf{Q} = (\boldsymbol{\pi}, \mathbf{0})$, and the SDW-phase, accompanied by an orbital ordering along the z axis (SDW+ferroorbital). For a weakly correlated case ($U/W \approx 0.29$), it was demonstrated that the SDW+ferro orbital phase has lower energy than the conventional SDW phase. Such an ordering leads to an additional orthorhombic distortion.

Influence of ferro-orbital ordering on magnetic structure of LaOFeAs is theoreticaly investigated in [502]. Such influence can explain giant anisotropy of exchange interaction in LaOFeAs.

5.9.3 *Spectrum of Magnetic Excitations*

The spin wave spectrum of the SDW phase coexisting with the orbital ordering was calculated in [500] in the classical pseudospin limit. In the static approximation for some degrees of freedom, the Hamiltonian of spin excitations has the same form as (5.8.3), where $A_{\mathbf{q}}$ and $B_{\mathbf{q}}$ the terms are added which are related to orbital

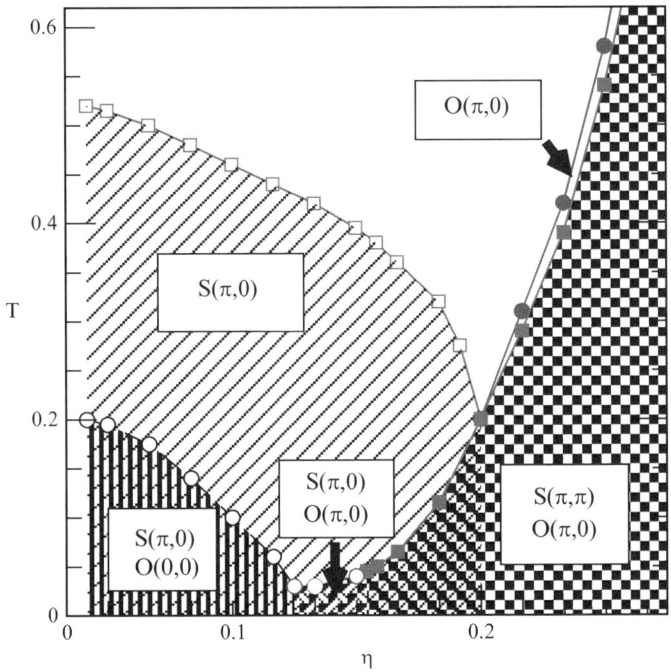

Fig. 5.73 $T - \eta$ phase diagram for $\delta = 0.01$, $\lambda = 0.7$ and $\gamma = -0.05$ [500]

static variables. The latter terms depend, in particular, on the components of the wave vector of orbital ordering $\mathbf{Q}$ and on the η parameter, on which the effective exchange interactions do depend: J_x, J_y – between the nearest neighbouring Fe atoms, and J_d – between the next-to-nearest Fe neighbours (along the diagonal of a square made by Fe atoms on the lattice). The calculated spectrum is depicted in Fig. 5.74. We see that with disordered orbitals, the spin wave spectrum has zeros at the wave vectors $\mathbf{q} = (\pi, 0)$ and $\mathbf{q} = (\pi, \pi)$. In case of the orbital-ferro-ordering on a wave vector, the gap is opening which decreases with the rise of η, and finally at $\eta = 0.15$ this gap disappears, whereas a character of the magnetic ordering undergoes a change: the $(\pi, 0)$-magnetic phase gets replaced by the (π, π)-phase. It is natural that in the spin wave spectrum, the gap decreases on the wave vector $\mathbf{q} = (\pi, \pi)$, and at $\mathbf{q} = (\pi, 0)$ the gap closes.

These peculiarities of the spin wave spectrum could have served as an indicator of orbital ordering, be the dispersion curves measurable in inelastic neutron scattering experiments. As is emphasized in [500], the orbital ordering in the FeAs-compounds can be directly measured in experiments on X-ray scattering.

The theory of orbital ordering suggested in [500] is based on analyzing the limit of strong Coulomb interaction that corresponds to a localized treatment of the system. In this approach, an itinerant character of the d-electrons of the system is fully neglected. Although the FeAs-compounds belong to the case of moderate

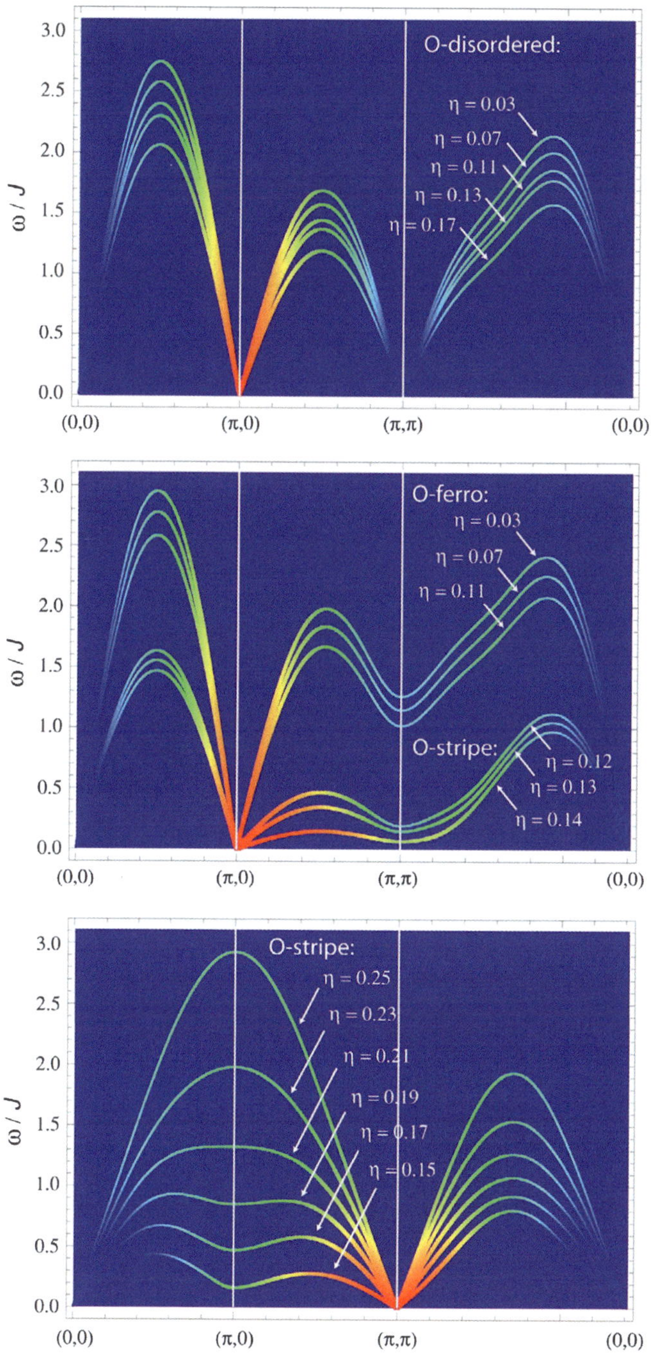

Fig. 5.74 Spin wave spectrum at different values of η. *Top panel*: $(\pi, 0)$-magnetism at disordered orbitals; *middle panel*: $(\pi, 0)$-magnetism at the orbital-ferro ordering; *bottom panel*: (π, π)-magnetism at the orbital-stripe ordering [500]

correlation ($U \simeq W$), the description of the phenomena traced back to spin-orbit coupling and obtained in the $U \gg W$ limit may be, nevertheless, qualitatively adequate for true FeAs-systems.

The theory [500] was developed for FeAs-compounds in their stoichiometric composition; it is less applicable for a description of doped compounds, because the degree of electron delocalization in them increases. We note, for a conclusion, an importance of a subject of frustration in discussing the problem of spin-orbit coupling in the FeAs-compounds. Within the model under discussion in this section, two types of frustration come about. The one is related to a presence of competing antiferromagnetic interactions between the nearest and next-to-nearest Fe atoms. The second one is related to a degeneracy of the $|xz\rangle$ and $|yz\rangle$ orbitals at the upper edge of the spectrum of a Fe ion. The result of a frustration of the first type is a competition between the $(\pi, 0)$ and (π, π) magnetic orderings. A frustration of the second type (with magnetic ordering given) gives rise to three possibilities of orbital ordering, shown in Fig. 5.71. At orbital-stripe- and orbital-antiferro-orderings, the x and y directions in the Fe lattice cease to be equivalent, and a transition into an orthorhombic phase occurs, leading to a magnetic transition. A close relation between all these transitions is typical for all FeAs-compounds.

Conclusion

The first years of the study of a new class of high-temperature superconductors resulted in quite impressive achievements. Several classes of FeAs-type compounds, which on doping turn superconducting, have been discovered. Despite differences in the chemical composition, all FeAs-type compounds exhibited similar properties. This is a consequence of the repetition in their crystal structures of the same motive – the planes made of quadratically arranged Fe atoms, sandwiched between the parallel above (and below) planes of As. The electronic states near the Fermi surface of any given compound of this family are dominated by the $3d$ electrons of Fe atoms. Because of a pronounced layered character of the crystal structure, the electronic states are quasi two-dimensional. The entire physics of these compounds is determined to what occurs in the FeAs planes.

In this aspect, the FeAs systems resemble cuprates, whose crystal structure is built out of CuO_2 planes, separated by other elements, due to which circumstance of the electronic structure is quasi two-dimensional, and the electronic states at the Fermi surface is primarily formed by the $3d$ electrons of Cu ions.

Another similarity between FeAs systems and cuprates is that they both, at stoichiometric composition, are antiferromagnets. A doping, in both cases, suppresses magnetism, and as soon as the long-range magnetic order disappears, in the compounds of both types a superconducting state is formed within a certain range of dopant concentration. Beyond this range the compounds become more metallic and exhibit Fermi-liquid-like properties. Broadly accepted is a concept that in both types of systems, the superconducting pairing occurs due to spin fluctuations which survive after the long-range magnetic ordering being suppressed.

However, there are important differences between cuprates and FeAs systems. First of all, stoichiometric cuprates are insulators, the magnetic moments in them are localized, and magnetic ordering can be well described by the Heisenberg model. On the contrary, stoichiometric FeAs-based compounds are metals; their magnetism is better consistent with an itinerant picture, so that the resulting magnetic ordering is conveniently called "spin density wave" (SDW), even if the wave vector of the latter corresponds to a simple doubling of the unit cell along one of the directions in the basal plane, sometimes combined with the doubling along the c axis. The mean magnetic moment on the Fe site is merely $\sim 0.3\,\mu_B$ in the 1111 compounds and $\sim 0.8\,\mu_B$ in the 122 compounds.

These differences can be traced to those in the strength of electron correlations. Namely, in cuprates the Coulomb interaction U between electrons on a Cu site is comparable with, or larger than, the band width W, and an undoped compound is a Mott insulator. The bulk of experimental data, especially those of X-ray spectroscopy, and numerical calculations within the LDA + DMFT formalism show that undoped FeAs-type compounds are far from the Mott metal-insulator transition, so that all FeAs compounds belong to the class of moderately correlated or weakly correlated electron systems, for which U is inferior to W. If comparing doped compounds of one and another type, one more difference could be pointed out: the density of states at the Fermi level is lower in cuprates than in the FeAs-based systems, and the carrier density is higher.

We now turn back to the FeAs systems. We have just pointed out that they are characterised by moderate electron correlation. Another question is – which model, the localized or itinerant one, is more appropriate for describing magnetism in such systems? In the theory chapter we outlined different approaches to this problem, the one based on the localized J_1–J_2-model and the other, entirely delocalized one, in which no exchange interaction parameters are a priori introduced, beyond the LDA band spectrum along with the Coulomb one-site interaction and the Hund-like single-atom exchange. Both approaches permit to explain the observed SDW-type magnetic ordering; it should be admitted however that the delocalized approach is more conclusive because it allows to relate the major feature of the magnetic ordering, its wave vector $\mathbf{Q}$, with the structure of the Fermi surface of the compound in question. This wave vector is close to the nesting vector which connect the electron and hole sheets of the Fermi surface.

It is remarkable that LDA calculations of the electronic structure of different FeAs-type systems show very close results in what concerns the shape of the Fermi surface: the latter is multi-sheet one, with two hole pockets at the Brillouin zone centre and two electron pockets at the corners. Notably, the sizes of the hole and electron pockets are comparable, that is exactly what makes the nesting possible. The nesting deteriorates on doping, as the size of either hole or electron pockets does gradually grow, depending on the dopand's nature. Therefore, the itinerant model binds together the three basic features of the FeAs-type systems: the Fermi surface topology, the wave vector of magnetic ordering, and the destruction of the latter that opens a possibility for electron pairing. An important element is still missing in this logically complete triad, namely an experimental confirmation of fluctuations of the magnetic ordering in the existence range of the superconducting phase. The suggested mechanism of pairing in FeAs-like systems via spin fluctuations is based on an analogy with cuprates, in which it is broadly believed to have been experimentally confirmed. Another argument in favour of this mechanism is that the Eliashberg equations as formulated in different electron models of FeAs-type compounds, taking into account their actual shape of multi-sheet Fermi surfaces, permit solutions with different symmetry of the superconducting order parameter. Among well-known symmetries of s and d type, many models offer a solution of a new, so-called $s^{\pm}$ type [503]. From the point of view of the group theory, the couplings of the s and the $s^{\pm}$ types belong to the same irreducible representation of the symmetry group, but their physical nature is different. In case of the $s^{\pm}$ pairing, the superconducting gap on the hole surface is of opposite sign to that on the electron sheets. Such solution of the Eliashberg equations emerges as a result of pairing interaction via spin fluctuations in the FeAs-type systems, taking into account their real Fermi surface topology, and becomes possible in the range of (U, J) parameters close to their actual values in FeAs-type compounds.

In cuprates, the symmetry of superconducting order parameter is well established by different experimental means. In case of strong one-site Coulomb interaction U, the $d_{x^2-y^2}$ configuration of a pair minimizes the Coulomb repulsion, because the probability to find an electron of such a pair on the site equals to zero. When considering the $s^{\pm}$ symmetry in FeAs-type systems, we note a strong on-site Coulomb interaction for two electrons of a pair is not completely excluded, but its role is strongly attenuated. This fact, along with a specific topology of the Fermi surface, apparently makes the $s^{\pm}$ coupling more favourable than the s or $d_{x^2-y^2}$ types.

The results of experimental studies of superconducting order parameter, as has been outlined in corresponding parts of the book, are by now contradictory and do not allow to make definite conclusions about the symmetry of the order parameter. Hopefully decisive might be experiments probing the phase of the order parameter, like the measurements of the Josephson current in different contacts.

Summarizing the studies of FeAs-type compounds done within a year after the discovery of high-temperature superconductivity in them, it is more astonishing to see an amount of already understood in their physical properties than to anticipate what is yet left to find out. This is even more fascinating when comparing, in retrospective, with results obtained within the first year since the discovery of high-temperature superconductivity in cuprates. Obviously in the studies of the new type of superconductors, a large use has been made of experience accumulated in the research on cuprates, notable the by now well-established theory concepts and experimental approaches. The studies of the FeAs-type compounds followed from the very beginning paved the way of searching how the decay of magnetic ordering in cuprates is related to the onset of electron pairing via spin fluctuations. In global, two problems in the physics of FeAs-type systems remain yet unsolved: an experimental verification of sufficiently strong fluctuations of magnetic ordering in the domain of existence of superconducting state, and finding out the symmetry of the superconducting order parameter. By whatever outcome of these prospective researches, it remains beyond doubt that cuprates and FeAs-type compounds are not the only two groups of materials in which high-temperature superconductivity may occur. The ways of the most promising search strategies for further materials containing d or f elements and antiferromagnetic order, such that their doping suppresses this order and increases the metallicity of the initial compounds, seem now to be recognized.

References

1. Y. Kamihara, H. Hiramatsu, M. Hirano, R. Kawamura, H. Yanagi, T. Kamiya, H. Hosono, J. Am. Chem. Soc. **128**, 10012 (2006)
2. M.V. Sadovskii, Phys. Usp. **51**, 1201 (2008)
3. A.L. Ivanovskii, Phys. Usp. **51**, 1229 (2008)
4. Yu.A. Izyumov, E.Z. Kurmaev, Phys. Usp. **51**, 1261 (2008)
5. Special issue of Physica C on superconducting pnictides, Physica C **469**, 313 (2009)
6. Special issue of New Journal of Physics, Focus on Iron-Based Superconductors, New J. Phys. **11**, 025003
7. T. Watanabe, H. Yanagi, T. Kamiya et al., Inorg. Chem. **46**, 7719 (2007)
8. Y. Kamihara, T. Watanabe, M. Hirano, H. Hosono, J. Am. Chem. Soc. **130**, 3296 (2008)
9. G.F. Chen, Z. Li, D. Wu, G. Li, W.Z. Hu, J. Dong, P. Zheng, J.L. Luo, N.L. Wang, Phys. Rev. Lett. **100**, 247002 (2008)
10. Z.-A. Ren, J. Yang, W. Lu, W. Yi, H.-C. Che, X.-L. Dong, L.-L. Sun, Z.-X. Zhao, Mater. Res. Innovat. **12**, 1 (2008)
11. Z.-A. Ren, W. Lu, J. Yang, W. Yi, X.-L. Shen, Z.-C. Li, G.-C. Che, X.-L. Dong, L.-L. Sun, F. Zhou, Z.-X.C. Zhao, Phys. Lett. **25**, 2215 (2008)
12. J. Zhao, Q. Huang, C. de la Cruz, S. Li, J.W. Lynn, Y. Chen, M.A. Green, G.F. Chen, G. Li, Z. Li, J.L. Luo, N.L. Wang, P. Dai, Nat. Mater. **7**, 953 (2008)
13. M. Rotter, M. Tegel, D. Johrendt, Phys. Rev. Lett. **101**, 107006 (2008)
14. J.H. Tapp, Z. Tang, B. Lv, K. Sasmal, B. Lorenz, P.C.W. Chu, A.M. Guloy, Phys. Rev. B **78**, 060505 (2008)
15. X.C. Wang, Q.Q. Liu, Y.X. Lv, W.B. Gao, L.X. Yang, R.C. Yu, F.Y. Li, C.Q. Jin, arXiv: 0806.4688 (2008)
16. Y. Mizuguchi, F. Tomioka, S. Tsuda, T. Yamaguchi, Y. Takano, Appl. Phys. Lett. **93**, 152505 (2008)
17. L. Boeri, O.V. Dolgov, A.A. Golubov, Phys. Rev. Lett. **101**, 026403 (2008)
18. K. Haule, J.H. Shim, G. Kotliar, Phys. Rev. Lett. **100**, 226402 (2008)
19. P. Fulde, *Electron Correlations in Molecues and Snolids* (Springer, Berlin, 1995)
20. L. Hozoi, P. Fulde, Phys. Rev. Lett. **102**, 136405 (2009)
21. H. Yang, X. Zhu, L. Fang, G. Mu, H.-H. Wen, arXiv: 0803.0623 (2008)
22. X. Zhu, H. Yang, L. Fang, G. Mu, H.-H. Wen, Supercond. Sci. Technol. **21**, 105001 (2008)
23. A.S. Sefat, M.A. McGuire, B.C. Sales, R. Jin, J.Y. Howe, D. Mandrus, Phys. Rev. B **77**, 174503 (2008)
24. M. Tropeano, C. Fanciulli, C. Ferdeghini, D. Marre', A.S. Siri, M. Putti, A. Martinelli, M. Ferretti, A. Palenzona, M.R. Cimberle, C. Mirri, S. Lupi, R. Sopracase, P. Calvani, A. Perucchi, arXiv: 0809.3500 (2008)
25. P. Cheng, H. Yang, Y. Jia, L. Fang, X. Zhu, G. Mu, H.-H. Wen, Phys. Rev. B **78**, 134508 (2008)
26. H. Mukuda, N. Terasaki, H. Kinouchi, M. Yashima, Y. Kitaoka, S. Suzuki, S. Miyasaka, S. Tajima, K. Miyazawa, P.M. Shirage, H. Kito, H. Eisaki, A. Iyo, J. Phys. Soc. Jpn. **77**, 093704 (2008)

27. W. Lu, X.-L. Shen, J. Yang, Z.-C. Li, W. Yi, Z.-A. Ren, X.-L. Dong, G.-C. Che, L.-L. Sun, F. Zhou, Z.-X. Zhao, Solid State Commun. **148**, 168 (2008)
28. Z.-A. Ren, J. Yang, W. Lu, W. Yi, X.-L. Shen, Z.-C. Li, G.-C. Che, X.-L. Dong, L.-L. Sun, F. Zhou, Z.-X. Zhao, Europhys. Lett. **82**, 57002 (2008)
29. J. Yang, Z.-C. Li, W. Lu, W. Yi, X.-L. Shen, Z.-A. Ren, G.-C. Che, X.-L. Dong, L.-L. Sun, F. Zhou, Z.-X. Zhao, Supercond. Sci. Technol. **21**, 082001 (2008)
30. C. Wang, L. Li, S. Chi, Z. Zhu, Z. Ren, Y. Li, Y. Wang, X. Lin, Y. Luo, S. Jiang, X. Xu, G. Cao, Z. Xu, Europhys. Lett. **83**, 67006 (2008)
31. L.-J. Li, Y.-K. Li, Z. Ren, Y.-K. Luo, X. Lin, M. He, Q. Tao, Z.-W. Zhu, G.-H. Cao, Z.-A. Xu, Phys. Rev. B **78**, 132506 (2008)
32. H.H. Wen, G. Mu, L. Fang, H. Yang, X.Y. Zhu, Europhys. Lett. **82**, 17009 (2008)
33. G. Mu, L. Fang, H. Yang, X. Zhu, P. Che, arXiv: 0806.2104 (2008)
34. G. Mu, B. Zeng, X. Zhu, F. Han, P. Cheng, B. Shen, H.-H. Wen, Phys. Rev. B **78**, 104501 (2008)
35. K. Kasperkiewicz, J.-W.G. Bos, A.N. Fitch, K. Prassides, S. Margadonna, Chem. Commun. 707 (2009)
36. L. Fang, H. Yang, P. Cheng, X. Zhu, G. Mu, H.-H. Wen, Phys. Rev. B **78**, 104528(2008)
37. G. Xu, W. Ming, Y. Yao, X. Dai, S. Zhang, Z. Fang, Europhys. Lett. **82**, 67002 (2008)
38. A.S. Sefat, A. Huq, M.A. McGuire, R. Jin, B.C. Sales, D. Mandrus, Phys. Rev. B **78**, 104505 (2008)
39. G. Cao, C. Wang, Z. Zhu, S. Jiang, Y. Luo, S.C.Z. Ren, Q. Tao, Y. Wang, Z. Xu, Phys. Rev. B **79**, 054521 (2009)
40. A. Kawabata, S.C. Lee, T. Moyoshi, Y. Kobayashi, M. Sato, J. Phys. Soc. Jpn. **77**, 103704 (2009), arXiv: 0807.3480 (2008)
41. H. Li, J. Li, S. Zhang, W. Chu, D. Chen, Z. Wu, arXiv: 0807.3153 (2008)
42. Z.A. Ren, G.-C. Che, X.-L. Dong, J. Yang, W. Lu, W. Yi, X.-L. Shen, Z.-C. Li, L.-L. Sun, F. Zhou, Z.-X. Zhao, Europhys. Lett. **83**, 17002 (2008)
43. N.R. Werthamer, E. Helfand, P.C. Hohenberg, Phys. Rev. **147**, 295 (1966)
44. X.L. Wang, R. Ghorbani, G. Peleckis, S.X. Dou, Adv. Mater. **21**, 236 (2009), arXiv: 0806.0063 (2008)
45. Y. Jia, P. Cheng, L. Fang, H. Luo, H. Yang, C. Ren, L. Shan, C. Gu, H.-H. Wen, Appl. Phys. Lett. **93**, 032503 (2008)
46. L. Ding, C. He, J.K. Dong, T. Wu, R.H. Liu, X.H. Chen, S.Y. Li, Phys. Rev. B **77**, 180510 R (2008)
47. Y. Ma, Z. Gao, L. Wang, Y. Qi, D. Wang, X. Zhang, Chin. Phys. Lett. **26**, 037401 (2009)
48. Y.J. Jo, J. Jaroszynski, A. Yamamoto, A. Gurevich, S.C. Riggs, G.S. Boebinger, D. Larbalastiera, H.H. Wen, N.D. Zhigadlo, S. Katrych, Z. Bukowski, J. Karpinski, R.H. Liu, H. Chen, Chen X.H., L. Balicas, Physica C **468**, 566 (2009), arXiv: 0902.0532 (2009)
49. H. Takahashi, K. Igawa, K. Avii, Y. Kamihara, M. Hirano, H. Hosono, Nature **453**, 376 (2008)
50. H. Okada, K. Igawa, H. Takahashi, Y. Kamihara, M. Hirano, H. Hosono, K. Matsubashi, Y. Uwatoko, J. Phys. Soc. Jpn. **77**, 113712 (2008)
51. H. Takahashi, H. Okada, K. Igawa, K. Arii, Y. Kamihara, S. Matsuishi, M. Hirano, H. Hosono, K. Matsubayashi, Y. Uwatoko, J. Supercond. Novel Magn. **22**, 595 (2009)
52. W. Yi, C. Zhang, L. Sun, Z.-A. Ren, W. Lu, X. Dong, Z. Li, G. Che, J. Yang, X. Shen, X. Dai, Z. Fang, F. Zhou, Z. Zhao, Europhys. Lett. **84**, 67009 (2008), arXiv: 0809.4702 (2008)
53. G. Garbarino, P. Toulemonde, M. Alvarez-Murga, A. Sow, M. Mezouar, M. Nunez-Regueiro, Phys. Rev. B **78**, 100507 (2008)
54. K. Igawa, H. Okada, K. Arii, H. Takahashi, Y. Kamihara, M. Hirano, H. Hosono, S. Nakano, T. Kikegawa, arXiv: 0809.1239 (2008)
55. M. Tropeano, C. Fanciulli, F. Canepa, M.R. Cimberle, C. Ferdeghini, G. Lamura, A. Martinelli, M. Puttia, M. Vignolo, A. Palenzona, Phys. Rev. B **79**, 174523 (2009), arXiv: 0903.1825 (2009)
56. L. Porovskii, V. Vildosola, S. Biermann, A. Georges, Eur. Phys. Lett. **84**, 37006 (2009)
57. L. Sun, X. Dai, C. Zhang, W. Yi, L. Zheng, Z. Jiang, X. Wei, Y. Huang, J. Yang, Z. Ren, W. Lu, X. Dong X., G. Che, Q. Wu, H. Ding, J. Liu, T. Hu, Z. Zhao, arXiv: 0907.4212 (2009)

58. Y. Luo, Y. Li, S. Jiang, J. Dai, G. Cao, Z. Xu, Phys. Rev. B **81**, 134422 (2010), arXiv: 0907.2961 (2009)
59. C.R. dela Cruz, Q. Huang, J.W. Lynn, J. Li, W. Ratcliff II, J.L. Zarestky, H.A. Mook, G.F. Chen, J.L. Luo, N.L. Wang, P. Dai, Nature **453**, 899 (2008)
60. J. Dong, H.J. Zhang, G. Xu, G. Li, W.Z. Hu, D. Wu, G.F. Chen, X. Dai, J.L. Luo, Z. Fang, N.L. Wang, Europhys. Lett. **83** 27006 (2008)
61. Y. Chen, J.W. Lynn, J. Li, G. Li, G.F. Chen, J.L. Luo, N.L. Wang, P.-C. Dai, C. dela Cruz, H.A. Mook, Phys. Rev. B **78**, 064515 (2008)
62. C. Cao, P.J. Hirschfeld, H.-P. Cheng, Phys. Rev. B **77**, 220506(R) (2008)
63. F. Ma, Z.-Y. Lu, Phys. Rev. B **78** 033111 (2008)
64. R. Cimberle, C. Ferdeghini, F. Canepa, M. Ferretti, A. Martinelli, A. Palenzona, A.S. Siri, M. Tropeano, arXiv: 0807.1688 (2008)
65. T. Yildirim, Phys. Rev. Lett. **101**, 057010 (2008)
66. T. Yildirim, A.B. Harris, E.E. Shender, Phys. Rev. B **53**, 6455 (1996)
67. S.H. Lee, C. Broholm, T.H. Kim, W. Ratclaff, S.-W. Cheong, Phys. Rev. Lett. **84**, 3718 (2000)
68. H.-H. Klauss, H. Luetkens, R. Klingeler, C. Hess, F.J. Litterst, M. Kraken, M.M. Korshunov, I. Eremin, Phys. Rev. Lett. **101**, 077005 (2008)
69. I. Opahle, H.C. Kandpal, Y. Zhang, C. Gros, R. Valenti, Phys. Rev. B **78**, 024509 (2008)
70. G.-B. Liu, B.-G. Liu, J. Phys. Condens. Matter **21**, 195701 (2009), arXiv: 0905.2005 (2009)
71. H. Maeter, H. Luetkens, Yu.G. Pashkevich, A. Kwadrin, R. Khasanov, A. Amato, A.A. Gusev, K.V. Lamonova, D.A. Chervinskii, R. Klingeler, C. Hess, G. Behr, B. Buchner, H.-H. Klauss, Phys. Rev. B **80**, 094524 (2009), arXiv: 0904.1563 (2009)
72. Q. Huang, J. Zhao, J.W. Lynn, G.F. Chen, J.L. Luo, N.L. Wang, P. Dai, Phys. Rev. B **78**, 054529 (2008)
73. M.A. McGuire, A.D. Christianson, A.S. Sefat, B.C. Sales, M.D. Lumsden, R. Jin, E.A. Payzant, D. Mandrus, Y. Luan, V. Keppens, V. Varadarajan, J.W. Brill, R.P. Hermann, M.T. Sougrati, F. Grandjean, G.J. Long, Phys. Rev. B **78**, 094517 (2008)
74. H. Luetkens, H.-H. Klauss, M. Kraken, F.J. Litterst, T. Dellmann, R. Klingeler, C. Hess, R. Khasanov, A. Amato, C. Baines, J. Hamann-Borrero, N. Leps, A. Kondrat, G. Behr, J. Werner, B. Buechner, Nat. Mater. **8**, 305 (2008), arXiv: 0806.3533
75. R.H. Liu, G. Wu, T. Wu, D.F. Fang, H. Chen, S. Y. Li, K. Liu, Y.L. Xie, X.F. Wang, R.L. Yang, L. Ding, C. He, D.L. Feng, X.H. Chen, Phys. Rev. Lett. **101**, 087001 (2008)
76. S.C. Riggs, J.B. Kemper, Y. Jo, Z. Stegen, L. Balicas, G.S. Boebinger, F.F. Balakirev, A. Migliori, H. Chen, R.H. Liu, X.H. Chen, arXiv: 0806.4011 (2008)
77. S. Margadonna, Y. Takabayashi, M.T. McDonald, M. Brunelli, G. Wu, R.H. Liu, X.H. Chen, K. Prassides, Phys. Rev. B **79**, 014503 (2009)
78. S. Sanna, R. De Renzi, G. Lamura, C. Ferdeghini, A. Palenzona, M. Putti, M. Tropeano, T. Shiroka, Phys. Rev. B **80**, 052503 (2009), arXiv: 0902.2156 (2009)
79. R. Klingeler, N. Leps, I. Hellmann, A. Popa, C. Hess, A. Kondrat, J. Hamann-Borrero, G. Behr, V. Kataev, B. Buchner, arXiv: 0808.0708 (2008)
80. S. Lebeque, Phys. Rev. B **75**, 035110 (2007)
81. D.J. Singh, M.-H. Du, Phys. Rev. Lett **100**, 237003 (2008)
82. I.A. Nekrasov, Z.V. Pchelkina, M.V. Sadovskii, JETP Lett. **87**, 620 (2008)
83. K. Kuroki, S. Onari, R. Arita, H. Usui, Y. Tanaka, H. Kontani, H. Aoki, Phys. Rev. Lett. **101**, 087004 (2008)
84. I.I. Mazin, M.D. Johannes, L. Boeri, D. Singh, J. Phys. Rev. B **78**, 085104 (2008)
85. V. Vildosola, L. Pourovskii, R. Arita, S. Biermann, A. Georges, Phys. Rev. B **78**, 064518 (2008)
86. F. Ma, Z.-Yi Lu, Xiang Tao, Front. Phys. China **5**, 150 (2010), arXiv: 0806.3526 (2008)
87. Z.P. Yin, S. Lebeque, M.J. Han, B. Neal, D.Y. Savrasov, W.E. Pickett, Phys. Rev. Lett. **101**, 047001 (2008)
88. S. Lebegue, Z.P. Yin, W.E. Pickett, New J. Phys. **11**, 025004 (2009), arXiv: 0810.0376
89. B. Lv, M. Gooch, B. Lorenz, F. Chen, A.M. Guloy, C.W. Chu, Nev.J. Phys. **11**, 025013 (2009), arXiv: 0903.0013 (2009)

90. I.I. Mazin, D.J. Singh, M.D. Johannes, M.H. Du, Phys. Rev. Lett. **101**, 057003 (2008)
91. S. Ishibashi, K. Terakura, H. Hosono, J. Phys. Soc. Jpn. **77**, 053709 (2008)
92. D.H. Lu, M. Yi, S.-K. Mo, A.S. Erickson, J. Analytis, J.-H. Chu, D.J. Singh, Z. Hussain, T.H. Geballe, I.R. Fisher, Z.-X. Shen, Nature **455**, 81 (2008)
93. C. Liu, T. Kondo, M.E. Tillman, R. Gordon, G.D. Samolyuk, Y. Lee, C. Martin, J.L. McChesney, S. Bud'ko, M.A. Tanatar, E. Rotenberg, P.C. Canfield, R. Prozorov, B.N. Harmon, A. Kaminski, arXiv: 0806.2147 (2008)
94. A.I. Coldea, J.D. Fletcher, A. Carrington, J.G. Analytis, A.F. Bangura, J.-H. Chu, A.S. Erickson, I.R. Fisher, N.E. Hussey, R.D. McDonald, Phys. Rev. Lett. **101**, 216402 (2008)
95. A. Carrington, A.I. Coldea, J.D. Fletcher, N.E. Hussey, C.M.J. Andrew, A.F. Bangura, J.G. Analytis, J.-H. Chub, A.S. Erickson, I.R. Fisher, R.D. McDonald, Physica C, **469**, 459 (2009), arXiv: 0901.3976 (2009)
96. D.H. Lu, M. Yi, S.-K. Mo, J.G. Analytis, J.-H. Chu, A.S. Erickson, D.J. Singh, Z. Hussain, T.H. Geballe, I.R. Fisher, Z.-X. Shen, Physica C **469**, 452 (2009), arXiv: 0902.2503 (2009)
97. K. Matano, Z.A. Ren, X.L. Dong, L.L. Sun, Z.X. Zhao, G.-Q. Zheng, Europhys. Lett. **83**, 57001 (2008)
98. H.-J. Grafe, D. Paar, G. Lang, N.J. Curro, G. Behr, J. Werner, J. Hamann-Borrero, C. Hess, N. Leps, R. Klingeler, B. Buchner, Phys. Rev. Lett. **101**, 047003 (2008)
99. N. Terasaki, H. Mukuda, M. Yashima, Y. Kitaoka, K. Miyazawa, P. Shirage, H. Kito, H. Eisaki, A. Iyo, J. Phys. Soc. Jpn. **78**, 013701 (2009)
100. S. Kawasaki, K. Shimada, G.F. Chen, J.L. Luo, N.L. Wang, G.-Q. Zheng, Phys. Rev. B **78**, 220506(R) (2008)
101. K. Ahilan, F.L. Ning, T. Imai, A.S. Sefat, R. Jin, M.A. McGuire, B.C. Sales, D. Mandrus, Phys. Rev. B **78**, 100501(R) (2008)
102. Y. Nakai, S. Kitagawa, K. Ishida, Y. Kamihara, M. Hirano, H. Hosono, New J. Phys. **11**, 045004 (2009), arXiv: 0810.3569 (2008)
103. Y. Kobayashi, A. Kawabata, S.C. Lee, T. Moyoshi, M. Sato, J. Phys. Soc. Jpn. **78**, N 7 (2009), arXiv: 0901.2830 (2009)
104. L. Shan, Y. Wang, X. Zhu, G. Mu, L. Fang, H.-H. Wen, Europhys. Lett. **83**, 57004 (2008)
105. T.Y. Chen, Z. Tesanovic, R.H. Liu, X.H. Chen, C.L. Chien, Nature **453**, 1224 (2008)
106. T.Y. Chen, S.X. Huang, Z. Tesanovic, R.H. Liu, X.H. Chen, C.L. Chen, Physica C **469**, 521 (2009), arXiv: 0902.4008 (2009)
107. G. Martin, M.E. Tollman, H. Kim, M.A. Tanatar, S.K. Kim, A. Kreyssig, R.T. Gordon, arXiv: 0903.2220 (2009)
108. O. Millo, I. Asulin, O. Yuli, I. Felner, Z.-A. Ren, X.-L. Shen, G.-C. Che, Z.-X. Zhao, Phys. Rev. B **78**, 092505 (2008)
109. Y. Wang, L. Shan, L. Fang, P. Cheng, C. Ren, H.-H. Wen, Supercond. Sci. Technol. **22**, 015018 (2009)
110. M.H. Pan, X.B. He, G.R. Li, J.F. Wendelken, R. Jin, A.S. Sefat, M.A. McGuire, B.C. Sales, D. Mandrus, E.W. Plummer, arXiv: 0808.0895 (2008)
111. T. Kondo, A.F. Santander-Syro, O. Copie, C. Liu, M.E. Tillman, E.D. Mun, J. Schmalian, S.L. Bud'ko, M.A. Tanatar, P.C. Canfield, A. Kaminski, Phys. Rev. Lett. **101**, 147003 (2008)
112. T. Mertelj, V.V. Kabanov, C. Gadermaier, J. Karpinski, D. Mihailovic, arXiv: 0808.2772 (2008)
113. P. Samuely, P. Szabo, Z. Pribulova, M.E. Tillman, S. Bud'ko, P.C. Canfield, Supercond. Sci. Technol. **22**, 014003 (2009)
114. C. Martin, M.E. Tillman, H. Kim, M.A. Tanatar, S.K. Kim, A. Kreyssig, R.T. Gordon, M.D. Vannette, S. Nandi, V.G. Kogan, S.L. Bud'ko, P.C. Canfield, A.I. Goldman, R. Prozorov, Phys. Rev. Lett. **102**, 247002 (2009)
115. Y. Aiura, K. Sato, H. Iwasawa, Y. Nakashima, A. Ino, M. Arita, K. Shimada, H. Namatame, M. Taniguchi, I. Hase, K. Miyazawa, P.M. Shirage, H. Eisaki, H. Kito, A. Iyo, J. Phys. Soc. Jpn. **77**, 103712 (2008)
116. K. Hashimoto, T. Shibauchi, T. Kato, K. Ikada, R. Okazaki, H. Shishido, M. Ishikado, H. Kito, A. Iyo, H. Eisaki, S. Shamoto, Y. Matsuda, Phys. Rev. Lett. **102**, 017002 (2009)

117. K. Kuroki, S. Onari, R. Arita, H. Usui, Y. Tanaka, H. Kontani, H. Aoki, Phys. Rev. Lett. **101**, 087004 (2008)
118. G.E. Blonder, M. Tinkham, T.M. Klapwijk, Phys. Rev. B **25**, 4515 (1982)
119. R.C. Dynes, V. Naraynamurti, J.P. Garno, Phys. Rev. Lett. **41**, 1509 (1978)
120. H.-Y. Choi, Y. Bang, arXiv: 0807.4604 (2008)
121. P. Ghaemi, F. Wang, A. Vishwanath, Phys. Rev. Lett. **102**, 157002 (2009), arXiv: 0812.0015 (2008)
122. J. Linder, A. Sudbo, Phys. Rev. B **79**, 020501(R) (2009)
123. A.A. Golubov, A. Brinkman, O.V. Dolgov, I.I. Mazin, Y. Tanaka, Phys. Rev. Lett. **103**, 077003 (2009), arXiv: 0812.5057 (2008)
124. M.A.N. Araujo, P.D. Sacramento, Phys. Rev. B **79**, 174529 (2009), arXiv: 0901.0398 (2009)
125. R.S. Gonnelli, D. Daghero, M. Tortello, G.A. Ummarino, V.A. Stepanov, R.K. Kremer, J.S. Kim, N.D. Zhigadlo, J. Karpinski, Physica C **469**, 512 (2009), arXiv: 0902.3441 (2009)
126. R.S. Gonnelli, D. Daghero, M. Tortello, G.A. Ummarino, V.A. Stepanov, J.S. Kim, R.K. Kremer, Phys. Rev. B **79**, 184526 (2009), arXiv: 0807.3149 (2008)
127. K.A. Yates, K. Morrison, J.A. Rodgers, G.B.S. Penny, J.-W.G. Bos, J.P. Attfield, L.F. Cohen, New J. Phys. **11**, 025015 (2009)
128. R. Prozorov, M.E. Tillman, E.D. Mun, P.C. Canfield, New J. Phys. **11**, 035004 (2009)
129. G.F. Chen, Z. Li, G. Li, W.Z. Hu, J. Dong, X.D. Zhang, P. Zheng, N.L. Wang, J.L. Luo, Chin. Phys. Lett. **25**, 3403 (2008)
130. K. Sasmal, B. Lv, B. Lorenz, A. Guloy, F. Chen, Y. Xue, C.W. Chu, Phys. Rev. Lett. **101**, 107007 (2008)
131. I.R. Shein, A.L. Ivanovskii, arXiv: 0806.0750 (2008)
132. C. Krellner, N. Caroca-Canales, A. Jesche, H. Rosner, A. Ormeci, C. Geibel, Phys. Rev. B **78**, 100504(R) (2008)
133. I.A. Nekrasov, Z.V. Pchelkina, M.V. Sadovskii, JETP Lett. **88**, 144 (2008)
134. D.J. Singh, Phys. Rev. B **78**, 094511 (2008)
135. A. Subedi, D.J. Singh, Phys. Rev. B **78**, 132511 (2008)
136. Y. Singh, Y. Lee, S. Nandi, A. Kreyssig, A. Ellern, S. Das, R. Nath, B.N. Harmon, A.I. Goldman, D.C. Johnston, Phys. Rev. B **78**, 104512 (2008)
137. S.Q. Xia, C. Myers, S. Bobev, Eur. J. Inortg. Chem. **27**, 4262 (2008)
138. J. An, A.S. Sefat, D.J. Singh, M-H. Du, D. Mandrus, Phys. Rev. B **79**, 075120 (2009), arXiv: 0901.0272 (2008)
139. D.J. Singh, A.S. Sefat, D. Mandrus, Phys. Rev. B **79**, 094429 (2009), arXiv: 0902.0945 (2009)
140. D.J. Singh, Phys. Rev. B **79**, 153102 (2009), arXiv: 0902.2502 (2009)
141. T. Shimojima, K. Ishizaka, Y. Ishida, N. Katayama, K. Ohgushi, T. Kiss, M. Okawa, T. Togashi, X.-Y. Wang, C.-T. Chen, S. Watanabe, R. Kadota, T. Oguchi, A. Chainani, S. Shin, Phys. Rev. Lett. **104**, 057002 (2010), arXiv: 0904.1632 (2009)
142. S.E. Sebastian, J. Gillett, N. Harrison, P.H.C. Lau, D.J. Singh, C.H. Mielke, G.G. Lonzarich, J. Phys. Condens. Matter **20**, 422203 (2008)
143. J.A. Analytis, R.D. McDonald, J.-H. Chu, S.C. Riggs, A.F. Bangura, C. Kucharczyk, M. Johannes, I.R. Fisher, Phys. Rev. B **80**, 064507 (2009), arXiv: 0902.1172 (2009)
144. D.J. Singh, Phys. Rev. B **79**, 174520 (2009), arXiv: 0904.1607 (2009)
145. J.G. Analytis, C.J. Andrew, A.I. Coldea, A. McCollam, J.-H. Chu, R.D. McDonald, I.R. Fisher, A. Carrington, Phys. Rev. Lett. **103**, 076401 (2009), arXiv: 0904.2405 (2009)
146. P. Vilmercati, A. Fedorov, I. Vobornik, U. Manju, G. Panaccione, A. Goldoni, A.S. Sefat, M.A. McGuire, B.C. Sales, R. Jin, D. Mandrus, D.J. Singh, N. Mannell, Phys. Rev. B **79**, 220503(R) (2009), arXiv: 0902.0756 (2009)
147. C. Liu, G.D. Samolyuk, Y. Lee, N. Ni, T. Kondo, A.F. Santander-Syro, S.L. Bud'ko, J.L. McChesney, E. Rotenberg, T. Valla, A.V. Fedorov, P.C. Canfield, B.N. Harmon, A. Kaminski, Phys. Rev. Lett. **101**, 177005 (2008)
148. H. Liu, W. Zhang, L. Zhao, X. Jia, J. Meng, G. Liu, X. Dong, G.F. Chen, J.L. Luo, N.L. Wang, W. Lu, G. Wang, Y. Zhou, Y. Zhu, X. Wang, Z. Zhao, Z. Xu, C. Chen, X. Zhou, J. Phys. Rev. B **78**, 184514 (2008)

149. L. Zhang, D.J. Singh, Phys. Rev. B **79**, 094528 (2009), arXiv: 0902.0265 (2009)
150. J. Fink, S. Thirupathaiah, R. Ovsyannikov, H.A. Duerr, R. Follath, Y. Huang, S. de Jong, M.S. Golden, Yu-Zhong Zhang, H.O. Jeschke, R. Valenti, C. Felser, S. Dastjani Farahani, M. Rotter, D. Johrendt, arXiv: 0903.0967 (2009)
151. V.B. Zabolotnyy, D.S. Inosov, D.V. Evtushinsky, A. Koitzsch, A.A. Kordyuk, J.T. Park, D. Haug, V. Hinkov, A.V. Boris, D.L. Sun, G.L. Sun, C.T. Lin, B. Keimer, M. Knupfer, B. Buechner, A. Varykhalov, R. Follath, S.V. Borisenko, Nature **457**, 569 (2009)
152. T. Sato, K. Nakayama, Y. Sekiba, P. Richard, Y.-M. Xu, S. Souma, T. Takahashi, G.F. Chen, J.L. Luo, N.L. Wang, H. Ding, Phys. Rev. Lett. **103**, 047002 (2009), arXiv: 0810.3047 (2008)
153. N.Y.-M. Xu, P. Richard, K. Nakayama, T. Kawahara, Y. Sekiba, T. Qian, M. Neupane, S. Souma, T. Sato, T. Takahashi, H. Luo, H.-H. Wen, G.-F. Chen, N.-L. Wang, Z. Wang, Z. Fang, H. Dai, H. Ding, arXiv: 0905.4467 (2009)
154. L. Zhao, H. Liu, W. Zhang, J. Meng, X. Jia, G. Liu, X. Dong, G.F. Chen, J.L. Luo, N.L. Wang, G. Wang, Y. Zhou, Y. Zhu, X. Wang, Z. Zhao, Z. Xu, C. Chen, X.J. Zhou, Chin. Phys. Lett. **25**, 4402 (2008)
155. H. Ding, P. Richard, K. Nakayama, T. Sugawara, T. Arakane, Y. Sekiba, A. Takayama, S. Souma, T. Sato, T. Takahashi, Z. Wang, X. Dai, Z. Fang, G.F. Chen, J.L. Luo, N.L. Wang, Europhys. Lett. **83**, 47001 (2008)
156. V.B. Zabolotnyy, D.V. Evtushinsky, A.A. Kordyuk, D.S. Inosov, A. Koitzsch, A.V. Boris, G.L. Sun, C.T. Lin, M. Knupfer, B. Buchner, A. Varykhalov, R. Follath, S.V. Borisenko, Physica C **469**, 448 (2009), arXiv:0904.4337 (2009)
157. T. Kondo, R.M. Fernandes, R. Khasanov, C. Liu, A.D. Palczewski, N. Ni, M. Shi, A. Bostwick, E. Rotenberg, J. Schmalian, S.L. Bud'ko, P.C. Canfield, A. Kaminski, Phys. Rev. B **81**, 060507(R) (2010), arXiv: 0905.0271 (2009)
158. M. Yi, D.H. Lu, J.G. Analytis, J.-H. Chu, S.-K. Mo, R.-H. He, X.J. Zhou, G.F. Chen, J.L. Luo, N.L. Wang, Z. Hussain, D.J. Singh, I.R. Fisher, Z.-X. Shen, Phys. Rev. B **80**, 024515 (2009), arXiv: 0902.2628 (2009)
159. W. Malaeb, T. Yoshida, A. Fujimori, M. Kubota, K. Ono, K. Kohou, P.M. Shirage, H. Kito, A. Iyo, H. Eisaki, Y. Nakajima, T. Tamegai, R. Arita, J. Phys. Soc. Jpn. **78**, 123706 (2009), arXiv: 0906.1846 (2009)
160. C. Liu, T. Kondo, N. Ni, A.D. Palczewski, A. Bostwick, G.D. Samolyuk, R. Khasanov, M. Shi, E. Rotenberg, S.L. Bud'ko, P.C. Canfield, A. Kaminski, Phys. Rev. Lett. **102**, 167004 (2009); arViv: 0903.4388 (2009)
161. Y. Ran, F. Wang, H. Zhai, A. Vishwanath, D.-H. Le, Phys. Rev. B **79**, 0145505 (2009)
162. N. Harrison, R.D. McDonald, C.H. Mielke, E.D. Bauer, F. Ronning, J.D. Thompson, arXiv: 0902.1481 (2009)
163. T. Terashima, M. Kimata, H. Satsukawa, A. Harada, K. Hazama, M. Imai, S. Uji, H. Kito, A. Iyo, H. Eisaki, H. Harima, J. Phys. Soc. Jpn. **78**, 033706 (2009), arXiv: 0901.3394 (2009)
164. A.I. Coldea, C.M.J. Andrew, J.G. Analytis, R.D. McDonald, A.F. Bangura, J.-H. Chu, I.R. Fisher, A. Carrington, Phys. Rev. Lett. **103**, 026404 (2009), arXiv: 0905.3305 (2009)
165. X. Zhu, F. Han, G. Mu, B. Zeng, P. Cheng, B. Shen, H.-H. Wen, Phys. Rev. B **79**, 024516 (2009)
166. I.R. Shein, A.L. Ivanovskii, arXiv: 0811.4475 (2008)
167. H. Ogino, Y. Matsumura, Y. Katsura, K. Ushiyama, S. Horii, K. Kishio, J.-I. Shimoyama, Supercond. Sci. Technol. **22**, 075008 (2009), arXiv: 0903.3314 (2008)
168. I.R. Shein, A.L. Ivanovskii, arXiv: 0903.4038 (2009)
169. Y.L. Xie, R.H. Liu, T. Wu, G. Wu, Y.A. Song, D. Tan, X.F. Wang, H. Chen, J.J. Ying, Y.J. Yan, Q.L. Li, X.H. Chen, Europhys. Lett. **86**, 57007 (2009), arXiv: 0903.5484 (2009)
170. H. Ogino, Y. Katsura, S. Horii, K. Kishio, J.-I. Shimoyama, Supercond. Sci. Technol. **22**, 085001 (2009), arXiv: 0903.5124 (2009)
171. G.F. Chen, T.-L. Xia, P. Zheng, J.L. Luo, N.L. Wang, Supercond. Sci. Technol. **22**, 072001 (2009), arXiv: 0903.5273 (2009)
172. I.R. Shein, A.L. Ivanovskii, arXiv: 0904.0117 (2009)
173. X. Zhu, F. Han, G. Mu, P. Cheng, B. Shen, B. Zeng, H.-H. Wen, Sci. China, Ser. G **52**, 1876 (2009), arXiv: 0904.0972 (2009)

174. Y. Matsumura, H. Ogino, S. Horii, Y. Katsura, K. Kishio, J-I. Shimoyama, Appl. Phys. Express **2**, 063007 (2009), arXiv: 0904.0825 (2009)
175. M. Tegel, F. Hummel, S. Lackner, I. Schellenberg, R. Poettgen, D. Johrendt, Z. Anorg. Allg. Chem. **635**, 2242 (2009), arXiv: 0904.0479 (2009)
176. I.R. Shein, A.L. Ivanovskii, Phys. Rev. B **79**, 245115 (2009)
177. H. Kotegawa, T. Kawazoe, H. Tou, K. Murata, H. Ogino, K. Kishio, J.-I. Shimoyama, J. Phys. Soc. Jpn. **78**, 123707 (2009), arXiv: 0908.1469 (2009)
178. X. Zhu, F. Han, G. Mu, P. Cheng, B. Shen, B. Zeng, H.-H. Wen, Phys. Rev. B **79**, 220512(R) (2009), arXiv: 0904.1732 (2009)
179. M. Tegel, I. Schellenberg, F. Hummel, R. Pottgen, D. Johrendt, Z. Naturforsch. B **64B**, 1815 (2009), arXiv: 0905.0337 (2009)
180. A.L. Ivanovskii, Russ. Chem. Rev. **79**, 1 (2010)
181. H. Chen, Y. Ren, Y. Qiu, W. Bao, R.H. Liu, G. Wu, T. Wu, Y.L. Xie, X.F. Wang, Q. Huang, X.H. Chen, Europhys. Lett. **85**, 17006 (2009)
182. J.-H. Chu, J.G. Analytis, C. Kucharczyk, I.R. Fisher, Phys. Rev. B **79**, 014506 (2009), arXiv: 0811.2463 (2008)
183. N. Ni, M.E. Tillman, J.-O. Yan, A. Kracher, S.T. Hannahs, S.L. Bud'ko, P.C. Canfield, Phys. Rev. B **78**, 214515 (2008)
184. F.L. Ning, K. Ahilan, T. Imai, A.S. Sefat, R. Jin, M.A. McGuire, B.C. Sales, D. Mandrus, J. Phys. Soc. Jpn. **78**, 013711 (2009)
185. Y. Nakajima, T. Taen, T. Tamegai, arXiv: 0811.2621 (2008)
186. A.S. Sefat, R. Jin, M.A. McGuire, B.C. Sales, D.J. Singh, D. Mandrus, Phys. Rev. Lett. **101**, 117004 (2008)
187. F.L. Ning, K. Ahilan, T. Imai, A.S. Sefat, R. Jin, M.A. McGuire, B.C. Sales, D.J. Mandrus, Phys. Soc. Jpn. **77**, 103705 (2008)
188. Y. Qi, Z. Gao, L. Wang, D. Wang, X. Zhang, Y. Ma, New J. Phys. **10**, 123003 (2008)
189. A. Leithe-Jasper, W. Schnelle, C. Geibel, H. Rosner, Phys. Rev. Lett. **101**, 207004 (2008), arXiv: 0807.2223 (2008)
190. N. Kumar, R. Nagalakshmi, R. Kulkarni, P.L. Paulose, A.K. Nigam, S.K. Dhar, A. Thamizhavel, Phys. Rev. B **79**, 012504 (2009)
191. D. Kasinathan, A. Ormeci, K. Koch, U. Burkhardt, W. Schnelle, A. Leithe-Jasper, H. Rosner, New J. Phys. **11**, 025023 (2009)
192. S. Sharma, A. Bharathi, S. Chandra, R. Reddy, S. Paulraj, A.T. Satya, V.S. Sastry, A. Gupta, C.S. Sundar, Phys. Rev. B **81**, 174512 (2010), arXiv: 0902.2728 (2009)
193. Y. Qi, L. Wang, Z. Gao, D. Wang, X. Zhang, Y. Ma, arXiv: 0903.4967 (2009)
194. F. Han, X. Zhu, P. Cheng, B. Shen, H.-H. Wen, arXiv: 0903.1028 (2009); Phys. Rev. B **80**, 024506 (2009)
195. X. Zhu, F. Han, H.-H. Wen, Phys. Rev. B **80**, 024506 (2009), arXiv: 0903.0323 (2009)
196. A.S. Sefat, D.J. Singh, L.H. Van Bebber, M.A. McGuire, Yu. Mozharivskyj, R. Jin, B.C. Sales, V. Keppens, D. Madrus, Phys. Rev. B **79**, 224524 (2009), arXiv: 0903.5546 (2009)
197. F. Han, X. Zhu, Y. Jia, L. Fang, P. Cheng, H. Luo, B. Shen, H.-H. Wen, Phys. Rev. B **80**, 024506 (2009), arXiv: 0902.3957 (2009)
198. F. Han, X. Zhu, P. Cheng, G. Mu, Y. Jia, L. Fang, Y. Wang, H. Luo, B. Zeng, B. Shen, L. Shan, C. Ren, H.-H. Wen, Phys. Rev. B **80**, 024506 (2009), arXiv: 0906.0403 (2009)
199. F. Ronning, N. Kurita, E.D. Bauer, B.L. Scott, T. Park, T. Klimczuk, R. Movshovich, J.D. Thompson, J. Phys. Condens. Matter **20**, 342203 (2008)
200. R.H. Liu, G. Wu, H. Chen, Y.L. Xie, Y.J. Yan, T. Wu, X.F. Wang, J.J. Ying, Q.J. Li, B.C. Shi, X.H. Chen, arXiv: 0811.2569 (2008)
201. S. Kasahara, T. Shibauchi, K. Hashimoto, K. Ikada, S. Tonegawa, H. Ikeda, H. Takeya, K. Hirata, T. Terashima, Y. Matsuda, Phys. Rev. B **81**, 184519 (2010), arXiv: 0905.4427 (2009)
202. M.A. Tanatar, J.-Ph. Reid, H. Shakeripour, X.G. Luo, N. Doiron-Leyraud, N. Ni, S.L. Bud'ko, P.C. Canfield, R. Prozorov, L. Taillefer, Phys. Rev. Lett. **104**, 067002 (2010), arXiv: 0907.1276 (2009)

203. G. Mu, B. Zeng, P. Cheng, Z. Wang, L. Fang, B. Shen, L. Shan, C. Ren, H.-H. Wen, Chin. Phys. Lett. **27**, 037402 (2010), arXiv: 0906.4513 (2009)
204. A. Leithe-Jasper, W. Schnelle, C. Geibel, H. Rosner, Phys. Rev. Lett. **101**, 207004 (2008)
205. X.F. Wang, T. Wu, G. Wu, H. Chen, Y.L. Xie, J.J. Ying, Y.J. Yan, R.H. Liu, X.H. Chen, Phys. Rev. Lett. **102**, 117005 (2009)
206. S.R. Saha, N.P. Butch, K. Kirshenbaum, J. Paglione, Phys. Rev. B **79**, 224519 (2009), arXiv: 0905.0859 (2009)
207. M. Kano, Y. Kohama, D. Graf, F.F. Balakirev, A.S. Sefat, M.A. McGuire, B.C. Sales, D. Mandrus, S.W. Tozer, arXiv: 0904.1418 (2009)
208. A. Yamamoto, J. Jaroszynski, C. Tarantini, L. Balicas, J. Jiang, A. Gurevich, D.C. Larbalestier, R. Jin, A.S. Sefat, M.A. McGuire, B.C. Sales, D.K. Christen, D. Mandrus, Appl. Phys. Lett. **94**, 062511 (2009)
209. M. Rotter, M. Tegel, I. Schellenberg, F.M. Schappacher, R. Poettgen, J. Deisenhofer, A. Guenther, F. Schrettle, A. Loidl, D. Johrendt, New J. Phys. **11**, 025014 (2009), arXiv: 0812.2827 (2008)
210. J.-H. Chu, J.G. Analytis, C. Kucharczyk, I.R. Fisher, Phys. Rev. B **79**, 014506 (2009)
211. D.K. Pratt, W. Tian, A. Kreyssig, J.L. Zarestky, S. Nandi, N. Ni, S.L. Bud'ko, P.C. Canfield, A.I. Goldman, R.J. McQueeney, arXiv: 0903.2833 (2009)
212. F. Rullier-Albenque, D. Colson, A. Forget, H. Alloul, Phys. Rev. Lett. **103**, 057001 (2009), arXiv: 0903.5243 (2009)
213. Y. Laplace, J. Bobroff, F. Rullier-Albenque, D. Colson, A. Forget, Phys. Rev. B **80**, 140501 (2009), arXiv: 0906.2125 (2009)
214. C. Bernhard, A.J. Drew, L. Schulz, V.K. Malik, M. Roessle, Ch. Niedermayer, Th. Wolf, G.D. Varma, G. Mu, H.-H. Wen, H. Liu, G. Wu, X.H. Chen, New. J. Phys. **11**, 055050 (2009), arXiv: 0902.0859 (2009)
215. G.L. Sun, D.L. Sun, M. Konuma, P. Popovich, A. Boris, J.B. Peng, K.-Y. Choi, P. Lemmens, C.T. Lin, Nev. J. Phys. **11**, 055050 (2009), arXiv: 0901.2728 (2009)
216. K. Kitagawa, N. Katayama, K. Ohgushi, M. Takigawa, J. Phys. Soc. Jpn. **78**, 063706 (2009), arXiv: 0903.3193 (2009)
217. M. Kumar, M. Nicklas, A. Jesche, N. Caroca-Canales, M. Schmitt, M. Hanfland, D. Kasinathan, U. Schwarz, H. Rosner, C. Geibel, Phys. Rev. B **78**, 184516 (2008)
218. L. Alireza Patricia, J. Gillett, Y.T. Chris Ko, S.E. Sebastian, G.G. Lonzarich, J. Phys. Condens. Matter **21**, 012208 (2009)
219. M.S. Torikachvili, S.L. Bud'ko, N. Ni, P.C. Canfield, Phys. Rev. Lett. **101**, 057006 (2008)
220. A. Kreyssig, M.A. Green, Y. Lee, G.D. Samolyuk, P. Zajdel, J.W. Lynn, S.L. Bud'ko, M.S. Torikachvili, N. Ni, S. Nandi, J. Leao, S.J. Poulton, D.N. Argyriou, B.N. Harmon, P.C. Canfield, R.J. McQueeney, A.I. Goldman, Phys. Rev. B **78**, 184517 (2008)
221. A.I. Goldman, A. Kreyssig, K. Prokes, D.K. Pratt, D.N. Argyriou, J.W. Lynn, S. Nandi, S.A.J. Kimber, Y. Chen, Y.B. Lee, G. Samolyuk, J.B. Leao, S.J. Poulton, S.L. Bud'ko, N. Ni, P.C. Canfield, B.N. Harmon, R.J. McQueeney, Phys. Rev. B **79**, 024513 (2009)
222. H. Lee, E. Park, T. Park, F. Ronning, E.D. Bauer, J.D. Thompson, Phys. Rev. **80**, 024519 (2009), arXiv: 0809.3550 (2008)
223. R.J. McQueeney, S.O. Diallo, V.P. Antropov, G. Samolyuk, C. Broholm, N. Ni, S. Nandi, M. Yethiraj, J.L. Zarestky, J.J. Pulikkotil, A. Kreyssig, M.D. Lumsden, B.N. Harmon, P.C. Canfield, A.I. Goldman, arXiv: 0809.1410 (2008)
224. Y.-Z. Zhang, H.C. Randpal, I. Opahle, H.O. Jeschke, R. Valenti, Phys. Rev. B **80**, 094530 (2009), arXiv: 0812.2920 (2008)
225. M. Parrinello, A. Rahman, Phys. Rev. Lett. **45**, 1196 (1980)
226. D.A. Tompsett, G.G. Lonzarich, arXiv: 0902.4859 (2009)
227. A. Hackl, M. Vojta, New J. Phys. **11**, 055064 (2009), arXiv: 0812.3394 (2008)
228. F. Ronning, E.D. Bauer, T. Park, S.-H. Baek, H. Sakai, J.D. Thompson, Phys. Rev. B **79**, 134507 (2009), arXiv: 0902.0663
229. C.F. Miclea, M. Nicklas, H.S. Jeevan, D. Kasinathan, Z. Hossain, H. Rosner, P. Gegenwart, C. Geibel, F. Steglich, Phys. Rev. B **79**, 212509 (2009), arXiv: 0808.2026 (2008)

230. M. Ishikawa, O. Fischer, Solid State Commun. **23**, 37 (1977)
231. Anupam, P.L. Paulose, H.S. Jeevan, C. Geibel, Z. Hossain, J. Phys. Condens. Matter **21**, 265701 (2009), arXiv: 0812.1131 (2008)
232. M. Gooch, B. Lv, B. Lorenz, A.M. Guloy, C.-W. Chu, Phys. Rev. B **78**, 180508(R) (2008)
233. M.S. Torikachvili, S.L. Budko, N. Ni, P.C. Canfield, arXiv: 0810.0241 (2008)
234. K. Ahilan, J. Balasubramaniam, F.L. Ning, T. Imai, A.S. Sefat, R. Jin, M.A. McGuire, B.C. Sales, D. Mandrus, J. Phys. Condens. Matter **20**, 472201 (2008)
235. F. Hardy, P. Adelmann, H. Loehneysen, T. Wolf, C. Meingast, arXiv: 0812.4176 (2008)
236. K. Ahilan, F.L. Ning, T. Imai, A.S. Sefat, M.A. McGuire, B.C. Sales, D. Mandrus, Phys. Rev. B **79**, 214520 (2009), arXiv: 0904.2215 (2009)
237. H. Kotegawa, T. Kawazoe, H. Sugawara, K. Murata, H. Tou, J. Phys. Soc. Jpn. **8**, 4 (2009), arXiv: 0904.4631 (2009)
238. E. Colombier, S.L. Bud'ko, N. Ni, P.C. Canfield, Phys. Rev. B **79**, 224518 (2009), arXiv: 0904.4488 (2009)
239. T. Terashima, M. Kimata, H. Satsukawa, A. Harada, K. Hazama, S. Uji, H.S. Suzuki, T. Matsumoto, K. Murata, J. Phys. Soc. Jpn. **78**, 083701 (2009), arXiv: 0904.2618 (2009)
240. K. Matsubayashi, N. Katayama, K.Y. Ohgushi, K. Munakata, T. Matsumoto, Y. Uwatoko, arXiv: 0905.0968 (2009)
241. M.S. Torikachvili, S.L. Bud'ko, N. Ni, P.C. Canfield, S.T. Hannahs, Phys. Rev. B **80**, 014521 (2009), arXiv: 0905.4002 (2008)
242. G. Li, W.Z. Hu, J. Dong, Z. Li, P. Zheng, G.F. Chen, J.L. Luo, N.L. Wang, Phys. Rev. Lett. **101**, 107004 (2008)
243. K. Nakayama, T. Sato, P. Richard, Y.-M. Xu, Y. Sekiba, S. Souma, G.F. Chen, J.L. Luo, N.L. Wang, H. Ding, T. Takahashi, Europhys. Lett. **85**, 67002 (2009)
244. L. Wray, D. Qian, D. Hsieh, Y. Xia, L. Li, J.G. Checkelsky, A. Pasupathy, K.K. Gomes, A.V. Fedorov, G.F. Chen, J.L. Luo, A. Yazdan, N.P. Ong, N.L. Wang, M.Z. Hasan, arXiv: 0808.2185 (2008)
245. T. Kondo, A.F. Santander-Syro, O. Copie, C. Liu, M.E. Tillman, E.D. Mun, J. Schmalian, S.L. Bud'ko, M.A. Tanatar, P.C. Canfield, A. Kaminski, Phys. Rev. Lett. **101**, 147003 (2008)
246. D.V. Evtushinsky, D.S. Inosov, V.B. Zabolotnyy, A. Koitzsch, M. Knupfer, B. Buchner, G.L. Sun, V. Hinkov, A.V. Boris, C.T. Lin, B. Keimer, A. Varykhalov, A.A. Kordyuk, S.V. Borisenko, Phys. Rev. B **79**, 054517 (2009)
247. P. Szabo, Z. Pribulovra, G. Pristras, S.L. Bud'ko, P.C. Canfield, P. Samuely, Phys. Rev. B **79**, 012503 (2009)
248. G. Mu, H. Luo, Z. Wang, L. Shan, C. Ren, H.-H. Wen, Phys. Rev. B **79**, 174501 (2009), arXiv: 0808.2941 (2008)
249. R. Khasanov, A. Amato, H.-H. Klauss, H. Luetkens, Ch. Niedermayer, G.L. Sun, C.T. Lin, J.T. Park, D.S. Inosov, V. Hinkov, Phys. Rev. Lett. **102**, 187005 (2009)
250. D.V. Evtushinsky, D.S. Inosov, V.B. Zabolotnyy, M.S. Viazovska, R. Khasanov, A. Amato, H.-H. Klauss, H. Luetkens, Ch. Niedermayer, G.L. Sun, V. Hinkov, C.T. Lin, A. Varykhalov, A. Koitzsch, M. Knupfer, B. Bchner, A.A. Kordyuk, S.V. Borisenko, New J. Phys. **11**, 055069 (2009), arXiv: 0903.4362 (2009)
251. H. Fukazawa, T. Yamazaki, K. Kondo, Y. Kohori, N. Takeshita, P.M. Shirage, K. Kihou, K. Miyazawa, H. Kito, H. Eisaki, A. Iyo, J. Phys. Soc. Jpn. **78**, 033704 (2009), arXiv: 0901.0177 (2009)
252. K.A. Yates, L.F. Cohen, Z.-A. Ren, J. Yang, W. Lu, X.-L. Dong, Z.-X. Zhao, Supercond. Sci. Technol. **21**, 092003 (2008)
253. K. Hashimoto, T. Shibauchi, S. Kasahara, K. Ikada, T. Kato, R. Okazaki, C.J. van der Beek, M. Konczykowski, H. Takeya, K. Hirata, T. Terashima, Y. Matsuda, Phys. Rev. Lett. **102**, 207001 (2009), arXiv: 0810.3506 (2008)
254. M. Hiraishi, R. Kadono, S. Takeshita, M. Miyazaki, A. Koda, H. Okabe, J. Akimitsu, J. Phys. Soc. Jpn. **78**, 023710 (2009)
255. G. Mu, H. Luo, Z. Wang, Z. Ren, L. Shan, C. Ren, H.-H. Wen, Phys. Rev. B **79**, 174501 (2009), arXiv: 0812.1188 (2008)

256. M. Gooch, B. Lv, B. Lorenz, A.M. Guloy, C.-W. Chu, Phys. Rev. B **79**, 104504 (2009)
257. L. Wray, D. Qian, D. Hsieh, Y. Xia, L. Li, J.G. Checkelsky, A. Pasupathy, K.K. Gomes, C.V. Parker, A.V. Fedorov, G.F. Chen, J.L. Luo, A. Yazdani, N.P. Ong, N.L. Wang, M.Z. Hasan, Phys. Rev. B **78**, 184508 (2008)
258. K. Hashimoto, M. Yamashita, S. Kasahara, Y. Senshu, N. Nakata, S. Tonegawa, K. Ikada, A. Serafin, A. Carrington, T. Terashima, H. Ikeda, T. Shibauchi, Y. Matsuda, Phys. Rev. B **81**, 220501(R) (2010), arXiv: 0907.4399 (2009)
259. Y.-R. Zhou, Y.-R. Li, J.-W. Zuo, R.-Y. Liu, S.-K. Su, G.F. Chen, J.L. Lu, N.L. Wang, Y.-P. Wang, arXiv: 0812.3295 (2008)
260. D.J. Van Harlingen, Rev. Mod. Phys. **67**, 515 (1995)
261. X. Zhang, Y.S. Oh, Y. Liu, L. Yan, K.H. Kim, R.L. Greene, I. Takeuchi, Phys. Rev. Lett. **102**, 147002 (2009)
262. X. Zhang, S. Saha, N.P. Butch, K. Kirshenbaum, J. Paglione, R.L. Greene, Appl. Phys. Lett. **95**, 062510 (2009), arXiv: 0907.4456 (2009)
263. M. Altarawneh, K. Collar, C.H. Mielke, N. Ni, S.L. Bud'ko, P.C. Canfield, Phys. Rev. B **78**, 220505 (2008)
264. Z.-S. Wang, H.-Q. Luo, C. Ren, H.-H. Wen, Phys. Rev. B **78**, 140501(R) (2008)
265. Y.J. Uemura, L.P. Le, G.M. Luke, B.J. Sternlieb, W.D. Wu, J.H. Brewer, T.M. Riseman, C.L. Seaman, M.B. Maple, M. Ishikawa, D.G. Hinks, J.D. Jorgensen, G. Saito, H. Yamochi, Phys. Rev. Lett. **66**, 2665 (1991)
266. C. Ren, Z.-S. Wang, H.-Q. Luo, H. Yang, L. Shan, H.-H. Wen, Phys. Rev. Lett. **101**, 257006 (2008)
267. A.J. Drew, F.L. Pratt, T. Lancaster, S.J. Blundell, P.J. Baker, R.H. Liu, G. Wu, X.H. Chen, I. Watanabe, V.K. Malik, A. Dubroka, K.W. Kim, M. Roessl, C. Bernhard, Phys. Rev. Lett. **101**, 097010 (2008)
268. A.A. Aczel, E. Baggio-Saitovitch, S.L. Bud'ko, P.C. Canfield, J.P. Carlo, G.F. Chen, P. Dai, T. Goko, W.Z. Hu, G.M. Luke, J.L. Luo, N. Ni, D.R. Sanchez-Candela, F.F. Tafti, N.L. Wang, T.J. Williams, W. Yu, Y.J. Uemura, Phys. Rev. B **78**, 214503 (2008)
269. T. Goko, A.A. Aczel, E. Baggio-Saitovitch, S.L. Bud'ko, P.C. Canfield, J.P. Carlo, G.F. Chen, P. Dai, A.C. Hamann, W.Z. Hu, H. Kageyama, G.M. Luke, J.L. Luo, B. Nachumi, N. Ni, D. Reznik, D.R. Sanchez-Candela, A.T. Savici, K.J. Sikes, N.L. Wang, C.R. Wiebe, T.J. Williams, T. Yamamoto, W. Yu, Y.J. Uemura, Phys. Rev. Cond. Mat. B **80**, 024508 (2009), arXiv: 0808.1425 (2008)
270. U. Welp, R. Xie, A.E. Koshelev, W.K. Kwok, H.Q. Luo, Z.S. Wang, G. Mu, H.H. Wen, arXiv: 0810.1944 (2008)
271. H.Q. Yuan, J. Singleton, F.F. Balakirev, S.A. Baily, G.F. Chen, J.L. Luo, N.L. Wang, Nature **457**, 565 (2009)
272. M.R. Eskildsen, L.Ya. Vinnikov, T.D. Blasius, I.S. Veshchunov, T.M. Artemova, J.M. Densmore, C.D. Dewhurst, N. Ni, A. Kreyssig, S.L. Bud'ko, P.C. Canfield, A.I. Goldman, Phys. Rev. B **79**, 100501(R) (2009)
273. Y. Su, P. Link, A. Schneidewind, Th. Wolf, Y. Xiao, R. Mittal, M. Rotter, D. Johrendt, Th. Brueckel, M. Loewenhaupt, Phys. Rev. B **79**, 064504 (2009)
274. Q. Huang, Y. Qiu, W. Bao, J.W. Lynn, M.A. Green, Y.C. Gasparovic, T. Wu, G. Wu, X.H. Chen, Phys. Rev. Lett. **101**, 257003 (2008)
275. A.I. Goldman, D.N. Argyriou, B. Ouladdiaf, T. Chatterji, A. Kreyssig, S. Nandi, N. Ni, S.L. Bud'ko, P.C. Canfield, R.J. McQueeney, Phys. Rev. B **78**, 100506(R) (2008)
276. J. Zhao, W. Ratcliff-II, J.W. Lynn, G.F. Chen, J.L. Luo, N.L. Wang, J. Hu, P. Dai, Phys. Rev. B **78**, 140504(R) (2008)
277. K. Kaneko, A. Hoser, N. Caroca-Canales, A. Jesche, C. Krellner, O. Stockert, C. Geibel, Phys. Rev. B **78**, 212502 (2008)
278. M. Kofu, Y. Qiu, W. Bao, S.-H. Lee, S. Chang, T. Wu, G. Wu, X.H. Chen, New.J. Phys. **11**, 055001 (2009), arXiv: 0901.0738 (2009)
279. S.D. Wilson, Z. Yamani, C.R. Rotundu, B. Freelon, E. Bourret-Courchesne, R.J. Birgeneau, Phys. Rev. B **79**, 184519 (2009), arXiv: 0903.2430 (2009)

280. C. Fang, H. Yao, W.-F. Tsai, J.-P. Hu, S.A. Kivelson, Phys. Rev. B **77**, 224509 (2008)
281. C. Xu, M. Muller, S. Sachdev, Phys. Rev. B **78**, 020501(R) (2008)
282. P. Becca, F. Mila, Phys. Rev. Lett. **89**, 037204 (2002)
283. D. Hou, Q.M. Zhang, Z.Y. Lu, J.H. Wei, arXiv: 0901.1525 (2009)
284. M. Tegel, M. Rotter, V. Weiss, F.M. Schappacher, R. Poettgen, D. Johrendt, J. Phys. Condens. Matter **20**, 452201 (2008)
285. H.S. Jeevan, Z. Hossain, C. Geibel, P. Gegenwart, Phys. Rev. B, **78**, 092406 (2008)
286. Z. Ren, Z. Zhu, S. Jiang, X. Xu, Q. Tao, C. Wang, C. Feng, G. Cao, Z. Xu, Phys. Rev. B, **78,** 052501 (2008)
287. T. Wu, G. Wu, H. Chen, Y.L. Xie, R.H. Liu, X.F. Wang, X.H. Chen, arXiv: 0808.2247 (2008)
288. Q.J. Zheng, Y. He, T. Wu, G. Wu, H. Chen, J.J. Ying, R.H. Liu, X.F. Wang, Y.L. Xie, Y.J. Yan, Q.J. Li, X.H. Chen, arXiv: 0907.5547 (2009)
289. Z. Ren, X. Lin, Q. Tao, S. Jiang, Z. Zhu, C. Wang, G. Cao, Z. Xu, Phys. Rev. B **79**, 094426 (2009)
290. S. Jiang, Y. Luo, Z. Ren, Z. Zhu, C. Wang, X. Xu, Q. Tao, G. Cao, Z. Xu, New J. Phys. **11**, 025007 (2009)
291. X.F. Wang, T. Wu, G. Wu, R.H. Liu, H. Chen, Y.L. Xie, X.H. Chen, New.J. Phys. **11** 045003 (2009), arXiv: 0811.2920 (2008)
292. Y. Singh, Y. Lee, S. Nandi, A. Kreyssig, A. Ellern, S. Das, R. Nath, B.N. Harmon, A.I. Goldman, D.C. Johnston, Phys. Rev. B **77**, 014410 (2008)
293. N. Ni, A. Thaler, A. Kracher, J.Q. Yan, S.L. Bud'ko, P.C. Canfield, Phys. Rev. B **80**, 024511 (2009), arXiv: 0905.4894 (2009)
294. J. Zhao, D.-X. Yao, S. Li, T. Hong, Y. Chen, S. Chang, W. Ratcliff, J.W. Lynn, H.A. Mook, G.F. Chen, J. L. Luo, N.L. Wang, E.W. Carlson, J. Hu, P. Dai, Phys. Rev. Lett. **101**, 167203 (2008)
295. K. Matan, R. Morinaga, K. Iida, T.J. Sato, Phys. Rev. B **79**, 054526 (2008)
296. A.D. Christianson, E.A. Goremychkin, R. Osborn, S. Rosenkranz, M.D. Lumsden, C.D. Malliakas, L.S. Todorov, H. Claus, D.Y. Chung, M.G. Kanatzidis, R.I. Bewley, T. Guidi, Nature **456**, 930 (2009)
297. M.D. Lumsden, A.D. Christianson, D. Parshall, M.B. Stone, S.E. Nagler, G.J. MacDougall, H.A. Mook, K. Lokshin, T. Egami, D.L. Abernathy, E.A. Goremychkin, R. Osborn, M.A. McGuire, A.S. Sefat, R. Jin, B.C. Sales, D. Mandrus, Phys. Rev. Lett **102**, 107005 (2009)
298. L.W. Harriger, A. Schneidewind, S. Li, J. Zhao, Z. Li, W. Lu, X. Dong, F. Zhou, Z. Zhao, J. Hu, P. Dai, Phys. Rev. Lett. **103**, 087005 (2009), arXiv: 0904.3775 (2009)
299. S. Iikubo, M. Fujita, S. Niitaka, H. Takagi, arXiv: 0904.3824 (2009)
300. A.D. Christianson, M.D. Lumsden, S.E. Nagler, G.J. MacDougall, M.A. McGuire, A.S. Sefat, R. Jin, B.C. Sales, D. Mandrus, Phys. Rev. Lett. **103**, 087002 (2009), arXiv: 0904.0767 (2009)
301. J. Zhao, D.T. Adroja, D.-X. Yao, R. Bewley, S. Li, X.F. Wang, G. Wu, X.H. Chen, J. Hu, P. Dai, Nat. Phys. **5,** 555 (2009), arXiv: 0903.2686 (2009)
302. S.O. Diallo, V.P. Antropov, C. Broholm, T.G. Perring, J.J. Pulikkoti, N. Ni, S.L. Bud'ko, P.C. Canfield, A. Kreyssig, A.I. Goldman, R.J. McQueeney, Phys. Rev. Lett. **102**, 187206 (2009), arXiv: 0901.3784 (2009)
303. F.L. Ning, K. Ahilan, T. Imai, A.S. Sefat, M.A. McGuire, B.C. Sales, D. Mandrus, P. Cheng, B. Shen, H.-H Wen, Phys. Rev. Lett. **104**, 037001 (2010), arXiv: 0907.3875 (2009)
304. F.-C. Hsu, J.-Y. Luo, K.-W. Yeh, T.-K. Chen, T.-W. Huang, P.M. Wu, Y.-C. Lee, Y.-L. Huang, Y.-Y. Chu, D.-C. Yan, M.-K. Wu, arXiv: 0807.2369 (2008)
305. S. Margadonna, Y. Takabayashi, M.T. McDonald, K. Kasperkiewicz, Y. Mizuguchi, Y. Takano, A.N. Fitch, E. Suarde, K. Prassides, Chem. Commun. 5607 (2008)
306. M.H. Fang, L. Spinu, B. Qian, H.M. Pham, T.J. Liu, E.K. Vehstedt, Y. Liu, Z.Q. Mao, Phys. Rev. B **78**, 224503 (2008)
307. K.-W. Yeh, T.-W. Huang, Y.-L. Huang, T.-K. Chen, F.-C. Hsu, P.M. Wu, Y.-C. Lee, Y.-Y. Chu, C.-L. Chen, J.-Y. Luo, D.-C. Yan, M.-K. Wu, arXiv: 0808.0474 (2008)
308. S. Margadonna, Y. Takabayashi, Y. Ohishi, Y. Mizuguchi, Y. Takano, T. Kagayama, T. Nakagawa, M. Takata, K. Prassides, Phys. Rev. B **80**, 064506 (2009), arXiv: 0903.2204 (2009)

309. S. Medvedev, T.M. McQueen, I. Trojan, T. Palasyuk, M.I. Eremets, R.J. Cava, S. Naghavi, F. Casper, V. Ksenofontov, G. Wortmann, C. Felser, Nat. Mater. **8**, 630 (2009), arXiv: 0903.2143 (2009)
310. G. Garbarino, A. Sow, P. Lejay, A. Sulpice, P. Toulemonde, W. Crichton, M. Mezouar, M. Núñez-Regueiro, Europhys. Lett. **86**, 27001 (2009), arXiv: 0903.3888 (2009)
311. S. Masaki, H. Kotegawa, Y. Hara, H. Tou, K. Murata, Y. Mizuguchi, Y. Takano, J. Phys. Soc. Jpn. **78**, N 6 (2009), arXiv: 0903.2594 (2009)
312. V.G. Tissen, E.G. Ponyatovskii, M.V. Nefedova, A.N. Titov, V.V. Fedorenko, Phys. Rev. B **80**, 092507 (2009), arXiv: 0905.3289 (2009)
313. M.H. Fang, B. Qian, H.M. Pham, J.H. Yang, T.J. Liu, E.K. Vehstedt, L. Spinu, Z.Q. Mao, Phys. Rev. B **78**, 220510(R) (2008)
314. R. Khasanov, K. Conder, E. Pomjakushina, A. Amato, C. Baines, Z. Bukowski, J. Karpinski, S. Katrych, H.-H. Klauss, H. Luetkens, A. Shengelaya, N.D. Zhigadlo, Phys. Rev. B **78**, 220510(R) (2008), arXiv: 0810.1716 (2008)
315. Y. Mizuguchi, F. Tomioka, S. Tsuda, T. Yamaguchi, Y. Takano, Appl. Phys. Lett. **94**, 012503 (2009), arXiv: 0811.0711
316. T.M. McQueen, Q. Huang, V. Ksenofontov, C. Felser, Q. Xu, H.W. Zandbergen, Y.S. Hor, J. Allred, A.J. Williams, D. Qu, J. Checkelsky, N.P. Ong, R.J. Cava, Phys. Rev. B **79**, 014522 (2009)
317. B.C. Sales, A.S. Sefat, M.A. McGuire, R.Y. Jin, Y.D. Mandrus, Phys. Rev.B. **79**, 094521 (2009), arXiv: 0902.1519 (2009)
318. R. Hu, J.B. Warren, C. Petrovic, arXiv: 0903.4430 (2009)
319. L. Li, Z.R. Yang, M. Ge, L. Pi, J.T. Xu, B.S. Wang, Y.P. Sun, Y.H. Zhang, arXiv: 0809.0128 (2008)
320. W. Bao, Y. Qiu, Q. Huang, M.A. Green, P. Zajdel, M.R. Fitzsimmons, M. Zhernenkov, M. Fang, B. Qian, E.K. Vehstedt, J. Yang, H.M. Pham, L. Spinu, Z.Q. Mao, Phys. Rev. Lett. **102**, 247001 (2009), arXiv: 0809.2058 (2008)
321. J.J. Pulikkotil, M. van Schilfgaarde, V.P. Antropov, arXiv: 0809.0283 (2008)
322. S. Li, C. de la Cruz, Q. Huang, Y. Chen, J.W. Lynn, J. Hu, Y.-L. Huang, F.-C. Hsu, K.-W. Yeh, M.-K. Wu, P. Dai, Phys. Rev. B **79**, 054503 (2009)
323. M.J. Han, S.Y. Savrasov, Phys. Rev. Lett. **103**, 067001 (2009), arXiv: 0903.2896 (2009)
324. L. Zhang, D.J. Singh, M.H. Du, Phys. Rev. B **79**, 012506 (2009), arXiv: 0810.3274 (2008)
325. C. Zhang, W. Yi, L. Sun, W. Lu, X. Dong, L. Bai, J. Liu, G. Chen, N. Wang, Z. Zhao, Phys. Rev. B **80**, 144519 (2009), arXiv: 0905.3249 (2009)
326. A. Subedi, L. Zhang, D.J. Singh, M.H. Du, Phys. Rev. B **78**, 134514 (2008)
327. D. Phelan, J.N. Millican, E.L. Thomas, J.B. Leao, Y. Qiu, R. Paul, Phys. Rev. B **79**, 014519 (2009)
328. T.-L. Xia, D. Hou, S.C. Zhao, A.M. Zhang, G.F. Chen, J.L. Luo, N.L. Wang, J.H. Wei, Z.-Y. Lu, Q.M. Zhang, Phys. Rev. B **79**, 140510(R) (2009), arXiv: 0811.2350 (2008)
329. F. Ma, W. Ji, J. Hu, Z.-Y. Lu, T. Xiang, Phys. Rev. Lett. **102**, 177003 (2009), arXiv: 0809.4732 (2008)
330. Y. Xia, D. Qian, L. Wray, D. Hsieh, G.F. Chen, J.L. Luo, N.L. Wang, M.Z. Hasan, arXiv: 0901.1299 (2009)
331. Y. Xia, D. Qian, L. Wray, D. Hsieh, G.F. Chen, J.L. Luo, N.L. Wang, M.Z. Hasan, Phys. Rev. Lett. **103**, 037002 (2009), arXiv: 0906.5392 (2009)
332. K.-W. Lee, V. Pardo, W.E. Pickett, Phys. Rev. B **78**, 174502 (2008)
333. P.P. Singh, arXiv: 0908.0303 (2009)
334. K. Nakayama, T. Sato, P. Richard, T. Kawahara, Y. Sekiba, T. Qian, G.F. Chen, J.L. Luo, N.L. Wang, H. Ding, T. Takahashi, F. Rullier-Albenque, D. Colson, A. Forget, H. Alloul, arXiv: 0907.0763 (2009)
335. Y. Qiu, W. Bao, Y. Zhao, C. Broholm, V. Stanev, Z. Tesanovic, Y.C. Gasparovic, S. Chang, J. Hu, B. Qian, M. Fang, Z. Mao, Phys. Rev. Lett. **103**, 067008 (2009), arXiv: 0905.3559 (2009)

336. M.D. Lumsden, A.D. Christianson, E.A. Goremychkin, S.E. Nagler, H.A. Mook, M.B. Stone, D.L. Abernathy, T. Guidi, G.J. MacDougall, C. de la Cruz, A.S. Sefat, M.A. McGuire, B.C. Sales, D. Mandrus, Nat. Phys. **6**, 182 (2010), arXiv: 0907.2417 (2009)
337. J. Wen, G. Xu, Z. Xu, Z.W. Lin, Q. Li, W. Ratcliff, G. Gu, J.M. Tranquada, Phys. Rev. B **80**, 104506 (2009), arXiv: 0906.3774 (2009)
338. Y.-F. Li, L.-F. Zhu, S.-D. Guo, Y.-C. Xu, B.-G. Liu, J. Phys. Condens. Matter **21**, 115701 (2009), arXiv: 0905.3300 (2009)
339. A.M. Turner, F. Wang, A. Vishwanath, arXiv: 0905.3782 (2009)
340. K. Miyoshi, Y. Takaichi, E. Mutou, K. Fujiwara, J. Takeuchi, J. Phys. Soc. Jpn **78**, 093703 (2009), arXiv: 0906.1534 (2009)
341. J.K. Dong, T.Y. Guan, S.Y. Zhou, X. Qiu, L. Ding, C. Zhang, U. Patel, Z.L. Xiao, S.Y. Li, Phys. Rev. B **80**, 024518 (2009), arXiv: 0906.3852 (2009)
342. M.J. Pitcher, D.R. Parker, P. Adamson, S.J.C. Herkel-rath, A.T. Boothroyd, S.J. Clarke, Chem. Commun. 5918 (2008)
343. M. Gooch, B. Lv, J.H. Tapp, Z. Tang, B. Lorenz, A.M. Guloy, P.C.W. Chu, Europhys. Lett. **85**, 27005 (2009), arXiv: 0901.4468 (2009)
344. I.A. Nekrasov, Z.V. Pchelkina, M.V. Sadovskii, JETP Lett. **88**, 543 (2008)
345. R.A. Jishi, H.M. Alyahyaei, arXiv: 0812.1215 (2008)
346. E.Z. Kurmaev, J. McLeod, N.A. Skorikov, L.D. Finkelstein, A. Moewes, M.A. Korotin, Yu.A. Izyumov, S. Clarke, J. Phys. Condens. Matter **21**, 345701 (2009), arXiv: 0903.4901 (2009)
347. S. Deng, J. Köhler, A. Simon, arXiv: 0908.0902 (2009)
348. F.L. Pratt, P.J. Baker, S.J. Blundell, T. Lancaster, H.J. Lewtas, P. Adamson, M.J. Pitcher, D.R. Parker, S.J. Clarke, Phys. Rev. B **79**, 052508 (2009)
349. D.R. Parker, M.J. Pitcher, S.J. Clarke, Chem. Commun. 2189 (2009), arXiv: 0810.3214 (2008)
350. Y-F. Li, B.-G. Liu, Eur. Phys. J. B **72**, 153 (2009), arXiv: 0905.3303 (2009)
351. L. Zhang, A. Subedi, D.J. Singh, M.H. Du, Phys. Rev. B **78**, 174520 (2008)
352. G.F. Chen, W.Z. Hu, J.L. Luo, N.L. Wang, arXiv: 0902.1100 (2009)
353. M. Tegel, S. Johansson, V. Wei, I. Schellenberg, W. Hermes, R. Pottgen, D. Johrendt, Europhys. **84**, 67007 (2008)
354. F. Han, X. Zhu, C. Mu, P. Cheng, H.-H. Wen, Phys. Rev. B **78**, 180503(R) (2008)
355. S. Matsuishi, Y. Inoue, T. Nomura, Y. Kamihara, M. Hirano, H. Hosono, New J. Phys. **11**, 025012 (2009)
356. X. Zhu, F. Han, P. Cheng, G. Mu, B. Shen, H.-H. Wen, Europhys. Lett. **85**, 17011 (2009)
357. G. Wu, Y.L. Xie, H. Chen, M. Zhong, R.H. Liu, B.C. Shi, Q.J. Li, X.F. Wang, T. Wu, Y.J. Yan, J.J. Ying, X.H. Chen, J. Phys. Condens. Matter **21**, 142203 (2009)
358. S. Li, C. de la Cruz, Q. Huang, G.F. Chen, T.-l. Xia, J.L. Lou, N.L. Wang, P. Dai, Phys. Rev. B **80**, 020504(R) (2009), arXiv: 0905.0525 (2009)
359. I.A. Nekrasov, Z.V. Pchelkina, M.V. Sadovskii, JETP Lett. **88**, 679 (2008)
360. I.R. Shein, A.L. Ivanovskii, arXiv: 0810.3498 (2008)
361. J.W. Lynn, P. Dai, Physica C **469**, 469 (2009), arXiv: 0902.0091 (2008)
362. C.H. Lee, A. Iyo, H. Eisaki, H. Kito, M.T. Fernandez-Diaz, T. Ito, K. Kihou, H. Matsuhata, M. Braden, K. Yamada, J. Phys. Soc. Jpn. **77**, 083704 (2008)
363. M. Ishikado, R. Kajimoto, S. Shamoto, M. Arai, A. Iyo, K. Miyazawa, P.M. Shirage, H. Kito, H. Eisaki, S.-W. Kim, H. Hosono, T. Guidi T., R. Bewley, S.M. Bennington, arXiv: 0809.5128v2 (2008)
364. J. Zhao, Q. Huang, C. de la Cruz, J.W. Lynn, M.D. Lumsden, Z.A. Ren, J. Yang, X. Shen, X. Dong, Z. Zhao, P. Dai, Phys. Rev. B **78**, 132504 (2008)
365. S.A.J. Kimber, D.N. Argyriou, F. Yokaichiya, K. Habicht, S. Gerischer, T. Hansen, T. Chatterji, R. Klingeler, C. Hess, G. Behr, A. Kondrat, B. Buchner, Phys. Rev. B**78**, 140503 (2008)
366. Y. Qiu, W. Bao, Q. Huang, T. Yildirim, J.M. Simmons, M.A. Green, J.W. Lynn, Y.C. Gasparovic, J. Li, T. Wu, G. Wu, X.H. Chen, Phys. Rev. Lett. **101**, 257002 (2008)
367. A. Jesche, N. Caroca-Canales, H. Rosner, H. Borrmann, A. Ormeci, D. Kasinathan, H.H. Klauss, H. Luetkens, R. Khasanov, A. Amato, A. Hoser, K. Kaneko, C. Krellner, C. Geibel, Phys. Rev. B **78**, 180504 (2008)

368. D. Fruchart, P. Convert, P. Wolfers, R. Madar, J.P. Senateur, R. Fruchart, Mater. Res. Bull. **10**, 169 (1975)
369. I.R. Shein, A.L. Ivanovskii, Eur. J. Combin. **30**, 1497 (2009), arXiv: 0811.2475 (2008)
370. G. Xu, H. Zhang, X. Dai, Z. Fang, Europhys. Lett. **84**, 67015 (2008)
371. H. Ding, P. Richard, K. Nakayama, K. Sugawara, T. Arakane, Y. Sekiba, A. Takayama, S. Souma, T. Sato, T. Takahashi, Z. Wang, X. Dai, Z. Fang, G.F. Chen, J.L. Luo, N.L. Wang, Europhys. Lett. **83**, 47001 (2008)
372. D. Parker, O.V. Dolgov, M.M. Korshunov, A.A. Golubov, I.I. Mazin, Phys. Rev. B **78**, 134524 (2008)
373. G. Mu, X.Y. Zhu, L. Fang, L. Shan, C. Ren, H.-H. Wen, Chin. Phys. Lett. **25**, 2221 (2008)
374. C. Ren, Z.-S. Wang, H. Yang, X. Zhu, L. Fang, G. Mu, L. Shan, H.-H. Wen, arXiv: 0804.1726 (2008)
375. Y. Nakai, K. Ishida, Y. Kamihara, M. Hirano, H. Hosono, J. Phys. Soc. Jpn. **77**, 073701 (2008)
376. X.L. Wang, S.X. Dou, Z.-A. Ren, W. Yi, Z.-C. Li, Z.-X. Zhao, S.-IK. Lee, arXiv: 0808.3398 (2008)
377. J.K. Dong, L. Ding, H. Wang, X.F. Wang, T. Wu, X.H. Chen, S.Y. Li, New J. Phys. **10,** 123031 (2008)
378. C.-T. Chen, C.C. Tsuei, M.B. Ketchen, Z.-A. Ren, Z.X. Zhao, Nat. Phys. **6**, 260 (2010), arXiv: 0905.3571 (2009)
379. R.H. Liu, T. Wu, G. Wu, H. Chen, X.F. Wang, Y.L. Xie, J.J. Yin, Y.J. Yan, Q.J. Li, B.C. Shi, W.S. Chu, Z.Y. Wu, X.H. Chen, Phys. Rev. Lett. **102**, 070403 (2009), arXiv: 0810.2604 (2008)
380. P.M. Shirage, P.M. Kihoux, K. Miyazawa, C.-H. Lee, H. Kito, H. Eisaki, Y. Tanaka, A. Iyo, Phys. Rev. Lett. **103**, 257003 (2009), arXiv: 0903.3515 (2009)
381. H.-Y. Choi, J.H. Yun, Y. Bang, H.C. Lee, Phys. Rev. B **80**, 052505 (2009), arXiv: 0904.1864 (2009)
382. W. Metzner, D. Vollhardt, Phys. Rev. Lett. **62**, 3241 (1989)
383. Yu.A. Izyumov, E.Z. Kurmaev, Phys. Usp. **51**, 23 (2008)
384. K. Held, Adv. Phys. **56**, 829 (2007)
385. Yu.A. Izyumov, V.I. Anisimov, *Electronic Structure of Compounds with Strong Electron Correlations* (M.-Izhevsk, 2008)
386. A.A. Abrikosov, L.P. Gor'kov, I.E. Dzyaloshinskii, Quantum Field Theoretical Methods in Statistical Physics, 2edn. (Pergamon, NY, 1965)
387. G. Kotliar, S.Y. Savrasov, K. Haule, V.S. Oudovenko, O. Parcollet, C.A. Marianetti, Rev. Mod. Phys. **78**, 865 (2006)
388. A.O. Shorikov, M.A. Korotin, S.V. Streltsov, D.M. Korotin, V.I. Anisimov, S.L. Skornyakov, J. Sci. Comput. **31**, 1774 (2009)
389. J.E. Han, M. Jarrell, D.E. Kondakov, T.M. Rice, M. Sigrist, Eur. Phys. J. B **25**, 191 (2002)
390. L. Craco, M.S. Laad, S. Leoni, H. Rosner, Phys. Rev. B **78**, 134511 (2008)
391. M.S. Laad, L. Craco, S. Leoni, H. Rosner, arXiv: 0810.1607 (2008)
392. V.I. Anisimov, E.Z. Kurmaev, A. Moewes, Yu.A. Izyumov, Physica C **469**, 442 (2009)
393. S.L. Skornyakov, A.V. Efremov, N.A. Skorikov, M.A. Korotin, Yu.A. Izyumov, V.I. Anisimov, A.V. Kozhevnikov, D. Vollhardt, Phys. Rev. B **80**, 092501 (2009), arXiv: 0906.3218 (2009)
394. S. de Jong, Y. Huang, R. Huisman, F. Massee, S. Thirupathaiah, M. Gorgoi, R. Follath, J.B. Goedkoop, M.S. Golden, Phys. Rev. B **79**, 115125 (2009), arXiv: 0901.2691 (2009)
395. C. Liu, G.D. Samolyuk, Y. Lee, N. Ni, T. Kondo, A.F. Santander-Syro, S.L. Bud'ko, J.L. McChesney, E. Rotenberg, T. Valla, A.V. Fedorov, P.C. Canfield, B.N. Harmon, A. Kaminski, Phys. Rev. Lett. **101**, 177005 (2008)
396. E.Z. Kurmaev, R.G. Wilks, A. Moewes, N.A. Skorikov, Yu.A. Izyumov, L.D. Finkelstein, R.H. Li, X.H. Chen, Phys. Rev. B **78**, 220503(R) (2008)
397. E.Z. Kurmaev, J.A. McLeod, A. Buling, N.A. Skorikov, A. Moewes, M. Neumann, M.A. Korotin, Yu.A. Izyumov, N. Ni, P.C. Canfield, arXiv: 0902.1141 (2009)
398. W. Malaeb, T. Yoshida, T. Kataoka, A. Fujimori, M. Kubota, K. Ono, H. Usui, K. Kuroki, R. Arita, H. Aoki, Y. Kamihara, M. Hirano, H. Hosono, J. Phys. Soc. Jpn. **77**, 093714 (2008)

399. V.R. Galakhov, A.I. Poteryaev, E.Z. Kurmaev, V.I. Anisimov, St. Bartkowski, M. Neumann, Z.W. Lu, B.M. Klein, T.-R. Zhao, Phys. Rev. B **56**, 4584 (1997)
400. A. Yamasaki, Imada, K. Takase, T. Muro, H. Kobori, A. Sugimura, N. Umeyama, H. Sato, Y. Hara, N. Miyagawa, S.I. Ikeda, arXiv: 0902.3314 (2009)
401. E.Z. Kurmaev, A.L. Ankudinov, J.J. Rehr, L.D. Finkelstein, P.F. Karimov, A. Moewes, J. Electr. Spectr. Relat. Phenom. **148**, 1 (2005)
402. M. Aichhorn, L. Pourovskii, V. Vildosola, M. Ferrero, O. Parcollet, T. Miyake, A. Georges, S. Biermann, Phys. Rev. B **80**, 085101 (2009), arXiv: 0906.3735 (2009)
403. X. Gao, D. Qi, S.C. Tan, A.T.S. Wee, X. Yu, H.O. Moser, J. Electr. Spectr. Relat. Phenom. **151**, 199 (2006)
404. S. Raghu, X.-L. Qi, C.-X. Liu, D. Scalapino, S.-C. Zhang, Phys. Rev. B **77**, 220503(R) (2008)
405. E. Dagotto, T. Hotte, A. Moreo, Phys. Rep. **344**, 1 (2001)
406. R. Yu, K.T. Trinh, A. Moreo, M. Daghofer, J.A. Riera, S. Haas, E. Dagotto, Phys. Rev. B **79**, 104510 (2009)
407. M. Daghofer, A. Moreo, J.A. Riera, A.E. Arrigoni, D.J. Scalapino, E. Dagotto, Phys. Rev. Lett. **101**, 237004 (2008)
408. E. Dagotto, Rev. Mod. Phys. **66**, 763 (1994)
409. M. Potthoff, M. Aichhorn, C. Dahnken, Phys. Rev. Lett. **91**, 206402 (2003)
410. A. Moreo, M. Daghofer, J.A. Riera, E. Dagotto, Phys. Rev. B **79**, 134502 (2009), arXiv: 0901.3544 (2008)
411. Y. Wan, Q.-H. Wang, arXiv: 0805.0923 (2008)
412. N. Bulut, D.J. Scalapino, R.T. Scaletter, Phys. Rev. B **45**, 5577 (1992)
413. K. Kuroki, R. Arita, Phys. Rev. B **64**, 024501 (2008)
414. S. Onari, K. Kuroki, R. Arita, H. Aoki, Phys. Rev. B **65**, 184525 (2002)
415. T. Takimoto, T. Hotta, K. Ueda, Phys. Rev. B **69**, 104504 (2004)
416. K. Kuroki, H. Aoki, Physica C **469**, 635 (2009), arXiv: 0902.3691 (2009)
417. N.E. Bickers, D.J. Scalapino, S.R. White, Phys. Rev. Lett. **62**, 961 (1989)
418. K. Kuroki, New J. Phys. **11**, 025017 (2009)
419. Z.-J. Yao, J.-X. Li, Z.D. Wang, New J. Phys. **11**, 025009 (2009)
420. H.H. Suhl, B.T. Matthias, L.R. Walker, Phys. Rev. Lett. **3**, 552 (1959)
421. D. Inotani, Y. Ohashi, arXiv: 0901.1718 (2008)
422. Y. Senga, H. Kontani, J. Phys. Soc. Jpn. **77**, 113710 (2008)
423. A.V. Chubukov, D. Efremov, I. Eremin, Phys. Rev. B **78**, 134512 (2008)
424. D. Parker, O.V. Dolgov, M.M. Korshunov, A.A. Golubov, I.I. Mazin, Phys. Rev. B **78**, 134524 (2008)
425. V. Stanev, J. Kang, Z. Tesanovic, Phys. Rev. B **78**, 184509 (2008), arXiv: 0809.0014 (2008)
426. Y. Bang, H.-Y. Choi, Phys. Rev. B **78**, 134523 (2008)
427. Y. Yanagi, Y. Yamakawa, Y. Ōno, J. Phys. Soc. Jpn. **77**, Suppl. C, 149 (2008), arXiv: 0808.1192 (2008)
428. S.-L. Yu, J. Kang, J.-X. Li, Phys. Rev. B **79**, 064517 (2009), arXiv: 0901.0821 (2009)
429. P.A. Lee, X.-G. Wen, Phys. Rev. B **78**, 144517 (2008), arXiv: 0804.1739 (2008)
430. S. Graser, T.A. Maier, P.J. Hirschfeld, D.J. Scalapino, New J. Phys. **11**, 025016 (2009)
431. M.J. Calderon, B. Valenzuela, E. Bascones, Phys. Rev. B **80**, 094531 (2009), arXiv: 0907.1259 (2009)
432. H. Eschrig, K. Koepernik, arXiv: 0905.4844 (2009)
433. G.-M. Zhang, Y.-H. Su, Z.-Y. Lu, Z.-Y. Weng, D.-H. Lee, T. Xiang, Europhys. Lett. **86**, 37006 (2009), arXiv: 0809.3874 (2008)
434. K. Kuroki, H. Usui, S. Onari, R. Arita, H. Aoki, Phys. Rev. Lett. **101**, 087004 (2008), arXiv: 0904.2612 (2009)
435. L. Benfatto, M. Capone, S. Caprara, C. Castellani, C. Di Castro, Phys. Rev. B **78**, 140502(R) (2008)
436. E.Z. Kuchinskii, M.V. Sadovskii, JETP Lett. **89**, 156 (2009), arXiv: 0901.0164 (2009)
437. G.A. Ummarino, M. Tortello, D. Daghero, R.S. Gonnelli, arXiv: 0904.1808 (2009)
438. F. Wang, H. Zhai, Y. Ran, A. Vishwanath, D.-H. Lee, Phys. Rev. Lett. **102**, 047005 (2009)
439. D. Zhang, T. Zhou, C.S. Ting, arXiv: 0904.3708 (2009)

440. T. Zhou, D. Zhang, X. Hu, C.S. Ting, arXiv: 0904.4273 (2009)
441. W.-F. Tsai, Y.-Y. Zhang, C. Fang, J. Hu, Phys. Rev. B **80**, 064513 (2009), arXiv: 0905.0734 (2009)
442. X. Hu, C.S. Ting, J.-X. Zhu, arXiv: 0905.0014 (2009)
443. K. Seo, B.A. Bernevig, J. Hu, Phys. Rev. Lett. **101**, 206404 (2008)
444. H. Zhai, F. Wang, D.-H. Lee, Phys. Rev. B **80**, 064517 (2009), arXiv: 0905.1711 (2009)
445. M. Guidry, Y. Sun, C.-L. Wu, arXiv: 0908.1147 (2009)
446. A.T. Zheleznyak, V.M. Yakovenko, I.E. Dzyaloahinskii, Phys. Rev. B **55**, 3200 (1997)
447. A. Chubukov, D. Efremov, I. Eremin, Phys. Rev. B **78**, 134512 (2008)
448. A.V. Chubukov, arXiv: 0902.4188 (2009)
449. C. Platt, C. Honerkamp, W. Hanke, arXiv: 0903.1963 (2009)
450. V. Barzykin, L.P. Gorkov, JETP Lett. **88**, 131 (2008)
451. A.V. Chubukov, M.G. Vavilov, A.B. Vorontsov, Phys. Rev. B **80**, 140515(R) (2009), arXiv: 0903.5547 (2009)
452. T.A. Maier, S. Graser, D.J. Scalapino, P.J. Hirschfeld, Phys. Rev. B **79**, 224510 (2009), arXiv: 0903.5216 (2009)
453. R. Thomale, C. Platt, J. Hu, C. Honerkamp, B.A. Bernevig, Phys. Rev. B **80**, 180505(R) (2009), arXiv: 0906.4475 (2009)
454. A.V. Vorontsov, M.G. Vavilov, A.V. Chubukov, Phys. Rev. B **79**, 060508(R) (2009)
455. P. Fulde, R. Ferrell, Phys. Rev. **135**A, 550 (1964)
456. A.I. Larkin, Y.N. Ovchinnikov, JETP **47**, 1136 (1964)
457. R. Khasanov, M. Bendele, A. Amato, P. Babkevich, A.T. Boothroyd, A. Cervellino, K. Conder, S.N. Gvasaliya, H. Keller, H.-H. Klauss, H. Luetkens, E. Pomjakushina, B. Roessli, Phys. Rev. B **80**, 140511(R) (2009), arXiv: 0907.3429 (2009)
458. P.L. Paulose, C.S. Yadav, K.M. Subhedar, Europhys. Lett. **90**, 27011 (2010), arXiv: 0907.3513
459. A.V. Vorontsov, M.G. Vavilov, A.V. Chubukov, Phys. Rev. B **79**, 140507(R) (2009), arXiv: 0901.0719 (2009)
460. R.T. Gordon, N. Ni, C. Martin, M.A. Tanatar, M.D. Vannette, H. Kim, G. Samolyuk, J. Schmalian, S. Nandi, A. Kreyssig, A.I. Goldman, J.Q. Yan, S.L. Bud'ko, P.C. Canfield, R. Prozorov, Phys. Rev. Lett **102**, 127004 (2009)
461. R.T. Gordon, C. Martin, H. Kim, N. Ni, M.A. Tanatar, J. Schmalian, I.I. Mazin, S.L. Bud'ko, P.C. Canfield, R. Prozorov, Phys. Rev. B **79**, 100506(R) (2009)
462. M.M. Parish, J. Hu, B.A. Bernevig, Phys. Rev. B, **78**, 144514 (2008)
463. P. Goswami, P. Nikolic, Q. Si, Europhys. Lett. **91**, 37006 (2010), arXiv: 0905.2634 (2009)
464. W. Chen, K.-Y. Yang, Y. Zhou, F.-C. Zhang, Phys. Rev. B **79**, 047006 (2009)
465. W.-L. Yau, S.-J. Gu, G.-S. Tian, H.-Q. Lin, Phys. Rev. B **79**, 014508 (2009)
466. Q. Han, Y. Chen, Z.D. Wang, Europhys. Lett. **82**, 37007 (2008)
467. M.M. Korshunov, I. Eremin, Phys. Rev. B **78**, 140509(R) (2008)
468. M.M. Korshunov, I. Eremin, Europhys. Lett. **83**, 67003 (2008)
469. F. Ma, Z.-Y. Lu, T. Xiang, Phys. Rev. B **78**, 224517 (2008)
470. Q. Si, E. Abrahams, Phys. Rev. Lett. **101**, 076401 (2008)
471. M.D. Johannes, I.I. Mazin, Phys. Rev. B **79**, 220510(R) (2009), arXiv: 0904.3857 (2009)
472. D.-X. Yao, E.W. Carlson, Phys. Rev. B **78**, 052507 (2008)
473. G.S. Uhrig, M. Holt, J. Oitmaa, O.P. Sushkov, R.R.P. Singh, Phys. Rev. B **79**, 092416 (2009)
474. R.A. Ewings, T.G. Perring, R.I. Bewley, T. Guidi, M.J. Pitcher, D.R. Parker, S.J. Clarke, A.T. Boothroyd, Phys. Rev. B **78**, 220501 (2008)
475. R.J. McQueeney, S.O. Diallo, V.P. Antropov, G.D. Samolyuk, C. Broholm, N. Ni, S. Nandi, M. Yethiraj, J.L. Zarestky, J.J. Pulikkotil, A. Kreyssig, M.D. Lumsden, B.N. Harmon, P.C. Canfield, A.I. Goldman, Phys. Rev. Lett. **101**, 227205 (2008)
476. S.D. Wilson, P. Dai, S.L. Li, S.X. Chi, H.J. Kang, J.W. Lynn, Nature **442**, 59 (2006)
477. R. Osborn, S. Rosenkranz, E.A. Goremychkin, A.D. Christianson, arXiv: 0902.3760 (2009)
478. S. Chi, A. Schneidewind, J. Zhao, L.W. Harriger, L. Li, Y. Luo, G. Cao, Z. Xu, M. Loewenhaupt, J. Hu, P. Dai, Phys. Rev. Lett. **102**, 107006 (2009)

479. D. Parshall, K.A. Lokshin, J. Niedziela, A.D. Christianson, M.D. Lumsden, H.A. Mook, S.E. Nagler, M.A. McGuire, M.B. Stone, D.L. Abernathy, A.S. Sefat, B.C. Sales, D.G. Mandrus, T. Egami, arXiv: 0903.4621 (2009)
480. H. Ding, P. Richard, K. Nakayama, T. Sugawara, T. Arakane, Y. Sekiba, A. Takayama, S. Souma, T. Sato, T. Takahashi, Z. Wang, X. Dai, Z. Fang, G.F. Chen, J.L. Luo, N.L. Wang, Europhys. Lett. **83**, 47001 (2008)
481. S. Hüfner, M.A. Hossain, A. Damascelli, G. Sawatsky, Rep. Progr. Phys. **71**, 2501 (2008)
482. M.S. Laad, L. Craco, Phys. Rev. Lett. **103**, 017002 (2009), arXiv: 0902.3400 (2009)
483. M.S. Laad, L. Craco, S. Leoni, H. Rosner, Phys. Rev. B **79**, 024515 (2009)
484. L. Craco, M.S. Laad, arXiv: 0903.1568 (2009)
485. M.S. Laad, L. Craco, arXiv: 0903.3732 (2009)
486. J. Zhao, L.-P. Regnault, C. Zhang, M. Wang, Z. Li, F. Zhou, Z. Zhao, P. Dai, Phys. Rev. B **81**, 104520 (2009), arXiv: 0908.1710 (2009)
487. T.A. Maier, S. Graser, D.J. Scalapino, P. Hirschfeld, Phys. Rev. B **79**, 134520 (2009), arXiv: 0903.0008 (2009)
488. D.S. Inosov, J.T. Park, P. Bourges, D.L. Sun, Y. Sidis, A. Schneidewind, K. Hradil, D. Haug, C.T. Lin, B. Keimer, V. Hinkov, Nat. Phys. **6**, 178 (2010), arXiv: 0907.3632 (2009)
489. M.D. Lumsden, A.D. Christianson, D. Parshall, M.B. Stone, S.E. Nagler, H.A. Mook, K. Lokshin, T. Egami, D.L. Abernathy, E.A. Goremychkin, R. Osborn, M.A. McGuire, A.S. Sefat, R. Jin, B.C. Sales, D. Mandrus, Phys. Rev. Lett. **102**, 107005 (2009)
490. M.M. Korshunov, I. Eremin, Phys. Rev. B **78**, 140509(R) (2008)
491. T.A. Maier, S. Graser, D.J. Scalapino, P. Hirschfeld, Phys. Rev. B **79**, 134520 (2009)
492. S.-P. Kou, T.-L. Li, Z.-Y. Weng, Europhys. Lett. **88**, 17010 (2009), arXiv: 0811.4111 (2008)
493. K. Haule, G. Kotliar, New J. Phys. **11**, 025021 (2009)
494. Th. Pruschke, R. Bulla, Eur. Phys. J. B **44**, 217 (2005)
495. J. Dai, Q. Si, J.-X. Zhu, E. Abrahams, Proc. Natl. Acad. Sci. **106**, 4118 (2009), arXiv: 0808.0305 (2008)
496. Q. Si, E. Abrahams, J. Dai, J.-X. Zhu, New. J. Phys. **11**, 045001 (2009), arXiv: 0901.4112 (2009)
497. K.I. Kugel, D.I. Khomskii, Sov. Phys. Usp. **136**, 621 (1982)
498. J. van den Brink, D.I. Khomskii, Phys. Rev. B **63**, 140416 (2001)
499. J. van den Brink, G. Khaliulin, D. Khomskii, Phys. Rev. Lett. **83**, 5118 (1999)
500. F. Kruger, S. Kumar, J. Zaanen, J. van den Brink, Phys. Rev. B **79**, 054504 (2009), arXiv: 0811.4104 (2008)
501. K. Kubo, P. Thalmeier, J. Phys. Soc. Jpn. **78**, 083704 (2009), arXiv: 0903.4064 (2009)
502. C.-C. Lee, W.-G. Yin, W. Ku, Phys. Rev. Lett. **103**, 267001 (2009), arXiv: 0905.2957 (2009)
503. I.I. Mazin, J. Schmalian, Physica C **469**, 614 (2009), arXiv:0901.4790 (2009)

Index

Printed by Printforce, the Netherlands